T0199481

SAMPLING AND SURVEYING RADIOLOGICAL ENVIRONMENTS

SAMPLING AND SURVEYING RADIOLOGICAL ENVIRONMENTS

Mark E. Byrnes

Contributors

David A. King
Susan F. Blackburn
Robert L. Johnson
Sebastian C. Tindall
Walter E. Remsen, Jr.
Samuel E. Stinnette
Nile A. Luedtke

CRC Press
Taylor & Francis Group
Boca Raton London New York

CRC Press is an imprint of the
Taylor & Francis Group, an **informa** business

CRC Press
Taylor & Francis Group
6000 Broken Sound Parkway NW, Suite 300
Boca Raton, FL 33487-2742

First issued in paperback 2019

No claim to original U.S. Government works

ISBN-13: 978-0-367-45547-7 (pbk)
ISBN-13: 978-1-56670-364-2 (hbk)

Visit the Taylor & Francis Web site at
http://www.taylorandfrancis.com

and the CRC Press Web site at
http://www.crcpress.com

Library of Congress Card Number 00-058777

Library of Congress Cataloging-in-Publication Data

Byrnes, Mark E.
Sampling and surveying radiological environments / Mark E. Byrnes.
p.   cm.
Includes bibliographical references and index.
ISBN 1-56670-364-6 (alk. paper)
1. Radioactive waste sites—Evaluation. 2. Radioactive pollution—Measurement. 3. Environmental sampling. 4. Radioactive waste disposal—Law and legislation—United States. I. Title.
TD898.155.E83 B95 2000
628.5'2—dc21

00-058777

Preface

The purpose of *Sampling and Surveying Radiological Environments* is to provide the environmental industry with guidance on how to design and implement defensible sampling programs in radiological environments, such as those found in the vicinity of uranium mine sites, nuclear weapons production facilities, nuclear reactors, radioactive waste storage and disposal facilities, and areas in the vicinity of nuclear accidents. This book presents many of the most effective radiological surveying and sampling methods for use in supporting:

- Environmental site characterization
- Building characterization
- Waste characterization
- Tank characterization
- Risk assessment
- Feasibility study
- Remedial design
- Postremediation site closeout
- Postdecontamination and decommissioning building closeout

Standard operating procedures have been provided for those sampling methods that do not require specialized training, such as:

- Swipe sampling
- Concrete sampling
- Paint sampling
- Soil sampling
- Sediment sampling
- Surface water sampling
- Groundwater sampling
- Drum sampling

For more specialized radiological investigative techniques (e.g., *in situ* gamma spectroscopy, downhole HPGe measurements, cone penetrometry), information has been provided to help the reader understand how the technique works and under what conditions it can be used most effectively.

Guidance is provided on how to use the Environmental Protection Agency (EPA) Data Quality Objectives (DQO) and Data Quality Assessment (DQA) process to support the design of defensible sampling programs and to ensure that the collected data are of adequate quality and quantity to meet the intended purpose. Templates have been provided to assist the user in going through the DQO Process and to assist in the writing of a DQO Summary Report and Sampling and Analysis Plan. These templates appear in Appendices A and B, and on the CD accompanying the book. The capabilities of multiple statistical sample design software packages are presented along with Web page addresses where copies of the software can be downloaded.

The book includes a summary of the major environmental laws and regulations that apply to radiological sites, including those that govern the actions of the U.S. Department of Energy (DOE) and Nuclear Regulatory Commission (NRC). Other major topics addressed by this book include radiation detection theory; sample preparation, documentation, and shipment; data verification and validation; data management; and equipment decontamination.

This book focuses on those methods and procedures that have proved themselves to be effective and/or are acknowledged by the EPA, DOE, NRC, and/or U.S. Department of Defense (DOD) as reputable techniques. The primary references used as guidance to support the preparation of this book include:

Byrnes, M.E., 1994, *Field Sampling Methods for Remedial Investigations*, Lewis Publishers, Ann Arbor, MI.
Driscoll, F.G., 1986, *Groundwater and Wells*, 2nd ed., Johnson Division, St. Paul, MN.
Environmental Protection Agency, 1987, A Compendium of Superfund Field Operations Methods, EPA/540/P-87/001a.
Environmental Protection Agency, 1988, Guidance for Conducting Remedial Investigations and Feasibility Studies under CERCLA, EPA/540/G-89/004.
Environmental Protection Agency, 1989, Methods for Evaluating the Attainment of Cleanup Standards, Volume 1: Soils and Solid Media, PB89-234959.
Environmental Protection Agency, 1991, Handbook of Suggested Practices for the Design and Installation of Ground-Water Monitoring Wells, EPA/600/4-89/034.
Environmental Protection Agency, 1992a, Statistical Methods for Evaluating the Attainment of Cleanup Standards, Volume 3: Reference-Based Standards for Soils and Solid Media, EPA 230-R-94-004.
Environmental Protection Agency, 1992b, RCRA Ground-Water Monitoring: Draft Technical Guidance, EPA/530-R-93-001.
Environmental Protection Agency, 1992c, Final Comprehensive State Ground Water Protection Program Guidance, 100-R-93-001.
Environmental Protection Agency, 1992d, Guide to Management of Investigation—Derived Waste, PB92-963353.
Environmental Protection Agency, 1994, Guidance for the Data Quality Objectives Process, EPA QA/G-4.
Environmental Protection Agency, 1996, Guidance for Data Quality Assessment, Practical Methods for Data Analysis, EPA QA/G-9.
Environmental Protection Agency, 1997, Multi-Agency Radiation Survey and Site Investigation Manual (MARSSIM), EPA 402-R-97-016.
Environmental Protection Agency, 1998, EPA Guidance for Quality Assurance Project Plans, EPA QA/G-5.
International Atomic Energy Agency, 1991, *Airborne Gamma Ray Spectrometer Surveying*, Vienna, Austria.
U.S. Department of Energy, 1994, Radiological Control Manual, DOE/EH-0256T, Rev. 1.
U.S. Department of Energy, 1996, Field Screening Technical Demonstration Evaluation Report, DOE/OR/21950-1012.

A number of commercially available scanning, direct measurement, and sampling methods are discussed in this book. While the author believes these methods should be considered as potentially appropriate methods for radiological investigations, he is by no means specifically endorsing or marketing these products. The author chose this approach over a more general discussion because he believes it will provide more valuable information to the reader.

The primary audiences for this book are U.S. and international government agencies and their contractors responsible for the remediation and/or decontamination and decommissioning of radiological sites and facilities. This book is also intended to be used as a university textbook to teach advanced undergraduate or graduate-level courses that deal with the practical elements of performing environmental investigations in radiological environments.

About the Author

Mark E. Byrnes is a senior data quality/ sampling specialist working for Science Applications International Corporation (SAIC), a 39,000-person, employee-owned science and engineering company. Mr. Byrnes works at the U.S. Department of Energy Hanford Nuclear Reservation, supporting environmental remediation and facility decontamination and decommissioning activities performed by Bechtel Hanford, Inc., and CH2M Hill under the U.S. Department of Energy Environmental Restoration Program.

Mr. Byrnes received his bachelor of arts degree in geology from the University of Colorado (Boulder), and his master of science degree in geology/geochemistry from Portland State University (Oregon). Mr. Byrnes is a registered professional geologist in the States of Tennessee and Kentucky, and is the author of the 1994 Lewis Publishers book titled, *Field Sampling Methods for Remedial Investigations*, which has been used as a textbook at Georgia Tech and many other major universities across the country.

About the Contributors

David A. King is a certified health physicist, working for SAIC. Mr. King supports environmental characterization and remediation of Formerly Utilized Sites Remedial Action Program (FUSRAP) sites operated by the U.S. Corps of Engineers. Mr. King specializes in dose/risk assessments for radiologically contaminated sites and radiological surveys using the EPA Multi-Agency Radiation Survey and Site Investigation Manual (MARSSIM). Mr. King received his bachelor of science degree in physics from Middle Tennessee State University, and his master of science degree in radiation protection engineering from the University of Tennessee (Knoxville). Mr. King received his certification from the American Board of Health Physics in 1999.

Susan F. Blackburn is employed by SAIC as a senior environmental statistician. She provides statistical support to a variety of environmental programs including the U.S. Department of Energy Environmental Restoration Program at the Hanford Nuclear Reservation, the Advanced Mixed Waste Treatment Facility at Idaho Falls, the U.S. Department of Energy River Protection Program at the Hanford Nuclear Reservation, and the Office of Civilian Radioactive Waste Management for the Yucca Mountain Project. Ms. Blackburn has a bachelor of science degree in mathematics, a master of science degree in human factors, and a master of science degree in quantitative methods (statistics) from the University of Illinois in Champaign/Urbana.

Robert L. Johnson is with the Environmental Assessment Division, Argonne National Laboratory. Dr. Johnson holds a master's degree in environmental systems from Johns Hopkins University, Baltimore, and a Ph.D. in soil and water resources from Cornell University, Ithaca, NY. Dr. Johnson's areas of expertise include adaptive sampling program design and environmental data management.

Sebastian C. Tindall is a senior environmental scientist with Bechtel Hanford, Inc., Richland, WA. Mr. Tindall works at the U.S. Department of Energy Hanford Nuclear Reservation, supporting environmental remediation and building decontamination and decommissioning activities performed by Bechtel Hanford, Inc., under the U.S. Department of Energy Environmental Restoration Program. Mr. Tindall received his bachelor of arts degree in chemistry and biology and his master of science degree in chemistry from the University of California at Santa Cruz. Mr. Tindall has taught chemistry and hazardous materials courses for over 15 years at the college and university level. He is now on the faculty at Washington State University. Mr. Tindall is a registered environmental assessor in the State of California and a certified hazardous materials manager (master level). Mr. Tindall is nationally recognized as an expert in systematic planning for environmental decision making based on the EPA Data Quality Objectives (DQO) process and has developed and delivered DQO training courses for the U.S. Department of Energy.

Walter E. Remsen, Jr., is a senior environmental scientist working for Bechtel Hanford, Inc., at the U.S. Department of Energy Hanford Nuclear Reservation. Mr. Remsen provides technical support for environmental and engineering activities performed by Bechtel Hanford, Inc., and CH2MHill under the U.S. Department of Energy Environmental Restoration Program. Mr. Remsen received his bachelor of science degree in oceanography from the University of California—Humboldt and master of science degree in geology from the University of California—Northridge.

Samuel E. Stinnette is a senior data analyst working for the SAIC office in Oak Ridge, TN. He has more than 15 years of experience in the field of statistical data analysis and statistical consulting, with more than 10 years of experience working with environmental data analysis. He has provided statistical and programming support to projects associated with the Comprehensive Environmental Response, Compensation, and Liability Act (CERCLA) and Resource Conservation and Recovery Act (RCRA) at sites across the United States, with a focus on human health risk assessment. He has provided varying levels of support to Remedial Investigation (RI) and RCRA Facility Investigation (RFI) site characterizations, removal actions, corrective measures studies, feasibility studies, and risk-based prioritization. Mr. Stinnette has a bachelor of science degree in mathematics and history from James Madison University, Harrisonburg, VA, and a master of science degree in statistics from Virginia Tech, Blacksburg, VA.

Nile A. Luedtke is a senior chemist and analytical laboratory coordinator, SAIC–Oak Ridge, TN. Mr. Luedtke's expertise encompasses analytical chemistry and quality assurance/quality control (QA/QC), spanning a variety of environmental areas. He has worked in oceanographic research, commercial laboratory operations, the nuclear power industry, laboratory oversight programs, and environmental project management. His career has included development and implementation programs in relation to analytical laboratory interfaces, project chemistry support, project data quality development, and project data quality assessment. Mr. Luedtke holds a bachelor's degree in chemistry from Hartwick College, Oneonta, NY, and a master's degree in analytical chemistry from the University of Rhode Island, Kingston.

Contents

Acronyms and Abbreviations

α	alpha error
β	beta error
Ac	actinium
AEC	Atomic Energy Act
Am	americium
Be	beryllium
Bi	bismuth
C	carbon
°C	degrees Celsius
CAA	Clean Air Act
CCD	charged coupled device
CERCLA	Comprehensive Environmental Response, Compensation, and Liability Act
CFR	Code of Federal Regulations
CLP	contract laboratory program
cm	centimeter
COC	contaminant of concern
cpm	counts per minute
CWA	Clean Water Act
CX	categorical exclusion
DISPIM	Decommissioning *In Situ* Plutonium Inventory Monitor
DNA	deoxyribonucleic acid
DNAPL	dense nonaqueous-phase liquid
DOD	U.S. Department of Defense
DOE	U.S. Department of Energy
DQI	data quality indicator
DQO	Data Quality Objectives
dpm	disintegrations per minute
EA	Environmental Assessment
EIC	electret ion chamber
EIS	Environmental Impact Statement
ELR	Environmental Law Reporter
EPA	Environmental Protection Agency
EPCRA	Emergency Planning and Community Right to Know Act
ERDA	Energy Research Development Administration
eV	electron volt
FONSI	Finding of No Significant Impact
fpm	feet per minute
fps	feet per second

ft	feet
GC/MS	gas chromatography/mass spectrometry
GIS	Geographical Information System
GM	Geiger–Mueller
gpm	gallons per minute
GPS	Global Positioning System
Gy	gray
h	hour
H	hydrogen
HPGe	high-purity germanium
in.	inch
J	joule
k	one thousand
kg	kilogram
K	potassium
keV	1000 electron volts
L	liter
LDR	Land Disposal Restrictions
LNAPL	light nonaqueous-phase liquid
MARSSIM	Multi-Agency Radiation Survey and Site Investigation Manual (EPA 402-R-97-016)
MCL	Maximum Contaminant Level
MCLG	Maximum Contaminant Level Goal
MeV	millions of electron volts
min	minute
mL	milliliter
mm	millimeter
mph	miles per hour
mrem	millirem
msec	millisecond
MSE	mean square error
mV	millivolt
Na	sodium
NaI	sodium iodide
NCRP	National Council on Radiation Protection and Measurements
NEPA	National Environmental Policy Act
NOI	Notice of Intent
NPDWR	National Primary Drinking Water Regulation
NRC	Nuclear Regulatory Commission
NTU	Nephelometric Turbidity Units
OPNA	Office of NEPA Project Assistance

P	percent recovery
PARCC	precision, accuracy, representativeness, comparability, completeness
Pa	protactinium
Pb	lead
PCBs	polychlorinated biphenols
PIC	pressurized ion chamber
PNNL	Pacific Northwest National Laboratory
Po	polonium
PRG	preliminary remediation goal
psi	pounds per square inch
PSQ	principal study question
QA	quality assurance
QA/QC	quality assurance/quality control
QAPjP	Quality Assurance Project Plan
QC	quality control
R	roentgen, Coulomb/(kg of air)
Ra	radium
rad	a conventional unit of absorbed dose (one rad equals 0.01 gray)
Rb	rubidium
RCRA	Resource Conservation and Recovery Act
rem	unit of exposure [erg/(g of tissue)]
RI/FS	Remedial Investigation/Feasibility Study
Rn	radon
RSD	relative standard deviation
sec	second
SAIC	Science Applications International Corporation
SARA	Superfund Amendments and Reauthorization Act
SDWA	Safe Drinking Water Act
SOP	standard operating procedure
TCLP	Toxic Characteristic Leaching Procedure
Th	thorium
Tl	thallium
TLD	thermoluminescence dosimeter
TPH	total petroleum hydrocarbon
TSCA	Toxic Substance Control Act
TSD	treatment, storage, and disposal
U	uranium
UST	underground storage tank
V	volt
WPI	Waste Policy Institute
ZnS	zinc sulfide

Acknowledgments

The author would like to acknowledge the SAIC Project Management Solutions Operation, led by Dr. Stephen Whitfield (Operation Manager) and Dennis Schmidt (Division Manager), which provided funding to support the preparation of portions of this document.

The author appreciates all of the efforts provided by the seven contributors to this book, including David King (SAIC), Susan Blackburn (SAIC), Robert L. Johnson (Argonne National Laboratory), Sebastian C. Tindall (Bechtel Hanford, Inc.), Walter Remsen (Bechtel Hanford, Inc.), Samuel Stinnette (SAIC), and Nile Luedtke (SAIC). These contributors provided technical expertise in the areas of radiation detection, statistics, data quality, data management, chemistry, environmental regulations, remediation, and decontamination and decommissioning.

Specific acknowledgment goes to the following staff who performed technical reviews on various sections of this book: Wendy Thompson (Bechtel Hanford, Inc.), Debbie Browning (SAIC), Thomas Rucker (SAIC), Patrick Ryan (SAIC), and Grant Ceffalo (Bechtel Hanford, Inc.). Their input is very much appreciated. Dr. Richard Gilbert (Pacific Northwest National Laboratory) is also acknowledged for giving me permission to include several statistical tables from his 1987 Van Nostrand Reinhold book titled, *Statistical Methods for Environmental Pollution Monitoring*.

The author appreciates the support and/or technical input provided by Merrick Blancq (Corps of Engineers), William Price (Bechtel Hanford, Inc.), Karl Fecht (Bechtel Hanford, Inc.), Roy Bauer (CH2M Hill), David Keefer (Parallax), Ronald Kirk (DOE-OR), Tracy Friend (SAIC), Eric Dysland (WPI), Dawn Standley (SAIC), Sharon Bailey (PNNL), Debbie McCallam (Northrop Grumman–Remotec), Tony Marlow (BNFL Instruments), Michael Pitts (BNFL Instruments), Frazier Bronson (Canberra), Thomas Kabis (Sibak Industries), Desia Anderson (Arts Manufacturing & Supply, Inc.), Judy Mangan (Landauer, Inc.), Suzanne D'Angelo (AIL Systems, Inc.), and RSI Research Ltd.

I appreciate the patience expressed by my wife, Karen, and children, Christine and Kathleen Byrnes, throughout the preparation of this book. I love them dearly. I would like to acknowledge my mother, Frieda Byrnes, for providing me encouragement throughout the preparation of this book.

This book is dedicated to my father, Francis J. Byrnes, who taught me to enjoy and appreciate the field of science/engineering, and who has provided me with guidance and encouragement throughout my professional career.

Introduction

This book has been written to provide the environmental industry with guidance on how to develop and implement defensible sampling and surveying programs in radiological environments. This book provides the reader with proven radiological surveying and sampling methods that can be used to support soil remediation, building decontamination and decommissioning, tank characterization, and surveys of highly radioactive environments using pipe crawling and other robotic devices.

The intent of this book is to provide the reader with all of the tools needed to develop and implement a cost-effective and defensible sampling and surveying program. The purpose of Chapter 1 is to provide the reader with background information about radiological contamination sources, impacted environmental media, contaminant migration pathways, routes of exposure, and definitions of radiological terminology.

Chapter 2 provides a summary of the major environmental laws and regulations that apply to radiological sites and provides Internet addresses where individual state environmental agency regulations can be obtained. Chapter 3 provides the reader with the fundamentals of radioactivity, radiation, and radiation detection. Chapter 4 provides guidance on how to develop a defensible sampling program. This chapter provides:

- A template to support the implementation of the scoping process
- Guidance on regulatory interfacing
- Details on how to implement the EPA seven-step DQO process
- Guidance on developing statistical sampling and survey designs
- Guidance on developing integrated sampling and surveying designs
- Information on capabilities of various statistical sampling design software packages
- Guidance on developing a Sampling and Analysis Plan
- Information on capabilities of various scanning and direct measurement methods
- Standard operating procedures for media sampling

Chapter 5 provides guidance on sample preparation, field documentation, and shipment of radiological samples to the laboratory for analysis. This chapter addresses issues such as bottle requirements, sample preservation, sample labeling, chain-of-custody, field and photographic logbooks, and field sampling forms. Chapter 6 provides guidance on data verification and validation. Chapter 7 addresses how radiological data should be managed. Chapter 8 provides guidance on implementing the EPA five-step data quality assessment (DQA) process. Chapter 9 provides radiological and chemical equipment decontamination procedures. The appendices and CD-ROM provide templates to assist the reader in implementing the EPA seven-step DQO procedure and developing a DQO Summary Report and Sampling and Analysis Plan. The appendices also provide statistics tables to support statistical calculations, a metric conversion chart, radiological decay chains, and sample container, preservation, and holding time requirements.

1.1 RADIOLOGICAL CONTAMINANT SOURCES

The primary sources of radiological contamination include uranium mine sites, uranium mill tailings, uranium processing plants, nuclear weapons production facilities, nuclear testing laboratories, nuclear reactors, and associated fuel storage and radioactive waste storage and disposal facilities.

Examples of various types of radioactive waste storage and disposal facilities include:

- Landfills
- Trenches
- Waste water and cooling water holding ponds
- Cribs
- French drains
- Aboveground or underground storage tanks
- Waste container storage yards

If radiological contamination migrates from any of these primary sources into the surrounding environmental media (e.g., soil, sediment, building material), the environmental media becomes a secondary source of contamination. For example, if radiologically contaminated cooling water migrates through cracks in a concrete holding pond and contaminates the underlying soil, the contaminated soil beneath the pond becomes a secondary source of contamination. Contamination from this secondary source can then migrate and contaminate other environmental media, such as groundwater (Figure 1.1).

The primary source of the radiological contamination will determine which specific isotopes should be considered contaminants of concern for a particular site. For example, the contaminants of concern for a uranium mine site often include Th-232, U-235, U-238, and the isotopes resulting from the decay of these parent isotopes, such as Ac-227, Pa-231, Ra-226, Ra-228, Th-230, U-234, etc. (see Appendix E). On the other hand, the contaminants of concern at a nuclear weapons production facility may include isotopes such as Co-60, Cs-137, Eu-152, Eu-154, Eu-155, Pu-239/Pu-240, Sr-90, etc.

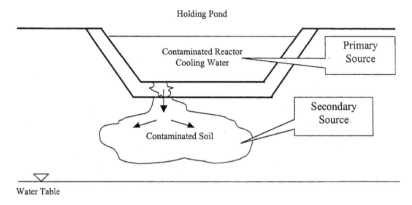

Figure 1.1 Example of a primary and secondary source of contamination.

As discussed in Section 4.1.1.5.1.6, the primary and secondary sources of contamination are essential components in the development of a Conceptual Site Model.

1.2 IMPACTED MEDIA

The media that may be impacted by radiological contamination will vary from site to site and will be influenced by the source(s) of contamination and the method by which the contamination is released (e.g., spill, leak, equipment malfunction) into the environment. For example, the impacted media for a leaking underground waste storage tank will be primarily deep soil and groundwater, while the impacted media from a surface spill within a plutonium processing plant may include air, concrete, paint, shallow soil, and surface water. The development of a Conceptual Site Model in Step 1 of the DQO procedure (Section 4.1.1.5.1.6), will assist in the identification of the impacted media.

Examples of the types of media that may be impacted at radiological sites include:

- Soil
- Sediment
- Sludge
- Surface water
- Groundwater
- Air
- Piping
- Ventilation ducts
- Concrete
- Asphalt
- Sheetrock
- Wood
- Roofing material
- Paint

1.3 CONTAMINANT MIGRATION PATHWAYS AND ROUTES OF EXPOSURE

The term *contaminant migration pathway* refers to the path by which contamination may spread into the surrounding environment. The three primary contamination migration pathways for chemical and radiological contaminants include:

- Air
- Surface water
- Groundwater

The pathway by which a contaminant may spread into the surrounding environment is dependent upon the location and type of media that contains the contaminant. For example, radiological contaminants contained in surface soils in an open field may be spread through both the air and/or surface water pathways. The surface water pathway may allow the contaminant to migrate both horizontally and vertically within the soil column. If surface water transports the radiological contaminants deep enough into the soil to reach groundwater, the groundwater will become a third pathway for contaminant migration. On the other hand, in the case of radiological contaminants being present on building wall surfaces, the primary pathway for contaminant migration is through the air, unless there are leaks in the building roof.

The term *route of exposure* refers to the path by which receptors (e.g., human and/or ecological population) may be exposed to the contamination. The routes by which human or ecological receptors may be exposed to radiological contamination include:

- Inhalation
- Ingestion
- Dermal contact
- Direct exposure

Receptors may be exposed to radiologically contaminated dust and soil particles through the inhalation, ingestion, or dermal contact routes. For example, a receptor walking through a radiologically contaminated area may kick up dust that is then inhaled. Contaminated dust or soil may also get into the receptor's food, which is then ingested, or may come in direct contact with the receptor's skin. Similarly, receptors may be exposed to radiologically contaminated surface water or groundwater either through the ingestion or dermal contact routes. Direct exposure simply occurs when receptors are in close enough proximity to radiological contamination to receive a dose.

Section 4.1.1.5.1.6 provides a detailed description of how the primary and secondary sources of contamination, migration pathways, and exposure routes are used to develop a Conceptual Site Model.

1.4 DEFINITIONS OF COMMON RADIOLOGICAL TERMS

This section provides definitions of terms that are commonly used in the radiological industry. These definitions were derived from 10 CFR 20 and DOE Order 5400.5.

Airborne radioactive material: Radioactive material dispersed in the air in the form of dusts, fumes, particulates, mists, vapors, or gases.

Annual limit on intake (ALI): The derived limit for the amount of radioactive material taken into the body of an adult worker by inhalation or ingestion in a year. ALI is the smaller value of intake of a given radionuclide in a year by the reference "man" that would result in a committed effective dose equivalent of 5 rems (0.05 Sv) or a committed dose equivalent of 50 rems (0.5 Sv) to any individual organ or tissue. (ALI values for intake by ingestion and by inhalation of selected radionuclides are given in 10 CFR 20 Table 1, Columns 1 and 2, of Appendix B to §§20.1001 to 20.2401.)

As low as reasonably achievable (ALARA): Phrase (acronym) used to describe an approach to radiation protection to control or manage exposures (both individual and collective to the workforce and the general public) and releases of radioactive material to the environment as low as social, technical, economic, practical, and public policy considerations permit. As used in U.S. Department of Energy (DOE) Order 5400.5, ALARA is not a dose limit, but rather it is a process that has as its objective the attainment of dose levels as far below the applicable limits of the Order as practical.

Background radiation: Radiation from cosmic sources; naturally occurring radioactive materials, including radon (except as a decay product of source or special nuclear material) and global fallout as it exists in the environment from the testing of nuclear explosive devices. "Background radiation" does not include radiation from source, by-product, or special nuclear materials regulated by the commission.

Best available technology (BAT): Phrase (acronym) that refers to the preferred technology for treating a particular process liquid waste, selected from among others after taking into account factors related to technology, economics, public policy, and other parameters. BAT is not a specific level of treatment, but rather the conclusion of a selection process that includes several treatment alternatives.

Bioassay (radiobioassay): The determination of kinds, quantities, or concentrations, and, in some cases, the locations of radioactive material in the human body, whether by direct measurement (*in vivo* counting) or by analysis and evaluation of materials excreted or removed from the human body.

By-product: (1) Any radioactive material (except special nuclear material) yielded in, or made radioactive by, exposure to the radiation incident to the process of producing or utilizing special nuclear material. (2) The tailings or wastes produced by the extraction or concentration of uranium or thorium from ore processed primarily for its source material content, including discrete surface wastes resulting from uranium solution extraction processes. Underground ore bodies depleted by these solution extraction operations do not constitute "by-product material" within this definition.

Controlled area: An area, outside of a restricted area but inside the site boundary, access to which can be limited by the licensee for any reason.

Derived air concentration (DAC): The concentration of a given radionuclide in air which, if breathed by the reference man for a working year of 2000 h under conditions of light work (inhalation rate 1.2 m³ of air/h), results in an intake of one ALI. (DAC values are given in 10 CFR 20 Table 1, Column 3, of Appendix B to §§20.1001 to 20.2401.)

Derived concentration guide (DCG): Phrase (acronym) that refers to the concentration of a radionuclide in air or water that, under conditions of continuous exposure for 1 year by one exposure mode (i.e., ingestion of water, inhalation), would exceed the allowable dose equivalent (i.e., 100 mrem). DCGs do not consider decay products when the parent radionuclide is the cause for the exposure.

Dose terms:

1. **Absorbed dose:** The energy imparted to matter by ionizing radiation per unit mass of irradiated material at the place of interest in that material. The absorbed dose is expressed in units of rad.
2. **Collective dose:** The sum of the individual doses received in a given period of time by a specified population from exposure to a specified source of radiation.
3. **Committed dose equivalent ($H_{T,50}$):** The dose equivalent to organs or tissues of reference (T) that will be received from an intake of radioactive material by an individual during the 50-year period following the intake.
4. **Committed effective dose equivalent ($H_{E,50}$):** The sum of the products of the weighting factors applicable to each of the body organs or tissues that are irradiated and the committed dose equivalent to these organs or tissues ($H_{E,50} = \Sigma \, W_T \, H_{T,50}$).
5. **Deep dose equivalent:** The dose equivalent in tissue at a depth of 1 cm deriving from external (penetrating) radiation.
6. **Dose equivalent:** The product of absorbed dose in rad in tissue and a quality factor. Dose equivalents are expressed in units of rem.
7. **External dose:** That portion of the dose equivalent received from radiation sources outside the body.
8. **Eye dose equivalent:** The external exposure of the lens of the eye taken as the dose equivalent at a tissue depth of 0.3 cm (300 mg/cm²).
9. **Internal dose:** That portion of the dose equivalent received from radioactive material taken into the body.

10. **Occupational dose:** The dose received by an individual in the course of employment in which the individual's assigned duties involve exposure to radiation or to radioactive material from licensed and unlicensed sources of radiation, whether in the possession of the licensee or another person. Occupational dose does not include dose received from background radiation, from any medical administration the individual has received, from exposure to individuals administered radioactive material and released in accordance with 10 CFR 20 §35.75, from voluntary participation in medical research programs, or as a member of the public.

11. **Public dose:** The dose received by members of the public from exposure to radiation and to radioactive material released by a facility or operation. It does not include dose received from occupational exposure, doses received from naturally occurring "background" radiation, doses received from medical practices, or doses received from consumer products.

12. **Quality factor:** The principal modifying factor used to calculate the dose equivalent from the absorbed dose. 10 CFR 20.1004 specifies the quality factors listed in Table 1.1. If sufficient information exists to estimate the approximate energy distribution of the neutrons, the licensee may use the fluence rate per unit dose equivalent or the appropriate Q value from Table 1.2 to convert a measured tissue dose in rads to dose equivalent in rems.

13. **Shallow-dose equivalent (Hs):** The external exposure of the skin or an extremity is taken as the dose equivalent at a tissue depth of 0.007 cm (7 mg/cm^2) averaged over an area of 1 cm^2.

14. **Total effective dose equivalent (TEDE):** The sum of the deep-dose equivalent (for external exposures) and the committed effective dose equivalent (for internal exposures).

15. **Weighting factor (W_T):** For an organ or tissue (T), the proportion of the risk of stochastic effects resulting from irradiation of that organ or tissue to the total risk of stochastic effects when the whole body is irradiated uniformly. For calculating the effective dose equivalent, the values of W_T are listed in Table 1.3.

Environmental surveillance: The collection and analysis of samples of air, water, soil, foodstuffs, biota, and other media and the measurement of external radiation for the purpose of demonstrating compliance with applicable standards, assessing radiation exposures to members of the public, and assessing any impacts to the local environment.

Exposure: Condition of being exposed to ionizing radiation or to radioactive material.

Gray (Gy): The SI unit of absorbed dose. One gray is equal to an absorbed dose of 1 J/kg (100 rad).

High radiation area: An area, accessible to individuals, in which radiation levels could result in an individual receiving a dose equivalent in excess of 0.1 rem (1 mSv) in 1 h at 30 cm from the radiation source or from any surface that the radiation penetrates.

Members of the public: Persons who are not occupationally associated with the facility or operation responsible for the radiation (i.e., persons whose assigned occupational duties do not require them to enter radioactive environments).

Monitoring (radiation monitoring, radiation protection monitoring): The measurement of radiation levels, concentrations, surface area concentrations, or quantities of radioactive material and the use of the results of these measurements to evaluate potential exposures and doses.

Nonstochastic effect: Health effects, the severity of which varies with the dose and for which a threshold is believed to exist. Radiation-induced cataract formation is an example of a nonstochastic effect (also called a deterministic effect).

Rad: The special unit of absorbed dose. One rad is equal to an absorbed dose of 100 erg/g or 0.01 J/kg (0.01 Gy).

Radiation (ionizing radiation): Alpha particles, beta particles, gamma rays, X rays, neutrons, high-speed electrons, high-speed protons, and other particles capable of producing ions. Radiation, as used in this part, does not include nonionizing radiation, such as radio- or microwaves, or visible, infrared, or ultraviolet light.

Radiation area: An area, accessible to individuals, in which radiation levels could result in an individual receiving a dose equivalent in excess of 0.005 rem (0.05 mSv) in 1 h at 30 cm from the radiation source or from any surface that the radiation penetrates.

Radioactivity: The property or characteristic of radioactive material to "disintegrate" spontaneously with the emission of energy in the form of radiation. The unit of radioactivity is the curie.

Rem: The special unit of any of the quantities expressed as dose equivalent. The dose equivalent in rems is equal to the absorbed dose in rads multiplied by the quality factor (1 rem = 0.01 Sv).

Restricted area: An area, access to which is limited by the licensee for the purpose of protecting individuals against undue risks from exposure to radiation and radioactive materials. Restricted area does not include areas used as residential quarters, but separate rooms in a residential building may be set apart as restricted areas.

Sievert (Sv): The SI unit of any of the quantities expressed as dose equivalent. The dose equivalent in sieverts is equal to the absorbed dose in grays multiplied by the quality factor (1 Sv = 100 rems).

Source material: (1) Uranium or thorium or any combination of uranium and thorium in any physical or chemical form; or (2) ores that contain, by weight, $1/20$ of 1% (0.05 percent), or more, of uranium, thorium, or any combination of uranium and thorium. Source material does not include special nuclear material.

Special nuclear material: (1) Plutonium, uranium-233, uranium enriched in the isotope 233 or in the isotope 235, and any other material that the Commission, pursuant to the provisions of Section 51 of the Act, determines to be special nuclear material, but does not include source material; or (2) any material artificially enriched by any of the foregoing, but does not include source material.

Stochastic effects: Health effects that occur randomly and for which the probability of the effect occurring, rather than its severity, is assumed to be a linear function of dose without threshold. Hereditary effects and cancer incidence are examples of stochastic effects.

Survey: An evaluation of the radiological conditions and potential hazards incident to the production, use, transfer, release, disposal, or presence of radioactive material or other sources of radiation. When appropriate, such an evaluation includes a physical survey of the location of radioactive material and measurements or calculations of levels of radiation, or concentrations or quantities of radioactive material present.

Unrestricted area: An area, access to which is neither limited nor controlled by the licensee.

Uranium fuel cycle: The operations of milling of uranium ore, chemical conversion of uranium, isotopic enrichment of uranium, fabrication of uranium fuel, generation of electricity by a light-water-cooled nuclear power plant using uranium fuel, and reprocessing of spent uranium fuel to the extent that these activities directly support the production of electrical power for public use. Uranium fuel cycle does not include mining operations, operations at waste disposal sites, transportation of radioactive material in support of these operations, and the reuse of recovered nonuranium special nuclear and by-product materials from the cycle.

Very high radiation area: An area, accessible to individuals, in which radiation levels could result in an individual receiving an absorbed dose in excess of 500 rads (5 Gy) in 1 h at 1 m from a radiation source or from any surface that the radiation penetrates.

Whole body: For purposes of external exposure, the head, trunk (including male gonads), arms above the elbow, or legs above the knee.

Working level (WL): Any combination of short-lived radon daughters (for radon-222: polonium-218, lead-214, bismuth-214, and polonium-214; and for radon-220: polonium-216, lead-212, bismuth-212, and polonium-212) in 1 L of air that will result in the ultimate emission of 1.3×10^5 MeV of potential alpha particle energy.

Working level month (WLM): An exposure to 1 working level for 170 h (2000 working hours per year/12 months per year = approximately 170 hours per month).

Table 1.1 Quality Factors for Various Types of Radiation

Radiation Type	Quality Factor	Absorbed Dose[a]
X-rays, gamma, or beta radiation	1	1
Alpha particles, multiple-charged particles, fission fragments and heavy particles of unknown charge	20	0.05
Neutrons of unknown energy	10	0.1
High-energy protons	10	0.1

[a] Equal to a unit dose equivalent; absorbed dose in rad equal to 1 rem.

Table 1.2 Mean Quality Factors (Q) and Fluence per Unit Dose Equivalent for Monoenergetic Neutrons

Neutron Energy (MeV)	Quality Factor[a] (Q)	Fluence per Unit Dose Equivalent[b] (neutrons cm^{-2} rem^{-1})
(Thermal)2.5×10^{-8}	2	980×10^6
1×10^{-7}	2	980×10^6
1×10^{-6}	2	810×10^6
1×10^{-5}	2	810×10^6
1×10^{-4}	2	840×10^6
1×10^{-3}	2	980×10^6
1×10^{-2}	2.5	1010×10^6
1×10^{-1}	7.5	170×10^6
5×10^{-1}	11	39×10^6
1	11	27×10^6
2.5	9	29×10^6
5	8	23×10^6
7	7	24×10^6
10	6.5	24×10^6
14	7.5	17×10^6
20	8	16×10^6
40	7	14×10^6
60	5.5	16×10^6
1×10^2	4	20×10^6
2×10^2	3.5	19×10^6
3×10^2	3.5	16×10^6
4×10^2	3.5	14×10^6

[a] Value of quality factor (Q) at the point where the dose equivalent is maximum in a 30-cm-diameter cylinder tissue-equivalent phantom.

[b] Monoenergetic neutrons incident normally on a 30-cm-diameter cylinder tissue-equivalent phantom.

Table 1.3 Weighting Factors (W_T) for Various Organs or Tissues

Organ or Tissue	Weighting Factor
Gonads	0.25
Breasts	0.15
Red bone marrow	0.12
Lungs	0.12
Thyroid	0.03
Bone surfaces	0.03
Remainder[a]	0.30
Whole body	1.00

[a] Remainder refers to the five other organs with the highest dose (e.g., liver, kidney, spleen, thymus, adrenal, pancreas, stomach, small intestine, or upper and lower large intestine, but excluding skin, lens of the eye, and extremities). The weighting factors for each of these organs is 0.06.

REFERENCES

10 CFR 20, *Code of Federal Regulations* (as amended).
U.S. Department of Energy, Order 5400.5.

CHAPTER 2

Environmental Laws and Regulations

The purpose of this chapter is to provide the reader with general guidance on the types of issues addressed under the major environmental laws and federal regulations, and to provide Internet addresses for state regulatory agencies from which their requirements can be obtained. This information is not intended to be cited as law, but rather is provided to give the reader general information about the issues addressed by each of the cited laws and regulations.

An environmental law is first introduced as a bill to the U.S. House of Representatives or the U.S. Senate. The bill is then passed on to a committee where it undergoes a detailed evaluation. As part of this evaluation, it is not uncommon for committee hearings to be held where expert witnesses are called to testify on the key technical aspects of the bill. If the bill is passed, it becomes an act that is then sent to the President of the United States to sign into law. Once an environmental law is passed, administrative agencies (e.g., U.S. Environmental Protection Agency) develop and promulgate regulations that are then enforced at the federal level. Individual states often choose to promulgate their own regulations, which are required either to meet or to exceed the federal standards.

2.1 ENVIRONMENTAL LAWS

This section discusses the major environmental laws that pertain to performing environmental studies at radiological and chemical sites. These laws include the following:

- Comprehensive Environmental Response, Compensation, and Liability Act (CERCLA)
- Superfund Amendments and Reauthorization Act (SARA)
- Resource Conservation and Recovery Act (RCRA)
- Toxic Substance Control Act (TSCA)
- National Environmental Policy Act (NEPA)
- Clean Water Act (CWA)
- Safe Drinking Water Act (SDWA)
- Clean Air Act (CAA)

Although each of these laws is different in terms of the types of materials or activities that it regulates, together they are our nation's means of controlling the quality of our environment (Figure 2.1).

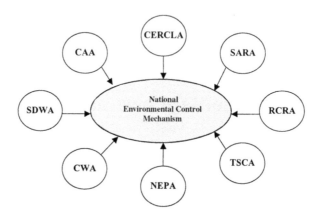

Figure 2.1 Major environmental laws protecting environment quality.

2.1.1 CERCLA Compliance

In 1980, Congress passed CERCLA (42 USC §§9601–9675, 40 CFR 300, 302), which regulates the cleanup of abandoned or closed waste sites across the country, in addition to providing requirements and guidance for response to unpermitted and uncontrolled releases of hazardous substances into the environment. CERCLA utilizes the Remedial Investigation & Feasibility Study (RI/FS) process to evaluate the environmental conditions at a site, and to select the final remedial alternative that will be implemented to remediate the site (EPA, 1988; Figure 2.2).

The RI/FS process consists of scoping, site characterization, development and screening of remedial alternatives, treatability investigations, and detailed evaluation of remedial alternatives. Scoping involves the detailed review of all pertinent historical documents, maps, photographs, and analytical data sets. Following the scoping process, the EPA's data quality objectives (DQO) process is used to support the development of a Sampling and Analysis Plan, which is composed of a Quality Assurance Project Plan and a Field Sampling Plan.

Site characterization involves conducting field investigations, performing laboratory analyses on environmental samples, and evaluating the results for the purpose of identifying the types, concentrations, sources, and extent of contamination at a site. These data are then used in the development of a Baseline Risk Assessment, and to evaluate potential remedial alternatives. Site characterization studies commonly use tools such as radiological scanning instruments, direct radiological measurements, surface and downhole geophysics, etc., in combination with the sampling of environmental media.

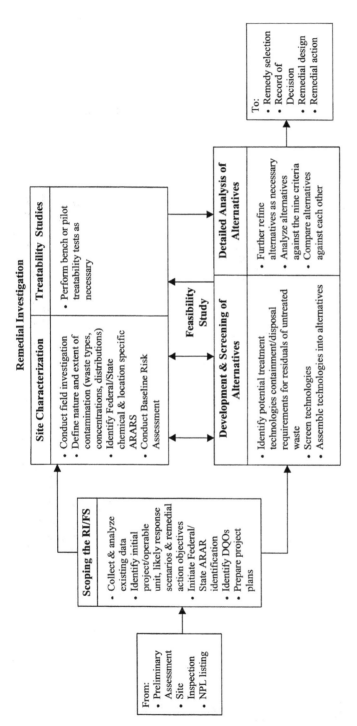

Figure 2.2 CERCLA RI/FS Process

Preliminary remedial alternatives should be developed when likely response scenarios are first identified. The development of alternatives requires the following:

- Identification of remedial action objectives;
- Identification of potential treatment, resource recovery, containment, and disposal technologies that will satisfy these objectives;
- Screening of technologies based on their effectiveness, implementability, and cost;
- Assembling of technologies and their associated containment or disposal requirements into alternatives for each contaminated media.

A range of treatment alternatives should be considered. The range should vary primarily in the extent to which the alternatives rely on long-term management. The upper bound of the range should be an alternative that would eliminate the need for any long-term management, while the lower bound should consist of an alternative that involves little or no treatment, such as no-action, or long-term monitoring.

Treatability studies provide sufficient data to develop and evaluate treatment alternatives to support the remedial design of the selected alternatives. They also reduce cost and performance uncertainties to help select a remedy.

If no treatability data are available for a site, treatability tests may be necessary to evaluate the effectiveness of a particular technology for that site. Treatability tests typically involve bench-scale testing, followed by pilot-scale testing if a technology appears to be feasible.

Once sufficient data are available, alternatives are evaluated in detail against the following nine evaluation criteria:

1. Overall protectiveness to human health and the environment;
2. Compliance with ARARs;
3. Long-term effectiveness;
4. Reduction of toxicity, mobility, or volume;
5. Short-term effectiveness;
6. Implementability;
7. Cost;
8. State acceptance;
9. Community acceptance.

Each alternative is analyzed against each of the above criteria and then compared against one another to determine the respective strengths and weaknesses.

2.1.2 SARA Compliance

In 1986, CERCLA (42 USC §§9601–9675, 40 CFR 300) was modified by the SARA. SARA did not change the basic structure of CERCLA, but rather modified many of the existing requirements, and added a few new ones.

SARA places a strong statutory preference on remedial alternatives that are highly reliable and provide long-term protectiveness. A preference is placed on alternatives that employ treatment that permanently and significantly reduces the volume, toxicity, or mobility of contaminants. Off-site transport and disposal of contaminated materials is not a favored alternative when practicable treatment or resource recovery technologies are available. After the initiation of an action, SARA requires a review of the remediation's effectiveness at least every 5 years for as long as contaminants may pose a threat to human health or the environment (ELR, 1990).

Before the passage of SARA, no laws specifically protected the health and safety of hazardous waste workers; consequently, workers were at the mercy of their employers' safety policies. SARA addresses the risk of exposure to hazardous wastes and the need to protect employees exposed to these materials under 29 CFR 1910.120. Workers who are covered by this regulation include those involved with cleanup, treatment, storage, disposal, and emergency response.

To ensure the protection of workers in hazardous environments, a health and safety program must be implemented. As part of this program, a health and safety plan must be written to identify, evaluate, and control health and safety hazards and provide for emergency response. The responsibility for developing this program and writing this plan rests on the employer of the hazardous waste workers.

As part of a health and safety program, medical surveillance is mandatory for all workers who (ELR, 1990):

- Are exposed to hazardous substances or health hazards at or above the permissible exposure limits;
- Wear a respirator for 30 days or more a year;
- Are injured, become ill, or develop signs or symptoms due to possible overexposure involving hazardous substances from an emergency response or hazardous waste operation;
- Belong to a hazardous material team.

SARA requires employers to establish a program to inform all personnel involved in a hazardous waste operation of the nature, level, and degree of exposure that can be anticipated as a result of their participation. Other key issues that SARA addresses include: the handling of drums and containers; procedures for decontamination; emergency response; illumination; and sanitation.

2.1.3 RCRA Compliance

In 1976, Congress passed the RCRA (40 CFR 260–280), which regulates solid and hazardous waste, prevents new uncontrolled hazardous sites from developing, and protects human health and the environment. The RCRA requires the "cradle-to-grave" management of hazardous waste.

Under RCRA, the term *waste* refers to any discarded material which is abandoned, disposed of, burned or incinerated, or stored in lieu of being abandoned. *Solid waste* materials include:

- Garbage;
- Refuse;
- Sludge from a waste treatment plant, water supply treatment plant, or air pollution control facility;
- Other discarded material.

Other discarded material may include solid, liquid, semisolid, and contained gaseous material resulting from industrial, commercial, mining, agricultural operations, and community activities. This material does not include solid or dissolved material in domestic sewage, solid or dissolved materials in irrigation return flows, industrial discharges, or special nuclear by-product material (ELR, 1990).

Once a material has been determined to be a "solid waste," it must next be determined whether or not it is a "hazardous waste." A solid waste may become a hazardous waste when, because of quantity, concentration, or physical, chemical, or infectious characteristics, it:

- Causes or contributes to an increase in mortality or increase in serious irreversible, or incapacitating reversible illness,
- Poses a substantial present or potential hazard to human health or environment if improperly treated, stored, transported, or disposed of.

Certain solid wastes are exempted from being considered hazardous wastes. In general, these materials include:

- Household, farming, mining, and fly ash wastes;
- Drilling fluids used in oil and gas exploration;
- Waste failing the Toxicity Characteristic Leaching Procedure (TCLP) for chromium if several tests were run and only one sample failed;
- Solid waste from processing ores;
- Cement kiln dust;
- Discarded wood products failing TCLP for arsenic;
- Petroleum-contaminated media failing TCLP if remediated under underground storage tank (UST) rules.

If a waste does not qualify for an exemption, it will be considered hazardous if it is either listed by the EPA in 40 CFR Part 261, Subpart D or if it exhibits any of the four characteristics found in 40 CFR Part 261, Subpart C.

There are three hazardous waste lists. The first list contains hazardous wastes from nonspecific sources such as various spent solvents. The second list contains hazardous waste from specific sources such as bottom sediment from a wood-preserving facility. The third list contains commercial chemical products that are deemed acute hazardous wastes or toxic (and therefore hazardous) when discarded.

If a waste is not listed as hazardous, it may still be considered hazardous by RCRA if it displays one of the following characteristics:

- Ignitable: Flash point of 140°F or lower.
- Corrosive: pH of 2 or lower (acid); 12.5 or higher (base).
- Reactive: Unstable, capable of detonation, explosive.
- Toxic: As determined by TCLP.

Table 2.1 presents the regulatory levels for each of the various contaminants tested under TCLP. If the TCLP results for a soil exceed any of these levels, the soil is defined as an RCRA Hazardous Waste.

Table 2.1 Maximum Concentration of Contaminants for the Toxicity Characteristic

EPA HW No.[a]	Contaminant	CAS No.[b]	Regulatory Level (mg/L)
D004	Arsenic	7440-38-2	5.0
D005	Barium	7440-39-3	100.0
D018	Benzene	71-43-2	0.5
D006	Cadmium	7440-43-9	1.0
D019	Carbon tetrachloride	56-23-5	0.5
D020	Chlordane	57-74-9	0.03
D021	Chlorobenzene	108-90-7	100.0
D022	Chloroform	67-66-3	6.0
D007	Chromium	7440-47-3	5.0
D023	o-Cresol	95-48-7	200.0[c]
D024	m-Cresol	108-39-4	200.0[c]
D025	p-Cresol	106-44-5	200.0[c]
D026	Cresol	—	200.0[c]
D016	2,4-D	94-75-7	10.0
D027	1,4-Dichlorobenzene	106-46-7	7.5
D028	1,2-Dichloroethane	107-06-2	0.5
D029	1,1-Dichloroethylene	75-35-4	0.7
D030	2,4-Dinitrotoluene	121-14-2	0.13[d]
D012	Endrin	72-20-8	0.02
D031	Heptachlor (and its epoxide)	76-44-8	0.008
D032	Hexachlorobenzene	118-74-1	0.13[d]
D033	Hexachlorobutadiene	87-68-3	0.5
D034	Hexachloroethane	67-72-1	3.0
D008	Lead	7439-92-1	5.0
D013	Lindane	58-89-9	0.4
D009	Mercury	7439-97-6	0.2
D014	Methoxychlor	72-43-5	10.0
D035	Methyl ethyl ketone	78-93-3	200.0
D036	Nitrobenzene	98-95-3	2.0
D037	Pentachlorophenol	87-86-5	100.0
D038	Pyridine	110-86-1	5.0[d]
D010	Selenium	7782-49-2	1.0
D011	Silver	7440-22-4	5.0
D039	Tetrachloroethylene	127-18-4	0.7
D015	Toxaphene	8001-35-2	0.5
D040	Trichloroethylene	79-01-6	0.5
D041	2,4,5-Trichlorophenol	95-95-4	400.0
D042	2,4,6-Trichlorophenol	88-06-2	2.0
D017	2,4,5-TP (Silvex)	93-72-1	1.0
D043	Vinyl chloride	75-01-4	0.2

[a] Hazardous waste number.

[b] Chemical abstracts service number.

[c] If o-, m-, and p-cresol concentrations cannot be differentiated, the total cresol (D026) concentration is used. The regulatory level of total cresol is 200 mg/L.

[d] Quantitation limit is greater than the calculated regulatory level. The quantitation limit therefore becomes the regulatory level.

Source: 40 CFR Part 261.24, Table 1.

RCRA also regulates medical waste, tanks, tank systems, surface impoundments, waste piles, land treatment facilities, landfills, incinerators, generators, and other types of miscellaneous units such as boilers, industrial furnaces, and underground injection wells.

Some of the responsibilities of waste generators under RCRA include knowing what types of waste materials are being generated at a particular site through either knowledge or testing. The generator must obtain an EPA identification number after notifying it of the type and status of the waste being generated. All hazardous waste must be consolidated at a permitted hazardous waste storage unit, or a satellite accumulation point, where this waste can be temporarily stored for 90 days before transport to a treatment storage and disposal facility.

Prior to transporting the hazardous waste material, the EPA must be notified of the proposed activity, proper manifesting requirements must be met, and emergency response procedures must be in place. A copy of the manifest must be kept on file to document that waste was properly disposed of at a permitted facility. Any transportation of hazardous waste must comply with all Department of Transportation regulations outlined in 49 CFR 171 through 180.

All treatment, storage, and disposal (TSD) facilities must use Part A permit applications and must comply with general facility standards, interim-status technical standards, closure/postclosure standards, and notification requirements. The Part A permit application identifies the type of hazardous waste managed, estimates the annual quantities of material managed, provides details on methods of waste management, and includes a facility map.

Each TSD facility is required to develop and implement a waste analysis plan to test incoming hazardous waste, which serves to ensure that the waste material received matches the manifest. It must also provide appropriate security measures, conduct regular facility inspections, implement a groundwater-monitoring program, provide appropriate personnel training, and keep operating records. The operating records must include information such as a description and the quantity of waste received, results of waste analysis testing, inspection findings, summary reports of incidents, closure and postclosure cost estimates, land disposal restriction certifications, and notifications.

RCRA requires that closure of a hazardous waste unit must begin within 90 days of the last receipt of hazardous waste, or upon approval of the closure plan, whichever is later. Postclosure care of a unit is required if closure in place is used (ELR, 1990).

2.1.4 TSCA Compliance

The TSCA (15 USC §§2601–2671, 40 CFR 761) requires all manufacturers, processors, and distributors to maintain records of the hazards that each of their products pose to human health and the environment. It also requires the EPA to compile and publish a list of each chemical substance manufactured or processed in the United States. The statute authorizes the EPA to conduct limited inspections of areas where substances are processed or stored, and of conveyances used to transport the substances.

TSCA requires all manufacturers and processors of new substances, or substances that will be applied to a significant new use, to notify the administrator of the EPA that they intend to manufacture or process the substance. If analytical testing is required, the manufacturer or processor must provide the results along with the notification.

If the EPA finds the analytical testing to be insufficient, a proposed order is written to restrict the manufacturing of the substance until adequate testing is completed. If the testing data indicate that the substance may present a significant risk of cancer, gene mutations, or birth defects, the EPA will promulgate regulations concerning the distribution, handling, and labeling of the substance. In the case of an imminently hazardous substance, the EPA may commence a civil action for seizure of the substance, and possibly a recall and repurchase of the substance previously sold (ELR, 1990).

The requirements of this statute generally do not apply to toxic substances distributed for export unless they would cause an unreasonable risk of harm within the United States. On the other hand, imported substances are subject to the requirements of the statute, and any substances that do not comply will be refused entry into the United States. Violations of this statute can result in both civil and criminal penalties, and the violating substance may be seized.

Some important regulations under TSCA govern the manufacture, use, and disposal of polychlorinated biphenyls (PCBs). PCBs are found in many substances, such as oils, paints, and contaminated solvents. The regulations establish concentration limits and define acceptable methods of disposal. PCBs may now be used only in totally enclosed systems.

TSCA also requires that asbestos inspections be performed in school buildings to define the appropriate level of response actions. The statute also requires the implementation of maintenance and repair programs, and periodic surveillance of school buildings where asbestos is located, as well as prescribing standards for the transportation and disposal of this material. For those school buildings containing asbestos, local educational agencies are required to develop an asbestos management program, which must include plans for response actions, long-term surveillance, and use of warning labels for asbestos remaining in the buildings.

In an attempt to control radon contamination inside buildings, the EPA is required by this statute to publish a document titled "A Citizen's Guide to Radon," which includes information on the health risks associated with exposure to radon, the cost and technical feasibility of reducing radon concentrations, the relationship between long-term and short-term testing techniques, and outdoor radon levels around the country. This statute also requires the EPA to determine the extent of radon contamination in the nation's schools, develop model construction standards and techniques for controlling radon levels within new buildings, and make grants available to states to assist them in the development and implementation of their radon programs (ELR, 1990).

2.1.5 NEPA Compliance

NEPA (42 USC §§4321–4370a, 40 CFR 1500–1508) was passed in 1969, and was one of the first statutes directed specifically at protecting the environment. NEPA documentation is necessary when any "major Federal action" that may have a significant impact on the environment may be undertaken. The NEPA process places heavy emphasis on public involvement. Public notice must be provided for NEPA-related hearings, public meetings, and to announce the availability of environmental documents. In the case of a NEPA action of national concern, notice is included in the *Federal Register* and notice is made by mail to national organizations reasonably expected to be interested in the matter.

The primary documents prepared under the NEPA process are the Notice of Intent (NOI), Environmental Impact Statement (EIS), Environmental Assessment (EA), Finding of No Significant Impact (FONSI), and Categorical Exclusion (CX). Any environmental document in compliance with NEPA may be combined or integrated with any other agency document to reduce duplication and paperwork.

Before preparing an EIS, an NOI must be issued for public review. The NOI describes the proposed action and possible alternatives, describes the federal agency's proposed scoping process including whether, when, and where any public scoping meetings will be held, and, finally, states the name and address of a person within the agency who can answer questions about the proposed action.

The EIS serves as an action-forcing device to ensure that the policies and goals defined in NEPA are infused into the ongoing programs and actions of the federal government. The objective of the EIS is to provide a full and fair discussion of significant environmental impacts, and is used to inform decision makers and the public of the reasonable alternatives, which would avoid or minimize adverse impacts or enhance the quality of the human environment. The EIS is meant to serve as the means of assessing the environmental impact of proposed federal agency actions, but is not used to justify decisions that have already been made.

The EIS and other NEPA documents should be written so the public can readily understand them. Wherever there is incomplete or unavailable information, it is critical to state this overtly in the document. No decision on the proposed action shall be made or recorded under a federal agency until the later of the following dates: 90 days after publication of the notice for a draft EIS or 30 days after publication of the notice for a final EIS.

An EA is a concise public document that determines whether or not to prepare an EIS. If there are no significant impacts on the environment, a FONSI is published. An EA can facilitate preparation of an EIS when one is needed, but is not necessary if it is already known that there will be significant impacts, and an EIS must be prepared.

A FONSI is a document prepared by a federal agency to describe briefly the reasons an action will not have a significant effect on the human environment, and for which an EIS is not needed. This document includes the EA or a summary of this study, and notes any other environmental documents related to it. If the EA is included, the finding need not repeat any of the discussion in the assessment, but may incorporate it by reference.

A CX refers to a category of actions that do not individually or cumulatively have a significant effect on the human environment and which have been found to have no such effect in procedures adopted by a federal agency in implementation of these regulations. Consequently, there is no need for the preparation of an EA or an EIS.

Where emergency circumstances make it necessary to take an action with significant environmental impact without observing the provisions of these regulations, the federal agency taking the action should consult with the council about alternative arrangements. Agencies and the council will limit such arrangements to actions necessary to control the immediate impacts of the emergency. Other actions remain subject to NEPA review (ONPA, 1988).

2.1.6 CWA Compliance

The objective of the CWA (33 USC §§1251–1387, 40 CFR 122–131) is to restore and maintain the chemical, physical, and biological integrity of the nation's waters. This act established interim water quality goals, which provide for the protection of human health and the environment, and assure the propagation of fish, shellfish, and wildlife until the pollutants can be completely eliminated. The act seeks to eliminate the discharge of pollutants into navigable waters, in addition to addressing research and related programs, grants for the construction of treatment works, standards and how they are enforced, and permits.

One of the primary objectives in establishing the CWA was to develop national programs for the prevention, reduction, and elimination of pollution. To achieve this objective, research, investigations, and experiments are required to identify ways of preventing future contamination problems. Studies are required to identify the types and distribution of contaminants at waste sites, with the intent of their future reduction and/or elimination. Other studies are striving to better understand the causes and effects that contaminants have on human health and the environment.

This statute requires the federal government to work with individual states and other interested agencies, organizations, and persons to investigate pollution sources. Research fellowships at public or nonprofit private educational institutions or research organizations are required to be made available to conduct research on the harmful effects on the health and welfare of persons who are exposed to pollutants in water. Grants must also be made available to any state agency or interstate agency to assist projects that focus on developing new or improved methods of waste treatment or water purification. Water bodies that are specifically addressed in the CWA include the Hudson River, Chesapeake Bay, the Great Lakes, Long Island Sound, Lake Champlain, and Onondaga Lake (ELR, 1990).

Grants from the federal government are provided for the construction of treatment works to assist the development and implementation of waste treatment management plans and practices that help to achieve the goals of the CWA. These plans and practices are expected to use the best practicable waste treatment technologies available. To the extent practicable, waste treatment management is to be maintained on an area-wide basis, to provide control or treatment of all point and nonpoint sources of pollution. Preference is always placed on waste treatment management that results in the construction of revenue-producing facilities providing for the following:

- Recycling of potential sewage pollutants through the production of agriculture, silviculture, or aquaculture products;
- Confined and contained disposal of pollutants not recycled;
- Reclamation of wastewater;
- Ultimate disposal of sludge in a manner that will not result in environmental hazards.

The statute requires that preference be placed on waste treatment management which results in integrating facilities for sewage treatment and recycling with facilities to treat, dispose of, or utilize other industrial and municipal wastes such as solid waste, waste heat, and thermal discharges.

The statute seeks to control pollutants being discharged into water by two permit programs. The first is the National Pollutant Discharge Elimination System. Either the EPA or a state that has established its own program issues permits for the discharge of any pollutant or combination of pollutants on condition that the discharge will meet all applicable standards or requirements. The U.S. Army Corps of Engineers issues another type of permit under the CWA for discharging dredged or fill material into navigable waters.

The CWA also defines standards for measuring pollution in water. Unless a source discharges into a publicly owned treatment works, point-source effluents must achieve the best practicable control technology and must satisfy pretreatment requirements for toxic substances. In addition, water quality standards apply where the technology-based limitations fail to attain a level protective of the public health. The EPA also periodically issues new source performance standards, which must take cost factors and environmental impacts into consideration (ELR, 1990).

2.1.7   SDWA Compliance

The purpose of the SDWA (42 USC §§300f–300j-26, 40 CFR 141–149) is to protect supplies of public drinking water from contamination. Under this act, the EPA is required to publish Maximum Contaminant Level Goals (MCLGs) and to promulgate National Primary Drinking Water Regulations (NPDWRs) for certain contaminants in the public water system. Public water systems are either community water systems (those which have at least 15 connections or regularly serve at least 25 people) or noncommunity water systems (all others). The MCLG is set at the level at which there are no known or anticipated adverse human health effects, in addition to allowing for a margin of safety.

Since the MCLGs are in many cases not attainable, the NPDWRs are required to specify maximum contaminant levels (MCLs) that are as close as feasible to the MCLG using the best available technology. The EPA may promulgate an NPDWR that requires the use of a treatment technique in lieu of setting an MCL if it finds that it is not economically or technologically feasible to ascertain the level of the contaminant.

A state has the primary enforcement responsibility for public water systems if the EPA administrator determines that the state has

- Adopted drinking water regulations that are no less stringent than the NPDWRs;
- Adequate enforcement and record-keeping mechanisms;
- Provisions for variances and exemptions that conform to the requirements of the statute;
- An adequate emergency drinking water plan.

The statute authorizes courts to issue injunctions and assess civil penalties against violators unless they have an approved variance. Variances may be granted to public water systems if they have implemented the best available technology or treatment technique but cannot meet the applicable MCL requirements because of characteristics of reasonably available raw water sources (ELR, 1990).

Since June 19, 1986, this statute has required that any pipe, solder, or flux used in the installation or repair of any public water system or any plumbing providing water for human consumption that is connected to a public water system is required to be free of lead. In addition, each state is required to assist local educational agencies in testing for and remedying lead contamination in school drinking water.

The statute protects underground sources of drinking water by preventing underground injection that may endanger a drinking water source. To protect underground water sources further, states are required to establish programs to protect wellhead areas from contamination. Finally, the statute regulates drinking watercoolers with lead-lined tanks (ELR, 1990).

2.1.8 CAA Compliance

The primary objective of the CAA (42 USC §§7601–7671q, 40 CFR 50–96) is to protect the quality of our nation's air by:

- Requiring air quality monitoring stations in major urban areas and other appropriate areas throughout the United States;
- Providing daily analysis and reporting of air quality data using a uniform air quality index;
- Providing for record keeping of all monitoring data for analysis and reporting.

The major issues that the statute addresses are noise pollution, acid deposition control, stratospheric ozone protection, and permitting requirements. The EPA has established an Office of Noise Abatement and Control, which studies noise and its effect on public health and welfare. The objectives of this group are to:

- Identify and classify causes and sources of noise;
- Project future growth in noise levels in urban areas;
- Study the psychological and physiological effects of noise on humans;
- Determine the effect noise has on wildlife and property.

The CAA is working to control acid deposition in the 48 contiguous states and the District of Columbia through reducing the allowable emissions of sulfur dioxide and nitrogen oxides. These reductions are being made in a staged approach, where prescribed emission limitations must be met by specific deadlines. The statute encourages energy conservation, use of renewable and clean alternative technologies, and pollution prevention as a long-term strategy (ELR, 1990).

In an effort to protect Earth's stratospheric ozone layer, all known ozone-depleting substances have been classified to control their use. Class I substances have been determined to have an ozone depletion potential of 0.2 or more, and currently include various forms of chlorofluorocarbons, halons, in addition to carbon tetrachloride and methyl chloroform. Class II substances include various forms of hydrochlorofluorocarbons and have an ozone depletion potential of less than 0.2. As more information becomes available about the causes of ozone depletion, additional substances will be added to these lists. The only time a substance can be removed from the Class II list is if it is moved to the Class I category. No substances on the Class I list can be removed.

The statute has established a permitting program for the purpose of controlling the releases of damaging chemicals into the air. Permits contain information such as enforceable emission limitations and standards, schedules for compliance, reporting requirements to permitting authorities, and any other information that is necessary to assure compliance with applicable requirements of this act (ELR, 1990).

2.2 FEDERAL REGULATIONS

In general, environmental laws give an administrative agency (e.g., the EPA) the authority to develop and promulgate regulations. Proposed regulations developed by an administrative agency are first published in the *Federal Register*, where the public is provided the opportunity to provide comments. Public comments may be received either through public hearings or by submission of written comments. Once these comments have been addressed, the regulations are published in the Code of Federal Regulations.

The Code of Federal Regulations is a codification of the rules published in the *Federal Register* by the executive departments and agencies of the federal government. The Code of Federal Regulations is divided into 50 titles, which represent broad areas subject to federal regulation. Title 40 contains most of the federal regulations administered by the EPA that set standards for the protection of the environment. Other titles that may apply to certain aspects of an environmental investigation include Titles 10, 29, and 49. Title 10 identifies all of the regulations that apply to the Nuclear Regulatory Commission and the U.S. Department of Energy. Occupational Safety and Health regulations are addressed under Title 29, and regulations pertaining to the transportation of hazardous and/or radioactive materials are addressed under Title 49. Table 2.2 provides a summary of the regulations that may pertain to environmental investigations performed in radiological environments. Table 2.3 presents the surface contamination levels identified in 10 CFR 835, Appendix D.

The EPA has established ten regions across the United States to regulate federal environmental protection policy. The states that fall into each of these regions are presented in Table 2.4. Each region can be accessed through the EPA Web site at www.epa.gov.

The full text of the Code of Federal Regulations is available from the Government Printing Office, attn: New Orders, P.O. Box 371954, Pittsburgh, PA 15250-7954. Electronic copies of the code are available from Electronic Information Dissemination Services, U.S. Government Printing Office, at the following Internet address: www.access.gpo.gov/nara/.

Figure 2.3 shows the major environmental laws and the federal regulations that contain the provisions for implementation and codified requirements.

2.3 STATE REGULATIONS

Many of the federal statutes contain provisions allowing individual states to enact and enforce their own environmental protection programs as long as they meet or exceed federal standards. Since states are given some leeway in interpreting federal enforcement criteria, environmental laws vary to some degree from state to state. Specific information pertaining to the environmental laws of each state may be accessed through the Internet addresses shown in Table 2.5.

2.4 OTHER REGULATIONS

The Atomic Energy Act of 1954 (AEA) approved the formation of the Atomic Energy Commission (AEC). The Act was promoted by the Eisenhower Administration as the "peaceful use of the atom." Among its provisions were the roots of what later became the nuclear power industry.

In 1974, the AEC was reorganized primarily to end the self-regulation of a government agency composed of a relatively small group of people making decisions vital to the future of our nation. The Nuclear Regulatory Commission (NRC) took over regulatory and oversight duties and the Energy Research Development Administration (ERDA) took over energy research and production. ERDA was designed to support nuclear weapons production as well as the research and development of peaceful uses of nuclear and other energy sources. ERDA later became the Department of Energy (DOE), which focused on the production of nuclear-grade weapon materials for the Department of Defense and served as the promotional arm of the nuclear power industry.

The NRC and DOE took two separate pathways in promulgating the requirements of major environmental laws. The NRC followed the guidelines contained in the Administrative Procedures Act. Proposed regulations were published in the *Federal Register*, the public was invited to comment, and the final rules were published in the Code of Federal Regulations (primarily Title 10). The DOE did not follow the same process. Instead, DOE self-promulgated environmental protection policies based on interpretation of existing regulations and issued DOE directives to be followed at DOE sites. Although these DOE directives do not have the full effect of the law, the DOE closely monitors the work of government employees, subcontractors, and vendors at DOE facilities to ensure compliance. DOE directives (Orders, Guides, Policies, Notices, and Manuals) can be accessed through the DOE Web site at http://www.explorer.doe.gov:1776.

Table 2.2　Summary of Code of Federal Regulations Addressing Radionuclides

Title	Part	Agency	Summary
10	20	NRC	*Standards for Protection against Radiation:* Establishes standards for protection against ionizing radiation resulting from activities conducted under licenses issued by the NRC. Addresses occupational dose limits for various exposure pathways, dose limits for members of the public, surveying and monitoring requirements, etc.
	36	NRC	*Licenses and Radiation Safety Requirements for Irradiators:* Contains requirements for the issuance of a license authorizing the use of sealed sources containing radioactive materials in irradiators used to irradiate objects or materials using gamma radiation. This part also contains radiation safety requirements for operating irradiators.
	39	NRC	*Licenses and Radiation Safety Requirements for Well Logging:* Prescribes requirements for the issuance of a license authorizing the use of licensed materials including sealed sources, radioactive tracers, radioactive markers, and uranium sinker bars in well logging in a single well. Addresses radiation safety requirements for persons using licensed materials in these operations.
	40	NRC	*Domestic Licensing of Source Material:* Establishes procedures and criteria for the issuance of licenses to receive title to, receive, possess, use, transfer, or deliver source and by-product materials. These regulations provide for the disposal of by-product material and for the long-term care and custody of by-product material and residual radioactive material. They also establish certain requirements for the physical protection of import, export, and transient shipment of natural uranium.
	60	NRC	*Disposal of High-Level Radioactive Wastes in Geologic Repositories:* Establishes rules governing the licensing of DOE to receive and possess source, special nuclear, and by-product material at a geologic repository operations area sited, constructed, or operated in accordance with the Nuclear Waste Policy Act of 1982.
	61	NRC	*Licensing Requirements for Land Disposal of Radioactive Waste:* Establishes, for the disposal of radioactive waste, the procedures, criteria, and terms and conditions upon which the commission issues licenses for the disposal of radioactive wastes containing by-product, source, and special nuclear material received from other persons.
	71	NRC	*Packaging and Transportation of Radioactive Material:* Establishes requirements for packaging, preparation for shipment, and transportation of licensed material. Establishes procedures and standards for NRC approval of packaging and shipping procedures for fissile material.

Table 2.2 (continued) Summary of Code of Federal Regulations Addressing

Title	Part	Agency	Summary
10	72	NRC	*Licensing Requirements for the Independent Storage of Spent Nuclear Fuel and High-Level Radioactive Waste:* Establishes requirements, procedures, and criteria for the issuance of licenses to receive, transfer, and possess power reactor spent fuel and other radioactive materials associated with spent fuel storage in an independent spent fuel storage installation and the terms and conditions under which the commission will issue such licenses, including licenses to DOE.
	765	DOE	*Reimbursement for Costs of Remedial Action at Active Uranium and Thorium Processing Sites:* Establishes policies, criteria, and procedures governing reimbursement of certain costs of remedial action incurred by licensees at active uranium or thorium processing sites as a result of by-product material generated as an incident of sales to the United States.
	835	DOE	*Occupational Radiation Protection:* Establish radiation protection standards, limits, and program requirements for protecting individuals from ionizing radiation resulting from the conduct of DOE activities.
			Subpart B Management and Administrative Requirements: DOE activities shall be conducted in compliance with a documented radiation protection program as approved by the DOE.
			Subpart C Standards for Internal and External Exposure:
			10 CFR 835.202 *Occupational dose received by general employees:* Shall be controlled such that the following limits are not exceeded in 1 year: (1) a total effective dose equivalent of 5 rem; (2) the sum of the deep dose equivalent for external exposures and the committed dose equivalent to any organ or tissue other than the lens of the eye of 50 rem; (3) a lens of the eye dose equivalent of 15 rem; and (4) a shallow dose equivalent of 50 rem to the skin or to any extremity.
			10 CFR 835.203 *Combining internal and external dose equivalents:* The total effective dose equivalent during a year shall be determined by summing the effective dose equivalent from external exposures and the committed effective dose equivalent from intakes during the year.
			10 CFR 835.206 *Limits for the embryo/fetus:* The dose equivalent limit for the embryo/fetus from the period of conception to birth, as a result of occupational exposure of a declared pregnant worker, is 0.5 rem.
			10 CFR 835.207 *Occupational dose limits for minors:* The dose equivalent limits for minors occupationally exposed to radiation and/or radioactive materials at a DOE activity are 0.1 rem total effective dose equivalent in a year and 10% of the occupational dose limits specified in §835.202.
			10 CFR 835.208 *Limits for members of the public entering a controlled area:* The total effective dose equivalent limit for members of the public exposed to radiation and/or radioactive material during access to a controlled area is 0.1 rem in a year.

Table 2.2 (continued) Summary of Code of Federal Regulations Addressing

Title	Part	Agency	Summary
10	835	DOE	10 CFR 835.209 *Concentrations of radioactive material in air:* The derived air concentration (DAC) values given in appendix A and C (of the CFR reference) shall be used in the control of occupational exposures to airborne radioactive material. The estimation of internal dose shall be based on bioassay data rather than air concentration values unless bioassay data are unavailable, inadequate, or internal dose estimates based on air concentration values are demonstrated to be as or more accurate.

Subpart E Monitoring of Individuals and Areas:

10 CFR 835.402 *Individual monitoring:* For the purpose of monitoring individual exposures to external radiation, personnel dosimeters shall be provided to and used by (1) radiological workers who, under typical conditions, are likely to receive an effective dose equivalent to the whole body of 0.1 rem or more in a year; a shallow dose equivalent to the skin or to any extremity of 5 rem or more in a year; a lens of the eye dose equivalent of 1.5 rem or more in a year; (2) declared pregnant workers who are likely to receive from external sources a dose equivalent to the embryo/fetus in excess of 10% of the limit at §835.206(a); (3) occupationally exposed minors likely to receive a dose in excess of 50% of the applicable limits at §835.207 in a year from external sources; (4) members of the public entering a controlled area likely to receive a dose in excess of 50% of the limit at §835.208 in a year from external sources; (5) individuals entering a high or very high radiation area.

10 CFR 835.403 *Air monitoring:* Monitoring of airborne radioactivity shall be performed (1) where an individual is likely to receive an exposure of 40 or more DAC-hours in a year; (2) as necessary to characterize the airborne radioactivity hazard where respiratory protection devices for protection against airborne radionuclides have been prescribed.

Subpart F Entry Control Program: Personnel entry control shall be maintained for each radiological area. This subpart expounds upon how control shall be maintained.

Subpart G Posting and Labeling: Except as otherwise provided in this subpart, postings and labels required by this subpart shall include the standard radiation warning trefoil in black or magenta imposed upon a yellow background. This subpart expounds upon postings for "radiation areas," "high radiation areas," "very high radiation areas," etc., and addresses how containers need to be labeled.

Subpart J Radiation Safety Training: Each individual shall complete radiation safety training commensurate with the hazards in the area. Topics should include risks of exposure, basic radiological fundamentals and radiation protection concepts, measures implemented at facilities to manage doses, responsibilities, and exposure reports. Radiation safety training shall be provided at a minimum of every 24 months or when there is a significant change to radiation protection policies.

Table 2.2 (continued)   Summary of Code of Federal Regulations Addressing

Title	Part	Agency	Summary
10	835	DOE	*Subpart K Design and Control:* Measures shall be taken to keep radiation exposure in controlled areas as low as reasonably achievable.
			Subpart L Radioactive Contamination Control: Material and equipment shall not be released from a contamination (or high contamination) area if (1) removable surface contamination levels on accessible surfaces exceed the removable surface contamination values specified in 10 CFR 835 Appendix D (*see Table 2.3 below*); (2) prior use suggests that the removable surface contamination levels on inaccessible surfaces are likely to exceed the removable surface contamination levels in 10 CFR 835 Appendix D.
			Appendix A to Part 835 Derived Air Concentrations (DAC) for Controlling Radiation Exposure to Workers at DOE Facilities: Data presented in this appendix are to be used for controlling individual internal doses and identifying the need for posting of airborne radioactivity areas. The DAC values in this appendix are given for individual radionuclides. For known mixtures of radionuclides, determine the sum of the ratio of the observed concentration of a particular radionuclide and its corresponding DAC for all radionuclides in the mixture. If this sum exceeds unity (1), the DAC (lowest value) for those isotopes not known to be absent shall be used.
			Appendix D to Part 835 Surface Contamination Values: The data presented in this appendix are to be used in identifying the need for posting of contamination and high contamination areas. *See Table 2.3 for the surface contamination values.*
	961	DOE	*Standard Contract for Disposal of Spent Nuclear Fuel and/or High-Level Radioactive Waste:* This part establishes the contractual terms and conditions under which DOE will make available nuclear waste disposal services to the owners and generators of spent nuclear fuel and high-level radioactive waste as provided in section 302 of the Nuclear Waste Policy Act of 1982.
	1021	DOE	*National Environmental Policy Act Implementation Procedure:* The purpose of this part is to establish procedures that DOE shall use to comply with section 102(2) of NEPA of 1969 and the Council on Environmental Quality regulations for implementing the procedural provisions of NEPA.
40	122	EPA	122.2 *Pollutant.* Means dredged spoil, solid waste, incinerator residue, filter backwash, sewer, garbage, sewage sludge, munitions, chemical wastes, biological waste materials, radioactive materials, except those regulated under the Atomic Energy Act of 1954, as amended (42 U.S.C. 2011 et seq.). *Note:* Radioactive materials covered by the Atomic Energy Act are those encompassed in its definition of source, by-product, or special nuclear materials. Examples of materials not covered include radium and accelerator-produced isotopes.

Table 2.2 (continued) Summary of Code of Federal Regulations Addressing

Title	Part	Agency	Summary
40	141	EPA	141.15 *Maximum contaminant levels for Ra-226, Ra-228, and gross alpha particle radioactivity in community water systems.* The following are the maximum contaminant levels for Ra-226, Ra-228, and gross alpha particle radioactivity: (1) combined Ra-226 and Ra-228 = 5 pCi/L; (2) gross alpha particle activity (including Ra-226 but excluding radon and uranium) = 15 pCi/L. 141.16 *Maximum contaminant levels for beta particle and photon radioactivity from man-made radionuclides in community water systems:* (1) The average annual concentration of beta particle and photon radioactivity from man-made radionuclides in drinking water shall not produce an annual dose equivalent to the total body or any internal organ greater than 4 mrem/year. (2) Except for the radionuclides listed in 40 CFR 141.16 Table A, the concentration of man-made radionuclides causing 4 mrem total body or organ dose equivalents shall be calculated on the basis of a 2 L/day drinking water intake using the 168-h data listed in "Maximum Permissible Body Burdens and Maximum Permissible Concentration of Radionuclides in Air or Water for Occupational Exposure" NBS Handbook 69 as amended August 1963, U.S. Department of Commerce. If two or more radionuclides are present, the sum of their annual dose equivalent to the total body or to any organ shall not exceed 4 mrem/year. 141.16 *Table A, Average Annual Concentration Assumed to Produce a Total Body or Organ Dose of 4 mrem/year:* Tritium (total body) 20,000 pCi/L; Sr-90 (bone marrow) 8 pCi/L.
	261	EPA	*Identification and Listing of Hazardous Waste:* This part identifies those solid wastes that are subject to regulation as hazardous wastes. 40 CFR 261.4(a)(4) *Exclusions:* Source, special nuclear, or by-product material as defined by the Atomic Energy Act of 1954, as amended, 42 U.S.C. 2011 is not a solid waste. *Subpart A Program Scope and Interim Prohibition:* §280.10 *Applicability:* (c) *deferrals.* Subparts B, C, D, E, and G do not apply to any of the following types of UST systems: (2) Any UST systems containing radioactive material that are regulated under the Atomic Energy Act of 1954 (42 U.S.C.2011 and following). *Subpart B Criteria for Identifying the Characteristics of Hazardous Waste and for Listing Hazardous Waste:* The administrator shall list a solid waste as a hazardous waste only upon determining that the solid waste meets one of the following criteria: (1) It exhibits any of the characteristics of hazardous waste identified in Subpart C. (2) It has been found to be fatal to humans in low doses or, in the absence of data on human toxicity, it has been shown in studies to have an oral LD_{50} toxicity (rat) of less than 50 mg/kg, and inhalation LC_{50} toxicity (rat) of less than 2 mg/L, or a dermal LD_{50} toxicity (rabbit) of less than 200 mg/kg. (3) It contains any of the toxic constituents listed in 40 CFR 261 Appendix VIII.

Table 2.2 (continued) Summary of Code of Federal Regulations Addressing

Title	Part	Agency	Summary
40	261	EPA	*Subpart C Characteristics of Hazardous Waste:* A solid waste is a hazardous waste if it exhibits any of the characteristics identified in this subpart. These characteristics include ignitability, corrosivity, reactivity, and toxicity.
			Subpart D Lists of Hazardous Wastes: A solid waste is a hazardous waste if it is listed in this subpart. This subpart presents tables which identify the F, K, P, and U listed waste codes that apply to various types of solid wastes.
	262	EPA	*Standards Applicable to Generators of Hazardous Waste:* This part addresses manifest requirements, pretransportation requirements, record keeping, reporting, etc.
	263	EPA	*Standards Applicable to Transporters of Hazardous Waste:* This part addresses compliance with manifest system/record keeping, and hazardous waste discharges.
	264	EPA	*Standards for Owners and Operators of Hazardous Waste Treatment, Storage, and Disposal Facilities:* The purpose of this part is to establish the minimum national standards which define the acceptable management of hazardous waste. Subparts A through EE address issues, such as general facility standards; preparedness and prevention; contingency plans; emergency procedures; release from solid waste management units; surface impoundments; landfills; corrective action; etc.
	265	EPA	*Interim Status Standards for Owners and Operators of Hazardous Waste Treatment, Storage, and Disposal Facilities:* The purpose of this part is to establish minimum national standards that define the acceptable management of hazardous waste during the period of interim status and until certification of final closure or, if the facility is subject to postclosure requirements, until postclosure responsibilities are fulfilled.
49	171	DOT	*General Information, Regulations, and Definitions:* This part prescribes DOT requirements governing the transportation of hazardous materials interstate, intrastate, and foreign commerce by rail car, aircraft, motor vehicle, and vessel.
	172	DOT	*Hazardous Materials Table, Special Provisions, Hazardous Materials Communication, Emergency Response Information, and Training Requirements:* This part lists and classifies those materials that the DOT has designated as hazardous materials for purposes of transportation and prescribes the requirements for shipping papers, package marking, labeling, and transport vehicle placarding. The Hazardous Materials Table presented in 49 CFR 172.101 addresses cesium, strontium, uranium, and radioactive materials in general. Table 2 of 49 CFR 172.101 Appendix A lists radionuclides that are hazardous substances and their corresponding reportable quantities. 49 CFR 172.554 notes that the radioactive placard background color must be white in the lower portion with a yellow triangle in the

Table 2.2　(continued)　Summary of Code of Federal Regulations Addressing

Title	Part	Agency	Summary
49	172	DOT	upper portion. The base of the yellow triangle must be 1.1 in. (29 mm) above the placard horizontal centerline. The radioactive symbol, text, class number, and inner border must be black. 49 CFR 172 Appendix B provides details on the trefoil symbol required for radioactive labels and placards.
	173	DOT	*Shippers General Requirements for Shipments and Packagings:* This part provides the requirements to be observed in preparing hazardous materials for shipment by air, highway, rail, or water.
			Subpart I Class 7 (Radioactive) Materials: This subpart sets forth requirements for the packaging and transportation of Class 7 (radioactive) materials. This subpart presents tables that address subjects, such as allowable contents of uranium hexafluoride "Heels" in a specification 7A cylinder; authorized quantities of uranium hexafluoride; nonfixed external radioactive contamination wipe limits; etc.

Table 2.3   Surface Contamination Levels[a] in dpm/100 cm[2] (10 CFR 835 Appendix D)

Radionuclide	Removable Activity[b,c]	Total Activity (Fixed + Removable)[b,d]
U-nat, U-235, U-238, and associated decay products	1000[e]	5000[g]
Transuranics, Ra-226, Ra-228, Th-230, Th-228, Pa-231, Ac-227, I-125, I-129	20	500
Th-nat, Th-232, Sr-90, Ra-223, Ra-224, U-232, I-126, I-131, I-133	200	1000
Beta–gamma emitters (nuclides with decay modes other than alpha emission or spontaneous fission) except Sr-90 and others noted above[f]	1000	5000
Tritium and tritiated compounds[g]	10,000	NA

[a] The values presented in this table, with the exception noted in footnote f, apply to radioactive contamination deposited on, but not incorporated into the interior or matrix of the contaminated item. Where surface contamination by both alpha– and beta–gamma-emitting radionuclides exists, the limits established for alpha– and beta–gamma-emitting radionuclides apply independently.

[b] As used in this table, disintegrations per minute (dpm) means the rate of emission by radioactive material as determined by correcting the counts per minute observed by an appropriate detector for background, efficiency, and geometric factors associated with the instrumentation.

[c] The amount of removable radioactive material per 100 cm[2] of surface area should be determined by swiping the area with dry filter or soft absorbent paper, applying moderate pressure, then assessing the amount of radioactive material on the swipe with an appropriate instrument of known efficiency. (*Note:* The use of dry material may not be appropriate for tritium.) When removable contamination on objects of surface area less than 100 cm[2] is determined, the activity per unit area shall be based on the actual area and the entire surface shall be wiped. It is not necessary to use swiping techniques to measure removable contamination levels if direct scan surveys indicate that the total residual surface contamination levels are within the limits for removable contamination.

[d] The levels may be averaged over 1 m[2] provided the maximum surface activity in any area of 100 cm[2] is less than three times the value specified. For purposes of averaging, any square meter of surface shall be considered to be above the surface contamination value if (1) from measurements of a representative number of sections it is determined that the average contamination level exceeds the applicable value; or (2) it is determined that the sum of the activity of all isolated spots or particles in any 100 cm[2] area exceeds three times the applicable value.

[e] Alpha.

[f] This category of radionuclides includes mixed fission products, including the Sr-90 that is present in them. It does not apply to Sr-90 that has been separated from the other fission products or mixtures where the Sr-90 has been enriched.

[g] Tritium contamination may diffuse into the volume or matrix of materials. Evaluation of surface contamination shall consider the extent to which such contamination may migrate to the surface to ensure the surface contamination value provided in this table is not exceeded. Once this contamination migrates to the surface, it may be removed, not fixed; therefore, a "Total" value does not apply.

Table 2.4   States Falling under Each EPA Region

Region Number	States
I	Maine, Vermont, New Hampshire, Massachusetts, Rhode Island, Connecticut
II	New York, New Jersey, Virgin Islands, Puerto Rico
III	Pennsylvania, Maryland, Delaware, West Virginia, Virginia
IV	Kentucky, Tennessee, North Carolina, South Carolina, Georgia, Florida, Alabama, Mississippi
V	Minnesota, Wisconsin, Illinois, Michigan, Indiana, Ohio
VI	New Mexico, Texas, Oklahoma, Arkansas, Louisiana
VII	Nebraska, Iowa, Kansas, Missouri
VIII	Montana, Wyoming, Utah, Colorado, North Dakota, South Dakota
IX	California, Nevada, Arizona, Hawaii, Guam, Northern Mariana Islands, Pacific Island Governments, American Samoa
X	Oregon, Washington, Idaho, Alaska

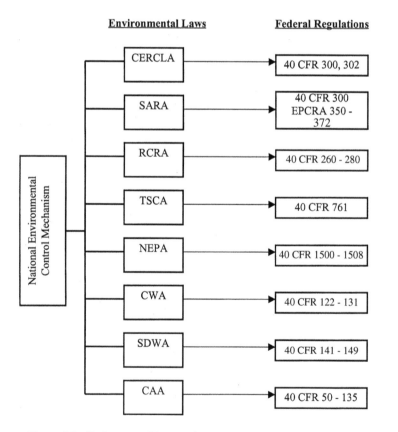

Figure 2.3 Environmental laws and corresponding federal regulations.

Table 2.5   Summary of State Environmental Protection Agencies

State	Environmental Agency	Internet Address
Alabama	Alabama Department of Environmental Management	http://www.adem.state.al.us
Alaska	Alaska Department of Environmental Conservation	http://www.state.ak.us/local/ akpages/ENV.CONSERV
Arizona	Arizona Department of Environmental Quality	http://www.adeq.state.az.us
Arkansas	Arkansas Department of Environmental Quality	http://www.adeq.state.ar.us
California	California Environmental Protection Agency	http://www.calepa.ca.gov
Colorado	Colorado Department of Public Health and Environment	http://www.cdphe.state.co.us
Connecticut	Connecticut Department of Environmental Protection	http://dep.state.ct.us
Delaware	Delaware Department of Natural Resources and Environmental Control	http://www.dnrec.state.de.us
Florida	Florida Department of Environmental Protection	http://www.dep.state.fl.us
Georgia	Georgia Department of Natural Resources	http://www.ganet.org/dnr
Hawaii	Hawaii Department of Health	http://www.state.hi.us/health/eh
Idaho	Idaho Division of Environmental Quality	http://www.state.id.us/deq
Illinois	Illinois Environmental Protection Agency	http://www.ipcb.state.il.us
Indiana	Indiana Department of Environmental Management	http://www.state.in.us/idem
Iowa	Iowa Department of Natural Resources	http://www.state.ia.us/ government/dnr
Kansas	Kansas Department of Health and Environment	http://www.kdhe.state.ks.us
Kentucky	Kentucky Natural Resources and Environmental Protection Cabinet	http://www.nr.state.ky.us
Louisiana	Louisiana Department of Environmental Quality	http://www.deq.state.la.us
Maine	Maine Department of Environmental Protection	http://www.state.me.us/dep
Maryland	Maryland Department of the Environment	http://www.mde.state.md.us
Massachusetts	Massachusetts Department of Environmental Protection	http://www.state.ma.us/dep
Michigan	Michigan Department of Environmental Quality	http://www.deq.state.mi.us
Minnesota	Minnesota Pollution Control Agency	http://www.pca.state.mn.us
Mississippi	Mississippi Department of Environmental Quality	http://www.deq.state.ms.us
Missouri	Missouri Department of Natural Resources	http://www.dnr.state.mo.us
Montana	Montana Department of Environmental Quality	http://www.deq.state.mt.us
Nebraska	Nebraska Department of Environmental Quality	http://www.deq.state.ne.us

Table 2.5 (continued) Summary of State Environmental Protection Agencies

State	Environmental Agency	Internet Address
Nevada	Nevada Department of Conservation and Natural Resources	http://www.state.nv.us/ndep
New Hampshire	New Hampshire Department of Environmental Services	http://www.des.state.nh.us
New Jersey	New Jersey Department of Environmental Protection	http://www.state.nj.us/dep
New Mexico	New Mexico Environment Department	http://www.nmenv.state.nm.us
New York	New York Department of Environmental Conservation	http://www.dec.state.ny.us
North Carolina	North Carolina Department of Environment and Natural Resources	http://www.enr.state.nc.us
North Dakota	North Dakota Department of Health	http://www.health.state.nd.us
Ohio	Ohio Environmental Protection Agency	http://www.epa.state.oh.us
Oklahoma	Oklahoma Department of Environmental Quality	http://www.deq.state.ok.us
Oregon	Oregon Department of Environmental Quality	http://www.deq.state.or.us
Pennsylvania	Pennsylvania Department of Environmental Protection	http://www.dep.state.pa.us
Rhode Island	Rhode Island Department of Environmental Management	http://www.dem.state.ri.us
South Carolina	South Carolina Department of Health and Environmental Control	http://www.state.sc.us/dhec
South Dakota	South Dakota Department of Environmental and Natural Resources	http://www.state.sd.us/denr
Tennessee	Tennessee Department of Environment and Conservation	http://www.state.tn.us/ environment
Texas	Texas Natural Resource Conservation Commission	http://www.tnrcc.state.tx.us
Utah	Utah Department of Environmental Quality	http://www.deq.state.ut.us
Vermont	Vermont Agency of Natural Resources	http://www.anr.state.vt.us
Virginia	Virginia Department of Environmental Quality	http://www.deq.state.va.us
Washington	Washington Department of Ecology	http://www.wa.gov/ecology
West Virginia	West Virginia Department of Environmental Protection	http://www.dep.state.wv.us
Wisconsin	Wisconsin Department of Natural Resources	http://www.dnr.state.wi.us
Wyoming	Wyoming Department of Environmental Quality	http://deq.state.wy.us

REFERENCES

10 CFR, Code of Federal Regulations (as amended).

40 CFR, Code of Federal Regulations (as amended).

49 CFR, Code of Federal Regulations (as amended).

ELR (Environmental Law Reporter), Statute Administrative Proceeding 001, Statute Binder, Environmental Law Institute, 1990.

EPA (Environmental Protection Agency), Guidance for Conducting Remedial Investigations and Feasibility Studies Under CERCLA, EPA 540/G-89/004, 1-7, 1988.

ONPA (Office of NEPA Project Assistance), NEPA Compliance Guide, I, 131–207, 1988.

CHAPTER 3

Radiation and Radioactivity

The purpose of this chapter is to provide the reader with the fundamentals of radioactivity, radiation, and radiation detection. Radiological contamination in the environment is of concern to human health because, if left uncontrolled, the contamination could lead to adverse health effects such as cancer. The interactions between radiation and the human body are essentially collisions between radiation "particles" and atoms. These collisions produce damage mostly by knocking electrons from their atomic orbit or leaving atoms in an energized state resulting in additional radioactivity. The trail of destruction produced by the radiation particle is on the atomic scale but may be sufficient to damage or kill cells in human tissue. The same interactions that produce adverse health effects may be used to locate and quantify radiological contamination in the environment. That is, collisions between a radiation particle and atoms could occur in a radiation detector leading to a response such as displacing a needle or producing an electronic pulse. The magnitude of cell damage or the characteristics of the detector response depend on the type and origin (source) of the radiation.

Radiation comes in many physical forms and from a range of sources. The types or forms of radiation of most interest originate as emission from an unstable nucleus or an excited atom. These emissions of radiation include various combinations of energetic electrons, protons, and neutrons (alpha particles and beta particles) and electromagnetic radiation (gamma rays and X rays). There are also more exotic radiation particles such as muons, pions, neutrinos, etc., that are less relevant when considering environmental contamination. Sources of radiation include rock and soil (primordial sources); nuclear reactors, high-energy particle accelerators, manufactured material, etc. (anthropogenic sources); and outer space (cosmic radiation). The type and source of radiation must be taken into consideration when planning environmental studies since they will influence the selection of the appropriate radiation detection instrumentation.

Radioactivity occurs when some part of an atom is unstable. The instability comes from having too many protons or too many neutrons in a nucleus, or when a proton or neutron is in an excited state (has too much energy). The type of radiation (alpha, beta, gamma, or X ray) that is emitted depends on the location of the instability. That is, alpha, beta, and gamma are only emitted from the nucleus, while X rays are only emitted from the electrons orbiting the nucleus. The energy of a radiation particle depends on the excited state of the nucleus or the orbiting electron. For example, a proton in a highly excited state may de-excite (lose the extra energy) by emitting a gamma particle that has a few million electronvolts of energy. An electron that is only slightly excited in its orbit around the nucleus may lose its extra energy by emitting an X ray with only a few electronvolts.

Radioactive materials may contain a number of discrete kinds of radioactive atoms. To categorize these atoms, the material is first broken into its elemental components (e.g., pure water is two parts hydrogen and one part oxygen). Once a particular element is identified, that element may be further categorized by isotope. Whereas an element is defined by the number of protons in its nucleus (all hydrogen atoms have one proton), an isotope of an element is defined by the number of neutrons in the nucleus. A cylinder full of pure hydrogen may contain atoms with zero, one, or two neutrons in the nucleus. The cylinder therefore contains three hydrogen isotopes. The isotopes that are radioactive are called radioisotopes. Hydrogen atoms with two neutrons in their nuclei are radioactive and are therefore radioisotopes. All radioisotopes that have the same number of protons and neutrons in the nucleus have identical physical properties. They are chemically identical, emit the same type of radiation, and emit the radiation at the same rate.

Radiation particles may be viewed as packets of energy or particles that carry energy. This energy is transferred during collisions with matter, producing tissue damage or a detector response. The unit often used to describe radiation energy is the electronvolt (eV), where 1 eV is defined as the amount of kinetic energy that an electron would gain if accelerated through 1 V of potential difference. A radiation particle may be very energetic with energies in the thousands of eV (keV) or millions of eV (MeV), or may have only fractions of an eV in energy. The more energetic particles are of most interest to an environmental study since these are the particles that produce the most damage in tissue and produce distinct detector responses. For example, consider a radiation particle with 1 MeV of energy. It takes an average of about 30 to 34 eV to knock an electron from its orbit around a nucleus. The 1 MeV alpha particle could potentially liberate approximately 30,000 electrons. In tissue, these electrons could disrupt cellular chemistry, break bonds in a DNA strand, and generally produce damage that could result in cell mutilation or cell death. In a radiation detector, the 30,000 electrons could be collected and used to characterize the radiation type and source. If a radiation particle has only a few electronvolts, there would be minimal tissue damage and little chance of a measurable detector response.

Because radioactivity results from instability in the atomic/nuclear structure, there is very little that can be done to change the radioactive properties. Changing the physical properties of a material by burning, dissolving, solidifying, etc., may change the chemistry of a material but does not change the structure of a nucleus or the radioactive properties. A material can be bombarded with neutrons or exposed in the beam of a high-energy particle accelerator to change the nuclear structure (and radioactive properties), but these methods are very expensive, creating new and possibly more hazardous materials, and are typically never considered in an environmental cleanup effort. The most reliable method to reduce the radioactivity of a material is to let time pass.

One property that all radioactive materials have in common is that the level of radioactivity decreases with time. Some materials may be radioactive for only a fraction of a second. These materials have relatively unstable nuclear structures that lose the excess energy quickly. Other materials can be radioactive for billions of years. These materials have slightly unstable nuclear structures that are not as anxious to lose the excess energy. The rate by which radioisotopes emit radiation or go through radioactive decay is defined by its half-life. A half-life is the amount of time it takes for one half of the radioactive atoms to decay. For example, if there are 1000 atoms of a radioisotope with a half-life of 1 year, about 500 will remain (and about 500 will have decayed) after 1 year. After another year, only about 250 will remain, about 125 in another year, etc., until all the atoms have decayed. By using this example, it is easy to see that a radioisotope with a half-life of 1 billion years will be around for a very long time. In fact, only a very small fraction of these atoms will undergo decay during a human's lifetime. On the other hand, a radioisotope with a half-life of a few minutes or less will be effectively gone in an hour.

Sometimes when a radioisotope decays, the remaining nucleus is also radioactive. The original radioisotope is called the parent and the remaining isotope is called the daughter or decay product. This first decay product can then decay into a second decay product, which may decay into a third, etc., until a nonradioactive (stable) decay product remains. Not all radioisotopes undergo a series of decays. For example, a carbon atom with six protons and eight neutrons (carbon-14) will emit a beta particle leaving a stable nitrogen atom. There are, however, three decay series found in nature that make up the radionuclides at most radioactively contaminated sites: the uranium series, the thorium series, and the actinium series. These series are shown in Tables 3.1, 3.2, and 3.3, respectively. The parent/daughter relationships, the modes of decay, energies of the radiation particles, and the half-lives presented for these series are always the same. When characterizing a site contaminated with uranium series radionuclides, the information presented in Table 3.1 should be used to select the proper field instrumentation, sampling procedures, laboratory analytical procedures, and health and safety procedures considering the degree to which equilibrium of the series is expected.

3.1 TYPES OF RADIATION

When considering environmental contamination, the most relevant forms of radiation include alpha particles, beta particles, X rays, and gamma rays. Each of these radiation particles has distinct physical characteristics that impact the way it interacts with matter, including human tissue or radiation detectors. Exotic forms of radiation and energetic neutrons may also be important under certain conditions, but rarely in an environmental setting. The following discussion describes the physical characteristics of the relevant radiation particles and corresponding effect the particle would have during collision interactions.

Table 3.1 Uranium Series

Nuclide	Historical Name	Half-Life	Major Radiation Energies (MeV) and Intensities[a]					
			Alpha		Beta		Gamma	
			MeV	%	MeV	%	MeV	%
$^{238}_{92}U$	Uranium I	4.468×10^9 year	4.15	22.9			0.0496	0.07
			4.20	76.8				
$^{234}_{90}Th$	Uranium X$_1$	24.1 days			0.076	2.7	0.0633	3.8
					0.095	6.2	0.0924	2.7
					0.096	18.6	0.0928	2.7
					0.1886	72.5	0.1128	0.24
$^{234m}_{91}Pa$	Uranium X$_2$	1.17 min			2.28	98.6	0.766	0.207
							1.001	0.59
$^{234}_{91}Pa$	Uranium Z	6.7 h			22 βs		0.132	19.7
					E Avg = 0.224		0.570	10.7
					E$_{max}$ = 1.26		0.883	11.8
							0.926	10.9
							0.946	12
$^{234}_{92}U$	Uranium II	244,500 year	4.72	27.4			0.053	0.12
			4.77	72.3			0.121	0.04
$^{230}_{90}Th$	Ionium	7.7×10^4 year	4.621	23.4			0.0677	0.37
			4.688	76.2			0.142	0.07
							0.144	0.045

99.87% 0.13%

$^{234m}_{91}Pa$ IT

Nuclide (decay scheme)	Name	Half-life	α energy (MeV)	α (%)	β energy (MeV)	β (%)	γ energy (MeV)	γ (%)
$^{226}_{88}$Ra →	Radium	1600 ± 7 year	4.60, 4.78	5.55, 94.4			0.186	3.28
$^{222}_{86}$Rn →	Emanation Radon (Rn)	3.823 days	5.49	99.9			0.510	0.078
$^{218}_{84}$Po 0.02%	Radium A	3.05 min	6.00	~100	0.33	0.02	0.837	0.0011
99.98% → $^{214}_{82}$Pb →	Radium B	26.8 min			0.67, 0.73, 1.03	48, 42.5, 6.3	0.2419, 0.295, 0.352, 0.786	7.5, 19.2, 37.1, 1.1
$^{218}_{85}$At →	Astatine	2 sec	6.65, 6.7, 6.757	6.4, 89.9, 3.6			0.053	6.6
$^{214}_{83}$Bi →	Radium C	19.9 min	5.45, 5.51	0.012, 0.008	1.42, 1.505, 1.54, 3.27	8.3, 17.6, 17.9, 17.7	0.609, 1.12, 1.765, 2.204	46.1, 15.0, 15.9, 5.0
99.979% 0.021%	Radium C′	164 μsec	7.687	100			0.7997	0.010
$^{214}_{84}$Po → $^{210}_{81}$Ti →	Radium C″	1.3 min			1.32, 1.87, 2.34	25, 56, 19	0.2918, 0.7997, 0.860, 1.110, 1.21, 1.310, 1.410, 2.010, 2.090	79.1, 99, 6.9, 6.9, 17, 21, 4.9, 6.9, 4.9

Table 3.1 (continued) Uranium Series

| Nuclide | Historical Name | Half-Life | Major Radiation Energies (MeV) and Intensities[a] | | | | | |
| | | | Alpha | | Beta | | Gamma | |
			MeV	%	MeV	%	MeV	%
$^{210}_{82}$Pb	Radium D	22.3 year	3.72	0.000002	0.016	80	0.0465	4
↓					0.063	20		
$^{210}_{83}$Bi	Radium E	5.01 days	4.65	0.00007	1.161	~100		
↓			4.69	0.00005				
~100% .00013%								
$^{210}_{84}$Po	Radium F	138.378 days	5.305	100			0.802	0.0011
↓ $^{206}_{81}$Tl	Radium E″	4.20 min			1.571	100	0.803	0.0055
↓								
$^{206}_{82}$Pb	Radium G	stable						

[a] Intensities refer to percentage of disintegrations of the nuclide itself, not to original parent of series. Gamma % in terms of observable emissions, not transitions.

Source: Shleien, *The Health Physics and Radiological Health Handbook, Scinta,* Incorporated, Silver Spring, MD, 1992.

Table 3.2 Thorium Series

Nuclide	Historical Name	Half-Life	Major Radiation Energies (MeV) and Intensities[a]					
			Alpha		Beta		Gamma	
			MeV	%	MeV	%	MeV	%
$^{232}_{90}$Th	Thorium	1.405×10^{10} year	3.83	0.2			0.059	0.19
			3.95	23			0.126	0.04
			4.01	76.8				
$\rightarrow$								
$^{228}_{88}$Ra	Mesothorium I	5.75 year			0.0389	100	0.0067	6×10^{-5}
$\rightarrow$								
$^{228}_{89}$Ac	Mesothorium II	6.13 h			0.983	7	0.338	11.4
					1.014	6.6	0.911	27.7
					1.115	3.4	0.969	16.6
					1.17	32	1.588	3.5
					1.74	12		
					2.08	8		
$\rightarrow$					(+ 33 more βs)			
$^{228}_{90}$Th	Radiothorium	1.913 year	5.34	26.7			0.84	1.19
			5.42	72.4			0.132	0.11
							0.166	0.08
							0.216	0.27
$\rightarrow$								
$^{224}_{88}$Ra	Thorium X	3.66 days	5.45	4.9			0.241	3.9
			5.686	95.1				
$\rightarrow$								
$^{220}_{86}$Rn	Emanation Thoron (Tn)	55.6 sec	6.288	99.9			0.55	0.07
$\rightarrow$								

Table 3.2 (continued)　Thorium Series

Nuclide	Historical Name	Half-Life	Alpha MeV	%	Beta MeV	%	Gamma MeV	%
$^{216}_{84}$Po	Thorium A	0.15 sec	6.78	100			0.128	0.002
$^{212}_{82}$Pb	Thorium B	10.64 h			0.158	5.2	0.239	44.6
					0.334	85.1	0.300	3.4
					0.573	9.9		
$^{212}_{83}$Bi	Thorium C	60.55 min	6.05	25	1.59	8	0.040	1.0
			6.09	9.6	2.246	48.4	0.727	11.8
							1.620	2.75
64.07% ↘　35.93% ↙								
$^{212}_{84}$Po	Thorium C′	305 nsec	8.785	100				
$^{208}_{81}$Tl	Thorium C″	3.07 min			1.28	25	0.277	6.8
					1.52	21	0.5108	21.6
					1.80	50	0.583	85.8
							0.860	12
$^{208}_{82}$Pb	Thorium D	Stable					2.614	100

[a] Intensities refer to percentage of disintegrations of the nuclide itself, not to original parent of series. Gamma % in terms of observable emissions, not transitions.

Source: Shleien, *The Health Physics and Radiological Health Handbook*, Scinta, Incorporated, Silver Spring, MD, 1992.

Table 3.3 Actinium Series

Nuclide	Historical Name	Half-Life	Major Radiation Energies (MeV) and Intensities[a]					
			Alpha		Beta		Gamma	
			MeV	%	MeV	%	MeV	%
$^{235}_{92}$U	Actinouranium	7.038×10^8 year	4.2–4.32	10.3			0.1438	10.5
			4.366	17.6			0.163	4.7
			4.398	56			0.1857	54
			4.5–4.6	11.3			0.205	4.7
$\rightarrow$								
$^{231}_{90}$Th	Uranium Y	25.5 h			0.205	15	0.0256	14.8
					0.287	49	0.0842	6.5
					0.304	35		
$\rightarrow$								
$^{231}_{91}$Pa	Protoactinium	3.276×10^4 year	4.95	23			0.0274	9.3
			5.01	25.6			0.2837	1.6
			5.029	20.2			0.300	2.3
			5.058	11.1			0.3027	4.6
$\rightarrow$							0.330	1.3
$^{227}_{89}$Ac	Actinium	21.77 year	4.94	0.53	0.019	10	0.070	0.017
			4.95	0.66	0.034	35	0.100	0.032
					0.044	44	0.160	0.019
98.62% 1.38%								
$^{237}_{90}$Th	Radioactinium	18.718 days	5.757	20.2			0.050	8.5
			5.978	23.3			0.236	11.2
			6.038	24.4			0.300	2.0
$\rightarrow$							0.304	1.1
							0.330	2.7
$^{223}_{87}$Fr	Actinium K	21.8 min	5.44	~0.006	1.15	~100	0.050	34
							0.0798	9.2

Table 3.3 (continued) Actinium Series

Nuclide	Historical Name	Half-Life	Alpha MeV	Alpha %	Beta MeV	Beta %	Gamma MeV	Gamma %
$^{223}_{88}\text{Ra}$	Actinium X	11.43 days	5.607	24.1			0.2349	3.4
			5.716	52.2				
			5.747	9.45				
$^{219}_{86}\text{Rn}$	Emanation Actinon (An)	3.96 sec	6.425	7.4	25 betas		0.144	3.3
			6.55	12.1	$E_{avg.} = 0.343$		0.154	5.6
			6.819	80.3	$E_{max} = 1.097$		0.269	13.6
							0.324	3.9
							0.338	2.8
							0.271	9.9
							0.4018	6.6
$^{215}_{84}\text{Po}$	Actinium A	1.78 msec	7.386	~100	0.74	~0.00023	0.4388	0.04
~100% 0.00023%								
$^{211}_{82}\text{Pb}$	Actinium B	36.1 min			0.26	4.8	0.405	3.0
					0.97	1.4	0.427	1.38
					1.37	92.9	0.832	2.8
$^{215}_{85}\text{At}$	Astatine	~0.1 msec	8.026	~100			0.404	0.047
$^{211}_{83}\text{Bi}$	Actinium C	2.14 min	6.28	16	0.579	0.27	0.351	12.7
			6.62	84				

Major Radiation Energies (MeV) and Intensities[a]

0.273%   99.73%
↓
$^{211}_{84}$Po   Actinium C′   0.516 sec   7.42   98.9

→
$^{207}_{81}$Tl   Actinium C″   4.77 min   1.42   99.8   0.570   0.54
                                                          0.898   0.52

→   0.897   0.24

$^{207}_{82}$Pb   Actinium D   Stable

a Intensities refer to percentage of disintegrations of the nuclide itself, not to original parent of series. Gamma % in terms of observable emissions, not transitions.

Source: Shleien, *The Health Physics and Radiological Health Handbook,* Scinta, Incorporated, Silver Spring, MD, 1992.

3.1.1 Alpha Particles

An alpha particle is basically an energetic helium nucleus, consisting of two protons and two neutrons. Alphas are emitted from the nucleus of an atom typically with energies in the million-electronvolt range. Because the energy levels are high, an alpha particle can ionize a large number of atoms when interacting with matter producing a relatively large amount of damage in tissue or creating a relatively large detector response. The two protons in an alpha particle create a +2 charge (neutrons have no charge) and the alpha particle is over 7000 times more massive than electrons with which it interacts. Both of these facts help limit the range an alpha particle travels. That is, because of the large mass and +2 charge (recall that electrons have a –1 charge), an alpha will undergo multiple collisions over a short track producing a high density of liberated electrons. In fact, a typical alpha particle will only travel a few inches in air, and cannot penetrate the dead layer of cells on the surface of skin. A common analogy used to describe the collisions between an alpha particle and electrons is to imagine throwing a bowling ball (symbolizing the alpha particle) through a room full of Ping-Pong balls (symbolizing the electrons). The bowling ball may easily displace many Ping-Pong balls at first, but will quickly lose its energy and come to rest after traveling a short distance.

Because the range of an alpha particle is short, detectors must be held close to the radiation source to make a measurement. Also because alpha particles are easily attenuated (shielded), a contaminated surface covered in dust, dirt, or paint may preclude alpha detection. Another important characteristic of alpha particles is that they are emitted from nuclei at discrete energies. For example, the radioisotope uranium-238 shown at the top of Table 3.1 emits an alpha particle at approximately 4.2 MeV. If a site is contaminated with uranium, field personnel could use detectors to look for the 4.2 MeV alpha to determine where uranium-238 levels are elevated.

3.1.2 Beta Particles

A beta particle is basically an energetic electron, but unlike electrons, beta particles can have a +1 or a –1 charge. Beta particles are emitted from the nucleus of an atom and can have energies in the million-electronvolt range. Unlike alpha particles that are emitted with discrete energies, beta particles are emitted with energies ranging from 0 eV to a maximum value characteristic of the radioisotope. Like alpha particles, the maximum beta energy may be used to identify the radio-nuclide. Beta particles have the same mass as an electron and a +1 or –1 charge and interact with electrons in matter (e.g., in tissue or detectors) more like Ping-Pong balls colliding with other Ping-Pong balls. Instead of producing a high density of liberated electrons over a short track, the beta bounces around changing directions many times until it loses all its energy through a series of collisions.

The range of a beta particle is energy dependent and it may travel several feet in air. Beta particles can penetrate the dead layer of cells on skin, but cannot penetrate through thin layers of paper, aluminum, wood, etc. To measure beta particles in the field it is best to hold the detector close to the contaminated surface. As with alpha particles, layers of dust, dirt, etc., can limit the ability to measure beta activity, but to a lesser extent than with alpha particles.

3.1.3   X Rays

An X ray occurs when an electron drops from a high energy level to a lower energy level as it orbits a nucleus. In this way an electron loses excess energy in the form of an energetic photon. An X ray (or any photon) has no mass and travels at the speed of light. In fact, an X ray is a kind of "light" or electromagnetic radiation—it simply has more energy than visible or ultraviolet light and interacts with matter more like an energetic particle than an electromagnetic wave. Radionuclides sometimes emit characteristic X rays, or X rays with discrete energies, that may be used to identify and quantify contamination. Field characterization is not often designed around measuring X rays as the X-ray radiation is usually not intense enough to measure in the field or other forms of radiation are more easily measured. X-ray measurements are more often made in a laboratory environment where they can be distinguished from other types of radiation.

X rays interact with matter in a couple of ways, but the mode relevant to environmental studies is called Compton scattering. During Compton scattering the X ray collides with an orbital electron transferring some fraction of the X-ray energy to the electron. The interaction is similar to collisions between two billiard balls where there could be a glancing blow or a head-on collision. The most energy is transferred during head-on collisions, and the postcollision energetic electron will likely break free from its orbit around a nucleus to produce damage in tissue or create a detector response. Instead of undergoing continuous interaction/collisions while passing through matter, X rays go through a discrete number of collisions. As a point of interest, the X ray may only transfer enough energy to bump the electron into a higher orbit or energy level. When this occurs, the electron will fall to a lower orbit by emitting an X ray from the absorbing material. In fact, interactions between all types of radiation (including alpha and beta particles) indirectly produce X rays in this manner. X rays can travel great distances in the air but have smaller ranges in dense materials like lead.

X-ray machines are used to generate high-energy photons for diagnostic, therapeutic, or research activities. However, these X rays are not relevant in an environmental setting.

3.1.4   Gamma Rays

Gamma rays, or gamma particles, are identical to X rays with two notable exceptions: gamma particles originate from the nucleus instead of the orbital electrons, and gamma particles are typically more energetic than X rays. A gamma particle is produced when a neutron or proton drops from a high energy level to a lower energy level from inside the nucleus. Gamma particle energies are characteristic of the radionuclide source, similar to X rays and alpha and beta particles. Many radionuclides emit energetic gamma particles well into the kilo- or megaelectronvolt range and at intensities that are a concern to human health. The billiard-ball-like collisions between a gamma particle and an electron are the same as the X-ray collision except the gamma energy and the energy transfer may be larger. These gamma particles may be measured in the field from a significant distance. In fact, gamma radiation surveys have been performed using detectors mounted on the bottom of helicopters flying hundreds of feet above the ground (see Section 4.2.1.1). Needless to say, gamma particles (or any energetic photon) can travel great distances in air, but like X rays have limited range in dense material such as lead. Field radiation surveys are often designed around measuring the gamma component of a radiological contaminant given the ease of detection compared with alpha and beta particles that can be shielded by thin layers of paint, soil, or other common barriers.

3.2 SOURCES OF RADIATION AND RADIOACTIVITY

3.2.1 Primordial Sources

Primordial sources of radiation contain radionuclides that remain from the creation of all matter billions of years ago. Theory has it that a wide range of radionuclides were created during the "Big Bang," but only those with very long half-lives remain today. Included in this group are two isotopes of uranium (U-235 and U-238) and one isotope of thorium (Th-232). These radionuclides are at the head of the decay series shown in Tables 3.1 through 3.3. Other radionuclides in these series would not be found in nature if they were not constantly reproduced by a long-lived parent. Other primordial radionuclides include K-40 with a half-life of 1.26 billion years and Rb-87 with a half-life of 48 billion years.

Primordial radionuclides are found in soil and rock, and are sometimes concentrated in ore. These radionuclides leach into water, are absorbed by plants, and are released into the air where they are consumed and inhaled by humans. Humans are also exposed to gamma-emitting primordial radionuclides while they are still bound to the soil and rock. The radiation that is emitted by primordial radionuclides make up approximately 90% of the average exposure to natural "background radiation" in the United States and 76% of the average exposure from all sources. The natural background concentrations in soil can be highly variable from one location to the next. For example, granite contains relatively high concentrations of uranium, and monazite sand contains relatively high concentrations of thorium. For a detailed discussion on background radiation and exposure to humans, see Ionizing Radiation Exposure of the Population in the United States (NCRP, 1987).

3.2.2 Cosmic Radiation

The other 10% of the average exposure to natural background radiation in the United States comes from cosmic radiation. Cosmic radiation originated from extra-terrestrial sources and consists mostly of highly energetic protons. These protons collide with atoms in the atmosphere creating high-energy muons, electrons, and a small number of neutrons (secondary particles) that cascade down to Earth's surface.

As the secondary particles travel through the atmosphere they lose energy through a series of collisions with atoms in the atmosphere or near Earth's surface until all energy is transferred. The exposure to cosmic radiation is highly dependent on altitude and overhead shielding. That is, cosmic radiation and the secondary particles have to travel a shorter distance from outer space to Denver, CO, than to reach sea level. Buildings provide shielding from cosmic radiation. A simple way to demonstrate shielding from cosmic radiation is to take an NaI detector outdoors. The detector response is larger (more counts or clicks per minute) outdoors than indoors because an overhead structure acts as a shield from cosmic radiation.

Cosmic radiation also produces cosmogenic radionuclides. These radionuclides are created when a cosmic radiation particle or two is absorbed by a nucleus (altering the nuclear structure), and the new nucleus becomes radioactive. The four most common cosmogenic radionuclides are H-3 (tritium), Be-7, C-14, and Na-22. Of these, H-3, C-14, and Na-22 are found naturally in the human body causing a relatively small amount of exposure compared to cosmic radiation and primordial radionuclides. Scientists often use C-14 to date very old organic material such as wood or bone. Scientists also take advantage of the fact that tritium is not produced far below Earth's surface. That is, tritium in deep groundwater may indicate communication between upper and lower groundwater aquifers.

3.2.3 Anthropogenic Sources

Anthropogenic sources make up 18% of an average U.S. citizen's radiation exposure. Humans produce a wide variety of radionuclides for use in manufactured goods and for energy production. Humans also produce radiation for research, medical diagnosis and therapy, etc. For example, Am-241 is used in smoke detectors, thorium is used in lantern mantles, uranium and thorium have been used to glaze pottery, and tobacco products concentrate polonium. There are many other examples of radionuclides in consumer products, each producing some exposure to humans. Even with the wide range of potential sources, consumer products only make up about 3% of the average exposure in the United States. While exposures during research activities make up a very small fraction of the average exposure in the United States, diagnostic and therapeutic X rays (from X-ray machines) and nuclear medicine make up approximately 15% of the average exposure.

Other anthropogenic sources of radiation that impact environmental studies include radionuclides resulting from the production of nuclear weapons, aboveground nuclear testing, and radionuclides that may be concentrated in building materials or fertilizers. Shallow soils often contain Cs-137, a radionuclide distributed across the globe as a result of past aboveground nuclear explosions. Cs-137 is often measured during environmental studies but is usually screened out as a non-site-related contaminant. Building materials such as brick often contain elevated concentrations of gamma-emitting Ra-226. During gamma radiation surveys, elevated readings may be collected near buildings or rubble piles. Because potassium is a major component of fertilizer, K-40 may be confused with other gamma-emitting radionuclides while surveying farmland or garden areas.

3.3 RADIATION DETECTION INSTRUMENTATION

Traditional radiation instruments consist of two components: a radiation detector and the power supply/display. However, radiation instruments come in a wide range of sizes and configurations ranging from very large and complex instruments (which will not be discussed here) to a simple plastic chip. This section identifies and very briefly describes the types of radiation detectors and associated display or recording equipment that are applicable to survey activities in support of environmental assessment or remedial action. For more information on radiation instrumentation, see Knoll (1989).

3.3.1 Radiation Detectors

The particular capabilities of a radiation detector will establish its potential applications in conducting a specific type of survey. Radiation detectors can be divided into four general classes based on the detector material or the application:

1. Gas-filled detectors
2. Scintillation detectors
3. Solid-state detectors
4. Passive integrating detectors

In most cases these detectors in some way measure the electrons that are liberated during collision interactions. The electrons may be:

- Collected continuously as an electrical current;
- Collected discretely as pulses;
- Used to produce light in a scintillation interaction;
- Collected on capacitors.

The collection and/or measurement of electrons is key in radiation detection instrumentation. In any case, the type of detector should be selected by the project health physicist to assure the selected detector is suitable for the site contaminants and is properly maintained.

3.3.1.1 Gas-Filled Detectors

Gas-filled detectors generally consist of a gas chamber where radiation interactions take place, some electronics to measure the radiation interactions, and a display for relaying relevant information to the detector user. Impinging radiation collides with gas particles knocking off electrons. The detector contains electronics used to create a voltage difference across the gas chamber. Liberated electrons with their negative charge are attracted to the positively charged anode and the now-ionized gas particles are attracted to the negatively charged cathode. The electrons and ionized gas particles (called ion pairs) can recombine at low voltages or accelerate and create other ion pairs at higher voltages (called gas multiplication).

Ion chambers or exposure rate meters operate at lower voltages (approximately 100 to 200 V) and can be made small enough to carry into the field for direct measurements. Ion chambers are most useful when measuring gamma radiation levels emanating from a contaminant. Ion chambers are also used to measure beta radiation levels. At higher voltages (approximately 250 to 500 V), gas multiplication occurs so that the number of ion pairs produced is proportional to the energy deposited by the radiation. Proportional counters may then be used to identify radiation particles (mainly beta and alpha particles) by their discrete energies, thus identifying the source of the radiation. Proportional counters are usually too large to be used in the field but may be very useful in a field office or laboratory. In the 600 to 800 V range, there is limited proportionality and a gas-filled detector is less useful. At still higher voltages (up to about 950 V), all proportionality is lost. Instead, radiation particles of all energies produce the same response. Handheld Geiger–Mueller (GM) counters operate in this voltage range and are commonly used to scan surfaces for beta and gamma contamination. Above the GM voltages, there is continuous discharge and no useful detector response.

The fill gases in these detectors vary, but the most common are air or argon with a small amount of organic methane (usually 10% methane by mass, referred to as P-10 gas). Others include argon or helium with a small amount of a halogen such as chlorine or bromine. Halogen fill gases must be replaced over time, and air-filled detectors are sometimes sensitive to humidity and pressure.

3.3.1.2 Scintillation Detectors

Some radiation detectors contain luminescent materials (called scintillators or phosphors). When a radiation particle loses energy in this material, the material releases low-energy photons that can be fed into a photomultiplier tube. The photomultiplier tube then amplifies the photon until an electrical pulse is produced. The pulse rate is proportional to the level of contamination. Common scintillators are sodium iodide (NaI) and zinc sulfide (ZnS). A NaI detector is efficient at measuring gamma radiation levels, and a detector with a 2×2-in. crystal can be used to locate radium contamination at about twice the background concentration. A scintillation detector with a ZnS foil can be effective for measuring alpha radiation on contaminated surfaces. These detectors may also be used to measure beta radiation, but are highly inefficient gamma radiation detectors. Note that the ZnS foil must be held close to the contaminated surface for the detector to work because alpha particles have a short range and may be shielded by thin layers of dust or moisture. Both the 2×2-in. NaI detector and the detector with a ZnS foil are handheld instruments that are commonly used in the field.

3.3.1.3 Solid-State Detectors

Solid-state detectors are detectors that contain semiconductor material such as high-purity germanium (HPGe) or NaI that are subjected to a potential difference. A radiation particle that undergoes collisions in the semiconductor material will liberate many electrons. The electrons are collected by the detector electronics to produce the detector response. Solid-state detectors are very useful for identifying the radiation sources as the number of electrons liberated is proportional to the energy deposited by the radiation particle. While solid-state detectors may be configured to detect beta radiation, the most common use is gamma and alpha detection. For example, the detector may identify gamma radiation with 1.33 MeV of energy. Because Co-60 has a 1.33-MeV gamma particle, it may be concluded that Co-60 is a contaminant at the site. The act of identifying the radiation source in this manner is called spectroscopy. The detector may also be calibrated to estimate the concentration of a contaminant. This is called spectrometry.

Spectrometry provides the means to discriminate among various radionuclides by separating radiation particles by energy. *In situ* gamma spectrometry (see Section 4.2.2.1.2.3) may be particularly effective in producing qualitative and quantitative data without waiting for laboratory reports. The availability of HPGe detectors permits measurement of low-abundance gamma emitters such as U-238 and Pu-239. NaI and other scintillation detectors may also be used *in situ*, but these systems are less sensitive than the HPGe system.

3.3.1.4 Passive Integrating Detectors

There is an additional class of instruments that consists of passive, integrating detectors and associated reading/analyzing instruments. This class includes thermoluminescence dosimeters (TLDs) and electret ion chambers (EICs). These detectors are often exposed for relatively long periods of time providing good sensitivity at low activity levels. Results from passive detectors are often compared with regulatory limits (e.g., fence-line measurements) or used for environmental surveillance. Passive integrating detectors are typically inexpensive and easy to operate. The ability to read and present data on site is also a useful feature, and such systems are comparable to direct reading instruments.

TLDs are essentially small semiconductor chips, but without the applied voltages used by spectrometers. When a radiation particle collides with the TLD, electrons are knocked into an excited state and, due to the special properties of the material, are trapped. After the TLD has been exposed to radiation for a period of time, the TLD is heated. The trapped electrons fall from their traps and while doing so emit low-energy photons. The number of photons counted is proportional to the energy absorbed from the radiation, thus producing measurements of the radiation levels at the measurement site. TLDs come in a large number of materials including:

- LiF
- CaF_2:Mn
- CaF_2:Dy
- $CaSO_4$:Mn
- $CaSO_4$:Dy
- Al_2O_3:C

TLDs can be used to measure alpha, beta, and gamma radiation levels.

The EIC consists of a very stable electret (a charged Teflon® disk) mounted inside a small air-filled chamber made of electrically charged plastic. The ion pairs produced by radiation interactions within the chamber are collected onto the electret, causing a reduction of its surface charge. The reduction in charge is a function of the total ionization during a specific monitoring period and the specific chamber volume. This change in electrical charge is measured with a surface potential voltmeter, and the voltage reading is compared with calibration information to indicate radiation levels. EICs are most often used to measure radon levels.

3.3.2 Instrument Inspection and Calibration

All instruments should be inspected and source-tested prior to use (Figure 3.1). Instrument inspection consists of inspecting the instrument for:

- Broken parts
- Loose or missing screws
- Loose or misaligned knobs
- Calibration potentiometers not aligned with access holes
- Circuit boards not secured
- Loose wires
- Loose connectors
- Loose components
- Testing of moving parts
- Making certain that batteries are fresh and properly installed.

The general operation of an instrument should be tested each time it is used. This consists of switching to check battery condition, verifying the set of the mechanical zero on the meter, testing the meter zero potentiometer, checking for switching transients, checking for zero drift on the meter, and checking for light sensitivity, if applicable.

Source tests consist of checking source response, geotropism, variability of readings, stability, temperature response, humidity response, and photon energy response. These tests should be performed on a random selection of 10% of the instrument batch or four instruments, whichever is larger. If one instrument in a sample of a large quantity fails the test, an additional 10% should be tested. An additional failure would require testing of the entire batch. It should be noted that temperature response of instruments can vary with components and that large changes can be observed at temperature extremes that may be in the recommended range for an acceptance test.

A formal instrument calibration (ANSI N323 (4.7.1), 10 CFR 835.401(c)(1)) "shall" be performed on each instrument at least annually. The calibration shall (ANSI N323(4)) include a precalibration inspection/test normally followed by a documented calibration over the entire range of the instrument. Calibration for ranges where the instrument is not intended to be used need not be conducted, as long as the specific limitations on instrument use are clearly marked on the instrument. The frequency of calibration should be adjusted to the use of the instrument and its durability.

For more detailed discussions on radiation and radioactivity, see Cember (1996), Turner (1992), the Health Physics Society Web site at www2.hps/org/hps/, or consult a certified health physicist.

Figure 3.1 Instrument calibration check.

REFERENCES

Cember, H., *Introduction to Health Physics*, McGraw-Hill, New York, 1996.

DOE (U.S. Department of Energy), Instrument Calibration for Portable Survey Instruments, G-10 CFR 835/E1–Rev. 1, November, 1994.

Knoll, G.F., *Radiation Detection and Measurement*, 2nd ed., John Wiley & Sons, New York, 1989.

NCRP (National Council on Radiation Protection and Measurements), Ionizing Radiation Exposure of the Population of the United States, Report No. 93, Bethesda, MD, 1987.

Turner, J.E., *Atoms, Radiation, and Radiation Protection*, McGraw-Hill, New York, 1992.

CHAPTER 4

Sampling and Surveying Radiological Environments

Environments that frequently require radiological sampling, direct measurements, and/or scanning include those associated with uranium mine sites, nuclear weapons production facilities, nuclear reactors, radioactive waste storage and disposal facilities, and areas in the vicinity of nuclear accidents. The primary objectives of sampling, direct measurements, and/or scanning performed in these environments include the following:

- Defining the nature and extent of contamination;
- Evaluating contaminant migration pathways;
- Predicting rates of contaminant migration;
- Assessing the risk to human health and the environment;
- Evaluating cleanup alternatives;
- Determining whether or not remedial action or decontamination and decommissioning objectives have been met;
- Dispositioning of the waste material.

One of the objectives of this chapter is to provide the reader with guidance on how to design cost-effective sampling programs that are both comprehensive and defensible. This guidance emphasizes the use of the EPA's Data Quality Objectives (DQO) process (EPA, 1994a) to assist in the development of defensible sampling programs that meet all the sampling objectives. The DQO process requires one to state the problem clearly, identify the decisions that need to be resolved, identify inputs needed to resolve those decisions, specify data quality requirements (e.g., precision, accuracy, detection limits), define the temporal and spatial boundaries that apply to the study, define error tolerances (e.g., false positive, false negative, width of gray region), and develop a sampling design that meets these requirements.

This chapter presents details on several useful statistical sample design software packages, guidance to assist the writing of a Sampling and Analysis Plan, and details on the most effective radiological scanning, direct measurements, and environmental media sampling methods available to support environmental studies.

4.1 DESIGNING A DEFENSIBLE SAMPLING PROGRAM

Each year, the EPA and the regulated community spend approximately $5 billion collecting environmental data for scientific research, regulatory decision making, and regulatory compliance. To ensure that these data are of sufficient quality and quantity to support defensible decisions, the process of collecting and analyzing data itself must be scientifically defensible.

When designing a sampling program for an environmental study, the goal should be to collect data of sufficient quality and quantity to resolve all of the decisions that need to be made to complete the entire study. Since substantial cost is incurred with the mobilization and demobilization of a field sampling team, every effort should be made to perform all of the required sampling and analysis under one mobilization.

Figure 4.1 identifies all of the key steps that are required to develop and implement a defensible sampling program that supports the environmental decision-making process. This life-cycle process was modified after the process developed and implemented by Bechtel Hanford, Inc., Department of Energy, and EPA (1997).

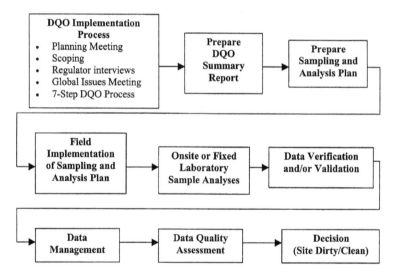

Figure 4.1 Data life cycle.

The following sections provide guidance on implementing each of the nine components of the data life cycle. If any one of the nine components is overlooked, the defensibility of the decision-making process will be severely impacted.

4.1.1 DQO Implementation Process

Prior to implementing the seven-step EPA DQO process, a number of preparatory steps must first be implemented. These steps include holding a project planning meeting, performing a thorough scoping effort, holding interviews with the regulators who will be involved in the decision-making process, and holding a Global Issues Meeting to resolve any disagreements with the requirements specified by the regulators, or disagreements between two or more regulatory agencies. If these preparatory steps are not implemented prior to beginning the seven-step process, the seven-step process will drag on for weeks or months because all of the required information needed to support the process will not be available.

4.1.1.1 *Planning Meeting*

The project manager should schedule and conduct a planning meeting with one or more technical advisors who have experience performing projects with a similar scope. The purpose of this meeting is to identify the project schedule, budget, staffing needs (DQO team), regulators, and procurement requirements. The size of the DQO team will vary between projects and is dependent upon the complexity of the problem. Examples of technical backgrounds that may be needed on the DQO team are provided in Section 4.1.5.1.3. The regulators are typically federal (e.g., EPA, NRC, DOE), state, and/or local regulators. Once the objectives of the planning meeting have been met, the project manager may begin the scoping process.

4.1.1.2 *Scoping*

An essential component to designing a defensible sampling program is scoping. Scoping involves the review and evaluation of all applicable historical documents, records, data sets, maps, diagrams, and photographs related to process operations, spills and releases, waste handling and disposal practices, and previous environmental investigations. The results from the scoping process are used in Step 1 (Section 4.1.1.5.1) of the DQO process to:

- Identify the contaminants of concern (COCs);
- Support the development of a conceptual site model;
- Develop a clear statement of the problem.

Since the results from the scoping process are used as the foundation upon which the sampling program will be designed, a project team should never attempt to rush through the scoping process in efforts to save money. Doing so could lead to the misidentification of the COCs, and the development of severely flawed conceptual site model and problem statement.

The scoping checklist presented in Table 4.1 identifies the key elements that should be researched during the scoping process. Table 4.1 is designed to assist the project manager in assigning scoping items to individual team members and documenting the results. A site visit should be scheduled following the completion of the scoping effort to familiarize the project team with the current site conditions. In addition, interviews should be scheduled and performed to verify that the information gathered during the scoping process is accurate, and to assist in filling in any informational data gaps. These interviews should include:

- Historical site workers/managers/owners;
- Federal, state, and/or local regulators;
- Potentially responsible parties (PRPs).

Interviews performed with federal, state, and/or local regulators should identify specific regulatory requirements that must be taken into consideration, and general concerns that they have related to the project. Examples of requirements and/or concerns expressed by regulatory agencies include:

- Cleanup guidelines
- Enforceable deadlines
- Waste classification and disposal requirements
- Preferred alternative actions for cleaning up the site
- Favored sampling and/or survey methods

Table 4.1 Scoping Checklist

Project Title:
Process Knowledge and Historical Information

1. Evaluation and Summary of Process Knowledge: Review historical records to identify the types of processes that were implemented at the site, the contaminants of concern, the types and estimated quantities of chemicals and radionuclides used, and any spills that may have occurred (note volume and type of chemical spilled).
    Person Assigned Responsibility:
    Summary:
2. Evaluation and Summary of Existing Information: Review all existing historical reports, analytical data, maps, diagrams, photographs, waste inventories, geophysical logs, drilling records, and other documents and/or records that could provide valuable information about the site under investigation.
    Person Assigned Responsibility:
    Summary:

Regulatory Issues

1. Identify All Applicable Regulatory Guidelines: Review regulations and agreements for the purpose of defining all applicable standards or risk-based guidelines for the various types of environmental media (e.g., groundwater, surface water, soil, air) that may be impacted at the site.
    Person Assigned Responsibility:
    Summary:
2. Identify all Agreements, or Regulatory/Statutory Obligations and Constraints: Identify all previously agreed upon milestones with regulatory agencies, waste acceptance criteria, and Applicable or Relevant and Appropriate Requirements (ARARs). Identify what regulatory pathway should be followed (e.g., RCRA, CERCLA, NEPA).
    Person Assigned Responsibility:
    Summary:
3. Identify All Regulatory Issues Pertaining to Waste Management: Identify Land Disposal Restrictions (LDR) that apply to waste material derived from the site. Identify waste acceptance criteria for potential waste disposal facilities.
    Person Assigned Responsibility:
    Summary:

Table 4.1 (continued) Scoping Checklist

Cultural Issues

1.   Identify All Cultural Issues That Must Be Taken into Consideration: Identify any Native American populations that may be impacted (e.g., burial grounds), or other cultural related issues that should be taken into consideration.
     Person Assigned Responsibility:
     Summary:

Ecological Risk Assessment Issues

1.   Identify All Ecological Issues That Must Be Taken into Consideration: Identify any threatened or endangered species of plants or animals that may be present in the vicinity of the site. Identify the potential contaminant sources, pathways, and receptors.
     Person Assigned Responsibility:
     Summary:

Human Health Risk Assessment Issues

1.   Identify All Human Health Concerns Associated with the Site: Use the identified existing information to perform a preliminary assessment of all potential chemical, radiological, and physical health hazards that may be encountered at the site. Identify the potential contaminant sources, pathways, and receptors.
     Person Assigned Responsibility:
     Summary:

Other Issues

1.   Identify Potential Data Uses: Identify all of the potential uses for existing or new analytical data. These uses may include defining the nature and extent of contamination, risk assessment, feasibility studies, treatability studies, remedial design, postremediation confirmation sampling, etc.
     Person Assigned Responsibility:
     Summary:
2.   Identify Sampling, Surveying, and/or Analytical Methods That Should Be Considered: Identify sample collection methods, field surveying methods, on-site laboratory methods, and laboratory methods that should be taken into consideration. Provide advantages and disadvantages to using each method, and general performance capabilities (e.g., detection limits, precision, accuracy).
     Person Assigned Responsibility:
     Summary:
3.   Identify Potential Risk Assessment Models That Should Be Considered: Provide the name of each model (e.g., RESRAD), the input requirements (e.g., Ra-226 activity levels in surface soil), and advantages and disadvantages of each.
     Person Assigned Responsibility:
     Summary:

4.1.1.3 Regulator Interviews

The regulators for a project may include federal (e.g., EPA, NRC, DOE), state, and/or local representatives who are responsible for making decisions on issues such as cleanup guidelines, deliverable milestone dates, waste classification and disposal requirements, preferred remedial alternatives, and/or favored sampling methods.

Prior to the commencement of the DQO process, the project manager should contact federal, state, and local regulatory agencies to identify the names of the representatives who will be involved with the project. Once the regulators have been identified, a 1- to 2-h interview should be scheduled with each of them individually. The purpose of the interview is to identify all of the key issues of concern that will need to be addressed by the DQO process. The project manager should consider bringing a few key technical experts to the regulator interviews to answer any technical questions that may arise.

4.1.1.4 Global Issues Meeting

A Global Issues Meeting is held whenever there are disagreements with any of the requirements specified by the regulators, or when the requirements from one regulator contradict the requirements of one of the other regulators. For example, the federal regulator may require the site under investigation to be remediated in accordance with the CERCLA process, while the state regulator may require the site to be remediated in accordance with the RCRA process. This meeting should be attended by the project manager, key technical project staff, and all of the regulators. All agreements made at the conclusion of the Global Issues Meeting should be carefully documented in a memorandum that is then entered into the document record.

4.1.1.5 Seven-Step DQO Process

The seven-step DQO process is a strategic planning approach developed by the EPA (EPA, 1994a) to prepare for data collection activities. This process provides a systematic procedure for defining the criteria that a data collection design should satisfy, including when/where/how to collect samples/measurements, tolerable limits on decision errors, and how many samples/measurements to collect. One of the advantages of the DQO process is that it enables data users and relevant technical experts to participate in the data collection planning process, where they can specify their specific data needs prior to data collection.

The DQO process should be implemented during the planning stage of an investigation prior to data collection. Using this process will ensure that the type, quantity, and quality of the environmental data used in the decision-making process will be appropriate for the intended application. The DQO process is intended to minimize expenditures related to data collection by eliminating unnecessary, duplicative, or overly precise data. In addition, the DQO process ensures that resources will not be committed to data collection efforts that do not support a defensible decision.

The DQO process consists of the seven steps identified in Figure 4.2. The output from each step influences the choices that will be made later in the process. Even though the DQO process is depicted as a linear sequence of steps, in practice, the process is iterative. In other words, the outputs from one step may lead to reconsideration of prior steps. This iterative approach should be encouraged because it will ultimately lead to a more efficient data collection design. During the first six steps of the process, the DQO team will develop the decision performance criteria, otherwise referred to as DQOs. The final step of the process involves developing the data collection design based on the DQOs.

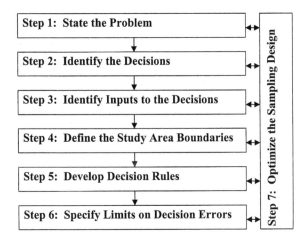

Figure 4.2 Seven steps that comprise the DQO process.

The first six steps must be completed prior to developing the data collection design. In Figure 4.2, the iterative link between DQO Steps 1 through 6 and DQO Step 7 "Optimize the Design" is illustrated by double arrows, which signify that it may be necessary to revisit any one or more of the first six steps to develop a feasible and appropriate data collection design. The DQO workbook template found in Appendix A and the accompanying CD-ROM should be used to help the user implement the DQO process. This workbook was developed as a joint effort between the author and Bechtel Hanford, Inc., technical staff. The results from the DQO process should then be used to support the preparation of a Sampling and Analysis Plan (see Section 4.1.1.7).

The key activities that are performed for each of the seven steps are summarized in Table 4.2. A more-detailed discussion of each of these steps is presented in the following sections.

4.1.1.5.1 Step 1: State the Problem

Objective: To define the problem clearly so that the focus of the project will be unambiguous.

Activities:

- Identify task objectives and assumptions.
- Identify members of the DQO team.
- Identify the regulators.
- Specify budget requirements and relevant deadlines.
- Identify the contaminants of concern.
- Develop conceptual site model.
- Develop a concise statement of the problem.

Table 4.2   DQO Seven-Step Process

DQO Step No.	Activities
Step 1: State the Problem	• Identify the contaminants of concern • Develop a conceptual site model • Formulate a concise problem statement
Step 2: Identify the Decisions	• Identify the principal study questions (PSQs) that the study will attempt to resolve • Identify the alternative actions that may result once each of the PSQs has been resolved • Join the PSQs and alternative actions to form decision statements
Step 3: Identify Inputs to the Decisions	• Identify the information needed to resolve each decision statement • Determine the source and level of quality for the information needed • Determine whether or not data of adequate quality already exist
Step 4: Define the Study Area Boundaries	• Define the population of interest and the geographic area/volume to which each decision statement applies • Divide the population into strata (statistical) that have relatively homogeneous characteristics • Define the temporal boundaries of the problem — Time frame to which each decision applies — When to collect the data • Define the scale of decision making
Step 5: Develop Decision Rules	• Define the statistical parameter of interest (e.g., mean) • Define the final action level • Develop decision rules which are "if … then …" statements that incorporate the parameter of interest, scale of decision making, action level, and alternative actions that would result from the resolution of the decision
Step 6: Specify Limits on Decision Errors	• Select between a statistical and nonstatistical sample design: — Define the expected concentration range for the parameter of interest — Identify the two types of decision error — Define the null hypothesis 1. Define boundaries of the gray region 2. Define tolerable limits for decision error
Step 7: Optimize the Sampling Design	*Nonstatistical Design:* • Provide summary of applicable surveying method alternatives • Provide summary of applicable sampling method alternatives • Develop an integrated sampling design *Statistical Design:* • Identify statistical sampling design alternatives (e.g., simple random, stratified random) and select the preferred alternative • Select the statistical hypothesis test for testing the null hypothesis • Evaluate multiple design options by varying the decision error criteria and width of the gray region • Select the preferred sampling design

Outputs:

- Administrative and logistical elements
- Concise statement of the problem

4.1.1.5.1.1 Background — In the process of defining the problem to be resolved, a combination of administrative and technical activities needs to be performed. The administrative activities include identifying the project objectives and assumptions, identifying the members of the DQO team and the regulators, and specifying budget requirements and relevant deadlines. The technical activities include identifying the contaminants of concern, developing a conceptual site model, and developing a concise statement of the problem.

4.1.1.5.1.2 Identify Task Objectives and Assumptions — This activity involves the development of a clear statement of the task objectives as they pertain to remedial activities. Initially identify the objectives on a large scale; then focus on the task-specific objectives. Identify all of the task-specific assumptions that have been made based on DQO team discussions and interviews with the regulators.

4.1.1.5.1.3 Identify Members of the DQO Team — The project manager identifies the members of the DQO team in the planning meeting (Section 4.1.1.1). The DQO team should be composed of technical staff members with a broad range of technical backgrounds. The number of members on the team should be directly related to the size and complexity of the problem. Complex tasks may require a team of ten or more members, while simpler tasks only require a few members. The required technical backgrounds for the DQO team members will vary depending on the scope of the project, but often include:

- Radiochemistry
- Hydrogeology
- Environmental engineering
- Radiation safety
- Statistics
- Groundwater/surface water/air modeling
- Quality assurance
- Waste management
- Risk assessment
- Remedial design

4.1.1.5.1.4 Identify the Regulators — The project manager identifies the regulators in the planning meeting (Section 4.1.1.1). Regulators are those who have authority over the study, and are responsible for making final decisions based on the recommendations of the DQO team. The regulators are often representatives of:

- Department of Energy
- Department of Defense
- Nuclear Regulatory Commission

- Environmental Protection Agency
- State regulatory agencies
- Local regulatory agencies

The project manager should encourage early regulator involvement since this will help ensure that the task stays on track.

4.1.1.5.1.5 Specify Budget Requirements and Relevant Deadlines — The project manager identifies the budget requirements and relevant deadlines in the planning meeting (Section 4.1.1.1). The specified budget requirement should include the cost for:

- Implementing the DQO process
- Preparing a DQO Summary Report
- Preparing a Sampling and Analysis Plan
- Implementing sampling activities
- Performing laboratory analyses
- Performing data verification/validation
- Performing data quality assessment
- Evaluating the resulting data

Identify all deadlines for completion of the study and any intermediate deadlines that may need to be met.

4.1.1.5.1.6 Develop a Conceptual Site Model — A conceptual site model is either a tabular or graphical depiction of the best understanding of the site conditions. The process of developing a conceptual site model helps one to identify any data gaps that may exist. The conceptual site model should identify:

- Primary and secondary sources of contamination
- Release mechanisms
- Pathways for contaminant migration
- Routes for exposure
- Receptors

In the example provided in Table 4.3 and Figure 4.3 the primary and secondary sources of contamination are the piles of chipped radiologically contaminated metal and the contaminated soils/sediment/water (surface water/groundwater) in the vicinity of the chipped metal piles, respectively. The contaminated metal chips were brought on site by railcar from another contaminated location. The contamination from the metal chips migrated into the surrounding soil, and was carried by surface water to a small nearby pond. Contamination was also transported by groundwater to a nearby drinking water well. The release mechanisms include wind, rain, pumping groundwater from the drinking water well, and/or human receptors walking through the contaminated area. The pathways for contaminant migration include air, surface water, and groundwater. The routes of exposure include the receptors breathing air containing radiologically contaminated dust particles, dermal contact with contaminated chipped metal and soil, and ingestion of contaminated soil, sediment, surface water, and/or groundwater. The receptors are humans living on the property, humans eating livestock/agricultural products from the property, and the surrounding ecological population.

The conceptual site model should be continuously refined throughout the implementation of the DQO process and sampling program.

4.1.1.5.1.7 Develop a Concise Statement of the Problem — Develop a concise problem statement that describes the problem as it is currently understood. The statement should be based on the conceptual site model described above.

4.1.1.5.2   Step 2: Identify the Decisions

Objective: Develop decision statements that address the concerns highlighted in the problem statement.

Activities:

- Identify the principal study questions (PSQs).
- Define the alternative actions.
- Join the PSQs and alternative actions into decision statements.

Outputs: Decision statements

4.1.1.5.2.1 Identify Principal Study Questions — The first activity to be performed under DQO Step 2 is identifying all of the PSQs. The PSQs are used to narrow the search for information needed to address the problem identified in DQO Step 1. The PSQs identify key unknown conditions or unresolved issues that reveal the solution to the problem. Note that only questions that require environmental data should be included as PSQs. For example, questions such as "Should a split-spoon or solid-tube sampler be used to collect the soil samples?" should not be included in the DQO process since this decision should be made based on experience and requires no analytical data to answer. The answers to the PSQs will provide the basis for determining what course of action should be taken to solve the problem.

4.1.1.5.2.2 Define Alternative Actions — Identify possible alternative actions that may be taken to solve the problem (including the alternative that requires no action). Alternative actions are taken only after the PSQ is resolved. Alternative actions are not taken to resolve the PSQ. Once the alternative actions have been identified, perform a qualitative assessment of the consequences of taking each alternative action if it is the wrong action. In other words, if one implements the no-action alternative when one should have implemented a soil removal action at a future day-care center, what are the potential consequences of this mistake?

Table 4.3 Example of a Tabular Conceptual Site Model

Primary Source of Contamination	Secondary Source of Contamination	Release Mechanisms	Migration Pathways	Exposure Routes	Receptors
Chips of radiologically contaminated metal brought onto the site by railcar	Contaminated soil immediately surrounding the chipped metal piles	Wind, rain, human receptors walking over soil	Air, surface water, groundwater	Inhalation, dermal contact, ingestion	Humans living on property, humans eating livestock/agricultural products, terrestrial ecological population, and birds of prey
	Contaminated sediment within drainage channels leading to a nearby pond	Wind, rain	Air, surface water, groundwater	Inhalation, dermal contact, ingestion	Humans living on property, humans eating livestock/agricultural products, terrestrial ecological population, and birds of prey
	Contaminated sediment at the base of a nearby pond	Percolation of pond water to underlying groundwater	Air, surface water, groundwater	Inhalation, dermal contact, ingestion	Humans living on property, humans eating livestock/agricultural products, and aquatic ecological population

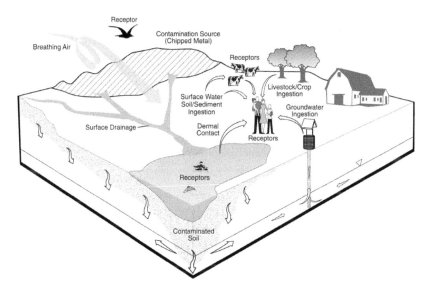

Figure 4.3 Graphical depiction of a conceptual site model

Express the consequences using the following terms:

- Low
- Moderate
- Severe

When assessing consequences, take the following aspects into consideration:

- Human health
- Environment (flora/fauna)
- Political
- Economic
- Legal

The results from this assessment will be used later in DQO Step 6 to assist in the selection between a statistical or nonstatistical sampling design.

4.1.1.5.2.3 Formulate Decision Statements — Decision statements are developed by linking PSQs with their corresponding alternative actions. The following format should be used when formulating the decision statements:

Determine whether or not [unknown environmental condition/issue/criterion from the problem statement] requires [taking alternative actions].

The following is an example of a correctly formatted decision statement:

Determine whether or not the sediment in the study area pond exceeds one or more cleanup guidelines and requires removal and disposal in a radiological landfill, or if no action is required.

4.1.1.5.3   Step 3: Identify Inputs to the Decisions

Objective: To identify the informational inputs that will be required to resolve the decision statements identified in DQO Step 2, and to determine which inputs require environmental measurements.

Activities:

- Identify information required to resolve each decision statement.
- Determine the source(s) for each item of information identified.
- Determine the level of quality required for the data.
- Evaluate the appropriateness of existing data.
- Identify the information needed to establish the action level.
- Confirm that appropriate analytical methods exist to provide the necessary data.

Outputs: Information needed to resolve decision statements

4.1.1.5.3.1 Identify Information Required to Resolve Each Decision Statement — Generally, all of the decisions identified in DQO Step 2 will be resolved by data (existing or new) from either environmental measurements or from scientific literature. Modeling may be used to resolve some decisions. However, all models require some input data (existing or new) to run the model.

For each decision statement, create a list of environmental variables of interest for which environmental measurements may be required. Examples of these variables include:

- U-238, Ra-226, Th-230, and Th-232 activity levels in shallow and deep soil;
- Metals concentrations in shallow and deep soil;
- U-238, Ra-226, Th-230, and Th-232 activity levels in surface water;
- Metals concentrations in surface water;
- U-238, Ra-226, Th-230, and Th-232 activity levels in groundwater;
- Metals concentrations in groundwater;
- Fixed and removable gross alpha and gross beta/gamma activity on building floor surfaces.

4.1.1.5.3.2 Determine the Source(s) for Each Item of Information Identified — Identify and list all of the potential sources of information that may be able to address each of the environmental variables identified in Section 4.1.1.5.3.1. Examples of potential data sources include:

- New data collection activities
- Previous data collection activities
- Historical records
- Scientific literature
- Regulatory guidance
- Professional judgment
- Modeling

4.1.1.5.3.3 Determine the Level of Quality Required for the Data — When determining the level of data quality required to resolve each decision, one should take into consideration human health, ecological, political, cost, and legal consequences of taking each of the alternative actions.

The four general levels of data quality presented from the lowest cost (highest detection limits) to highest cost (lowest detection limits) include:

- Field screening measurements
- On-site laboratory analyses
- Standard laboratory analyses
- CLP laboratory analyses

To minimize cost, one should consider collecting a larger number of low-cost field screening measurements and/or on-site laboratory analyses with 5 to 10% of the samples being sent to a fixed laboratory for confirmation analysis (standard or CLP analysis). This assumes that detection limit requirements can be met by the field screening or on-site laboratory analyses.

4.1.1.5.3.4 Evaluate the Appropriateness of Existing Data: Usability Assessment — To determine whether an existing data set is of adequate quality to resolve one or more decision statements, one must evaluate the results from the accompanying quality control data. Quality control data should include the results from the analysis of:

- Spike samples
- Duplicate samples
- Blank samples

The results from spike and duplicate samples are used to estimate the accuracy and precision of the analytical methods, respectively. Blank samples are analyzed to show the instrument was not contaminated from the analysis of previous samples.

When evaluating the appropriateness of existing data, one should also take into consideration:

- Instrument detection limits;
- Types of samples collected (e.g., grab, composite, integrated);
- Sample collection design (e.g., random, systematic, judgmental).

Remove data that are of poor quality, do not have low enough detection limits, or that are not representative of the population.

4.1.1.5.3.5 Identify the Information Needed to Establish the Action Level — The action level is the threshold value that provides the criterion for choosing between alternative actions. The action level may be based on:

- Regulatory thresholds or standards, e.g., maximum contaminant levels (MCLs), land disposal restrictions (LDRs);
- Problem-specific risk analysis, e.g., preliminary remediation goals (PRG).

Simply determine the criteria that will be used to set the numerical value for the action level. The actual numerical action level is set in DQO Step 5 (Section 4.1.1.5.5).

4.1.1.5.3.6 Confirm That Appropriate Analytical Methods Exist — For any new environmental measurements to be made, develop a comprehensive list of potentially appropriate measurement methods. Use the list of environmental variables of interest identified earlier in this step (Section 4.1.1.5.3.1).

4.1.1.5.4   Step 4: Define the Study Area Boundaries

Objective: To define the spatial and temporal boundaries that are covered by each decision statement.

Activities:

- Define the population of interest.
- Define the spatial boundaries of each decision statement.
- Define the temporal boundary of each decision statement.
- Define the scale of decision making.
- Identify any practical constraints on data collection.

Outputs: Definition of scale of decision making

4.1.1.5.4.1 Define the Population of Interest — It is difficult to make a decision with data that has not been drawn from a well-defined population. The term *population* refers to the total universe of objects to be studied, from which an estimate will be made. For example, if one is collecting surface soil samples from a football field to determine the U-238 concentration in the soil, the population is the total number of potential 1-kg soil samples that could be collected to a depth of 6 in. within the perimeter of the field.

Since it would be cost-prohibitive to sample and analyze every member of the population for U-238 concentrations, a statistical "sample" of the population is collected to provide an estimate of the U-238 concentrations in the population. Keep in mind this estimate will have error associated with it.

Other examples of the population of interest include:

- All subsurface soil samples (1 kg) within the area of interest to a depth of 15 ft;
- All surface water samples (1 L) within perimeter boundaries of the pond;
- All sediment samples (1 kg) from the top 0.5 ft of the lake bottom;
- All direct surface activity measurement areas (100 cm²) on the building floor and wall surfaces.

4.1.1.5.4.2 Define the Spatial Boundaries of Each Decision Statement — Define the geographic area to which each decision statement applies. The geographic area is a region distinctively marked by some physical feature, such as:

- Area (e.g., surface soil within the backyard of the Smith property);
- Volume (e.g., soil to a depth of 20 feet within the perimeter of the waste pit);
- Length (e.g., length of a pipeline);
- Some identifiable boundary (the natural habitat range of a particular animal/plant species).

It is often necessary to divide the population into strata (statistical) that have relatively homogeneous characteristics. Dividing the population into strata is desirable for the purpose of:

- Addressing subpopulations;
- Reducing variability;
- Reducing the complexity of the problem (breaking it into more manageable pieces).

Figures 4.4 and 4.5 provide examples of how two different sites may be stratified. The stratification approach presented in Figure 4.4 is based on current and past land use, while the stratification approach presented in Figure 4.5 is based on a site inspection or preliminary data results.

4.1.1.5.4.3 Define the Temporal Boundary of Each Decision Statement — Defining the temporal boundaries of a decision initially involves determining when to collect the data. In other words, one must determine when conditions will be most favorable for collecting data.

For example, a study to measure ambient airborne particulate matter may give misleading information if the sampling is conducted in the wetter winter months rather than the drier summer months. Several factors that should be considered when determining when to collect data include:

- Temperature
- Humidity
- Wind direction
- Rainfall
- Amount of sunlight
- Presence of receptors (e.g., site workers only present during work hours)

When defining the temporal boundaries, one should also take into consideration the time frame to which the decision applies. For example, if a residential property is determined to have radiologically contaminated soil in the backyard, the time frame to which a decision related to the risk condition of an average resident may be set at 8 years if that were the average length of residence for one family in the home.

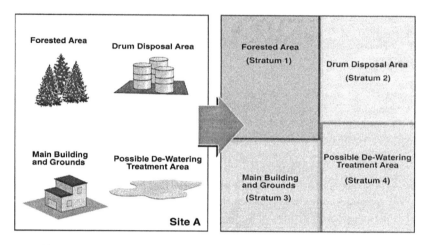

Figure 4.4 Example 1 of site stratification based on current/past land use.

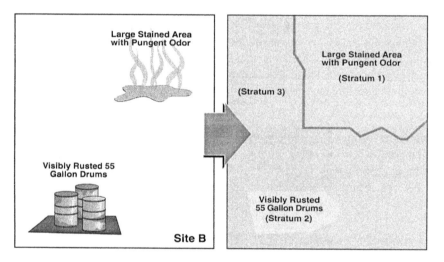

Figure 4.5 Example 2 of site stratification based on current/past land use.

4.1.1.5.4.4 Define the Scale of Decision Making — The scale of decision making merges the spatial and temporal boundaries described above. Two examples of scales of decision making are presented below:

1. Surface soil (top 0.5 ft) within the backyard of the Smith property over the next 8 years (sampling to be performed under clear weather conditions, preferably during summer months);
2. Soil to a depth of 20 ft within the perimeter of the waste pit over the next 6 months (sampling to be performed under clear weather conditions with minimal wind).

Figure 4.6 presents a graphical illustration of how the scale of decision making is defined.

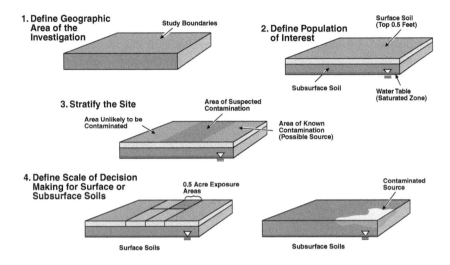

Figure 4.6 Defining the scale of decision making.

4.1.1.5.4.5 Identify Any Practical Constraints on Data Collection — Identify any constraints or obstacles that could potentially interfere with the full implementation of the data collection design. These should be taken into consideration in the sampling design and when developing implementation schedules. Examples of these constraints or obstacles include:

- Seasonal or meteorological conditions when sampling is not possible;
- Inability to gain site access or informed consent;
- Unavailability of personnel, time, or equipment;
- Presence of building or other structure that could prevent access to sampling locations;
- Security clearance requirements to access site.

4.1.1.5.5   Step 5: Develop Decision Rules

Objective: Combine the parameter of interest, scale of decision making, action level, and alternative actions to produce decision rules that provide a logical basis for choosing between alternative actions.

Activities:

- Specify the statistical parameter to characterize the population.
- Specify the action level.
- Develop a decision rule.

Outputs: If/then decision rule statements

4.1.1.5.5.1 Specify the Statistical Parameter to Characterize the Population — The statistical parameter of interest is a descriptive measure (e.g., mean, median) that specifies the characteristic or attribute that the regulators would like to know about the population. The statistical parameter of interest is used for comparison against the action level to determine whether or not the remedial action objectives have been met. Agreements made with regulatory agencies should specify which statistical parameter of interest should be used.

4.1.1.5.5.2 Specify the Action Level — Specify the numerical action levels for each of the contaminants of concern (COCs) for the purpose of allowing one to choose between alternative actions. The action levels may either be regulatory thresholds or standards (e.g., MCLs, LDRs), or they may be calculated based on a problem-specific risk analysis (e.g., PRG). The specified action levels will require regulator approval.

4.1.1.5.5.3 Develop a Decision Rule — For each of the decision statements identified in DQO Step 2, develop a decision rule as an "if ... then ..." statement that incorporates the parameter of interest, the scale of decision making, the action level, and the action(s) that would result from resolution of the decision.

The recommended decision rule format is as follows:

If the [parameter of interest] within the [scale of decision] is greater than the [action level], then take [alternative action A]; otherwise take [alternative action B].

Example 1

If the [mean activity of U-238 in the surface soil over the next 8 years] within [the perimeter of the backyard of the Smith property] is greater than [7 pCi/g], then [dispose of soil in a radiological landfill]; otherwise [leave the soil in place].

Example 2

If the mean concentration of cadmium in fly ash leachate over the next 6 months within [a container truck] is greater than [1 mg/kg], then [the ash waste will be considered hazardous and will be disposed of in an RCRA permitted landfill]; otherwise [the waste fly ash will be disposed of in a municipal landfill].

4.1.1.5.6   Step 6: Specify Limits on Decision Errors

Objective: Since analytical data can only provide an estimate of the true condition of a site, decisions that are based on such data could potentially be in error. The purpose of this step is to define tolerable limits on decision errors.

Activities:

- Select between a statistical and nonstatistical sample design.
- Determine the possible range of the parameter of interest.
- Identify the decision errors.
- Choose the null hypothesis.
- Specify the boundaries of the gray region.
- Assign tolerable limits on decision error.

Outputs:

- Boundaries of the gray region
- Decision error tolerances

4.1.1.5.6.1 Select between a Statistical and Nonstatistical Sample Design — Since analytical data can only estimate the true condition of a site or facility under investigation, decisions that are made based on measurement data could potentially be in error (i.e., decision error). For this reason, the primary objective of DQO Step 6 is to determine which decision statements require a statistical vs. a nonstatistical sample design. For those decision statements requiring a statistical design, DQO Step 6 defines tolerable limits on the probability of making a decision error. For those decision statements requiring a nonstatistical design, proceed to DQO Step 7 (Section 4.1.1.5.7).

The primary factors that should be taken into consideration in selecting a statistical vs. a nonstatistical design include:

- The qualitative consequences (low/moderate/severe) of decision error;
- The accessibility of the site or facility if resampling is required;
- The time frame over which each of the decision statements applies.

A statistically based sample design should be used whenever the consequences of decision error are moderate or severe. For example, if a child day-care center is proposed to be built on top of a site that has been remediated, the consequences of decision error (e.g., concluding the site is clean when it is contaminated) are severe since children would be exposed to contamination, and therefore a statistical sampling design is warranted.

The accessibility of the site if resampling is required will also impact the severity of the decision error. For example, if a remediated site is to be covered with 15-ft of backfill material after the site is pronounced clean followed by the construction of a building, the consequences of decision error are severe because the backfill material and potentially the building would need to be removed if it was later determined that the site was still contaminated and additional remediation was required. A statistical sample design is warranted for this situation.

The time frame for which a decision applies should be taken into consideration when defining the severity of decision error. For example, decisions that apply for only a few years have a less severe consequence than decisions that apply for 1000 years. For example, some decisions (e.g., those associated with defining the extent of contamination) apply only until the site is remediated (a few years). On the other hand, the decision that a remediated site is now clean may apply for thousands of years depending upon the half-life of the radionuclides involved.

A nonstatistical sampling design may be used whenever the consequences of decision error are relatively low. For example, if samples are being collected from a waste material simply to determine the most appropriate storage area for the waste while awaiting final characterization and disposition, a nonstatistical sampling design is adequate. The consequence of decision error in this case is just the inconvenience of moving the misidentified waste material after final characterization to the proper area to await disposal.

4.1.1.5.6.2 Nonstatistical Sample Designs — For those decision statements requiring a nonstatistical design, there is no need to define the range for the parameter of interest, types of decision errors, null hypothesis, boundaries of the gray region, or tolerable limits on decision error. These only apply to statistical sample designs. Proceed to DQO Step 7 (Section 4.1.1.5.7).

4.1.1.5.6.3 Statistical Sample Designs — The purpose in using a statistical sample design is to reduce the chances of making a decision error to a tolerable level. DQO Step 6 provides a mechanism for the regulator to define tolerable limits on the probability of making a decision error.

The two types of decision error that can occur include:

1. Walking away from a dirty site.
2. Cleaning up a clean site.

The decision error that causes one to walk away from a dirty site is the more serious of the two consequences since it could negatively impact human health and the environment. The decision error that causes one to clean up a clean site results in higher remediation costs.

Figure 4.7 illustrates the decision error of walking away from a dirty site. In this example, the true mean concentration (which only God knows) of an undefined isotope is above the action level of 100 pCi/g, while the sample mean (calculated based on a sample of the population) is below the action level. Since decisions are made based on the sample mean, the site is incorrectly determined to be clean. The action is to walk away from a dirty site. When the null hypothesis (Section 4.1.1.5.6.3.3) assumes the site to be contaminated until shown to be clean, the above error is referred to as an "α error" or a "false positive."

Figure 4.8 illustrates the decision error of cleaning up a clean site. In this example, the true mean concentration (which only God knows) of an undefined isotope is below the action level of 100 pCi/g, while the sample mean (calculated based on a sample of the population) is above the action level. Since decisions are made based on the sample mean, the site is incorrectly determined to be contaminated. The action is to clean up a clean site. When the null hypothesis assumes the

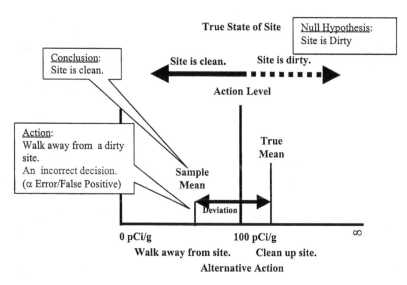

Figure 4.7 Decision error causing one to walk away from a dirty site.

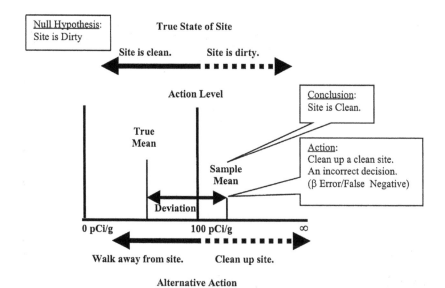

Figure 4.8 Decision error causing one to clean up a clean site.

site to be contaminated until shown to be clean, the above error is referred to as a "β error" or a "false negative."

To control the amount of decision error for a sample design it is necessary to define the range for the parameter of interest (e.g., mean, median), types of decision errors, null hypothesis, boundaries of the gray region, and tolerable limits on decision error. These criteria are needed to support statistical calculations performed in DQO Step 7 when determining the required number of samples/measurements to collect.

4.1.1.5.6.3.1 Determine the possible range of the parameter of interest — An initial step in the process of establishing a statistically based sample design is to define the expected range of the statistical parameter of interest (e.g., mean, median) for each COC. This should be defined using the results from historical analytical data. If no historical data are available, process knowledge should be used to estimate the expected range.

4.1.1.5.6.3.2 Identify the decision errors — The two types of decision error that can occur are walking away from a dirty site and cleaning up a clean site. For a site that is assumed to be contaminated until it is shown to be clean, the former is referred to as a "false positive" or "α error" (see Figure 4.7). The latter is referred to as a "false negative" or "β error" (see Figure 4.8). The α error has the more serious consequence since it results in contamination being left behind where human and ecological receptors could be impacted. The β error results in higher remediation costs because one is unnecessarily cleaning up soil that is below the action level.

4.1.1.5.6.3.3 Choose the null hypothesis — In the process of establishing a statistically based sample design, it is necessary to define the null hypothesis or baseline condition of the site. The two possible null hypotheses for environmental sites are

- The site is assumed to be *contaminated* until shown to be *clean.*
- The site is assumed to be *clean* until shown to be *contaminated.*

When selecting the null hypothesis, keep in mind that the null hypothesis should state the "opposite" of what the project eventually hopes to demonstrate. Since for an environmental site, the objective is almost always to show that a site is clean after remediation is complete, the null hypothesis should assume the site is contaminated until shown to be clean.

4.1.1.5.6.3.4 Specify the boundaries of the gray region — The gray region is a range of possible parameter values where the consequences of a decision error are relatively minor. It is bounded on one side by the action level, and on the other side by the parameter value where the consequences of decision error begin to be significant (Figure 4.9). It is necessary to specify the gray region because variability in the population and unavoidable imprecision in the measurement system combine to produce variability in the data such that a decision may be "too close to call" when the true parameter value is very near the action level.

In the example provided in Figure 4.9, the lower bound of the gray region is set at 80 pCi/g, and the upper bound of the gray region is set at the action level (100 pCi/g). In this example, the sample mean showed up above the action level when the true mean was actually within the gray region. This error is determined to be acceptable.

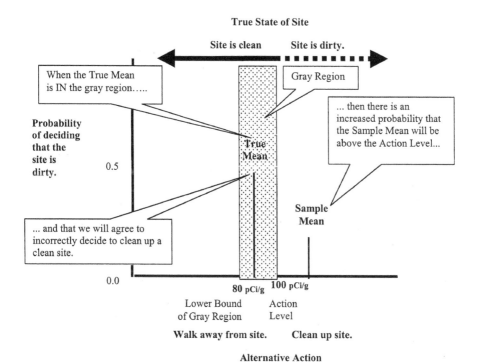

Figure 4.9 Boundaries of the gray region.

The lower bound of the gray region is one of many inputs used in statistical calculations to determine the total number of samples needed to resolve a decision. Setting a narrower gray region (e.g., 95 to 100 pCi/g) would result in lower uncertainty but higher sampling costs (larger number of samples required). On the other hand, a wider gray region (e.g., 70 to 100 pCi/g) would result in lower sampling costs but higher uncertainty (fewer samples required). While the default value for the lower bound of the gray region is typically set at 80% of the action level, it should be set on a project-by-project basis. When setting the lower bound of the gray region, keep in mind the consequences of decision error. For example, a less stringent (wider) lower bound of the gray region may be used to support waste disposition as opposed to site closeout sampling.

4.1.1.5.6.3.5 Assign tolerable limits on decision error — Assign probability values to points above and below the gray region that reflect the tolerable limits for making an incorrect decision. At a minimum, one should specify the tolerable decision error limits at the action level and at the lower bound of the gray region (Figure 4.10). The default value for α (false positive) errors is typically set at 5%, while the default value for β (false negative) is typically set at 20%.

One should consider evaluating the severity of the potential consequences of decision errors at different points within the domains of each type of decision error, since the severity of consequences may change as the parameter moves farther away

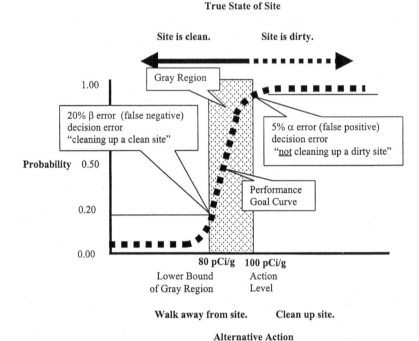

Figure 4.10 Example of decision performance goal diagram.

from the action level. This results in the creation of the performance goal curve shown in Figure 4.10.

4.1.1.5.7   Step 7: Optimize the Sampling Design

The objective of DQO Step 7 is to present alternative data collection designs that meet the minimum data quality requirements specified in DQO Steps 1 through 6. A selection process is then used to identify the most resource-effective data collection design. DQO Step 6 uses the severity of decision error consequences to differentiate between those decision statements that require a statistical sampling design from those that may be resolved using a nonstatistical design.

4.1.1.5.7.1 Nonstatistical Designs — Judgmental sampling is a nonstatistical sampling method that utilizes information gathered during the scoping process, or utilizes field screening instruments (e.g., gamma walkover surveys) to collect data that help focus the investigation on those areas that have the highest likelihood of being contaminated. Judgmental sampling designs provide data that represent the worst-case conditions for a site. For this reason, this type of sampling is most commonly used to support site characterization activities where the objective is to define the nature and extent of radiological contamination.

Judgmental sampling should not be used when collecting data to support post-remediation or postdecontamination and decommissioning site/facility closeout decisions since these data cannot be evaluated statistically. Even when performing site characterization activities, judgmental sampling should be combined with one or more statistical sampling approaches (e.g., simple random sampling, systematic sampling) since these data are often needed to support risk calculations, modeling studies, etc.

When a judgmental design is determined to be adequate to resolve one or more decision statements, the next step is to identify all potential surveying technologies and/or judgmental sampling methods that could potentially be used to provide the required data for each type of media (e.g., soil, concrete, paint). Sections 4.2 through 4.5 provide a number of scanning, direct measurement, and sampling methods that should be taken into consideration. The identified surveying and sampling methods should then be joined to form multiple alternative implementation designs.

Finally, the limitations and cost associated with each implementation design should be used to support the selection of the preferred implementation design. A Sampling and Analysis Plan is prepared following the completion of the DQO process in accordance with guidance provided in Section 4.1.1.7. Figure 4.11 provides a flowchart showing the implementation process for a judgmental sampling design.

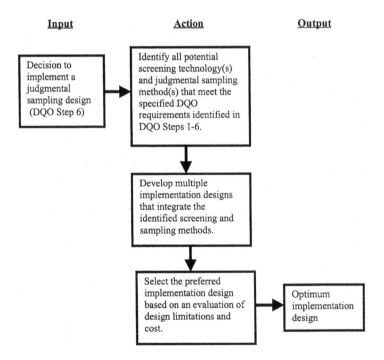

Figure 4.11 Implementation process for judgmental sampling design.

4.1.1.5.7.2 Statistical Designs — The purpose of this section is to provide general information on statistical sampling concepts. Reference is made to commonly used statistical hypothesis tests and formulae for calculating the number of samples required and confidence limits. However, this section is not intended as a technical discussion of these topics. Nor is the intent to provide all formulae needed to perform various statistical hypothesis tests. Rather, the purpose is to make the reader aware of what is commonly used, what is available, and where to find more-detailed information on topics of interest.

A statistical sampling design should be considered whenever the consequences of decision error are moderate or severe. Several commonly used statistical sampling designs are described below. However, before discussing particular sampling schemes, it is important to understand what happens, in general, when sampling occurs.

Suppose the concentration of U-238 needs to be determined for the surface soil present at a site that measures 1000×1000 ft. Further assume that the site is divided into 1-ft^2 sections for sampling purposes. That is, each 1 ft^2 is considered to be one sampling unit. In this example there are 1,000,000 possible surface soil samples that could be taken from this site. These 1,000,000 samples comprise the population of interest. Since it is cost- and time-prohibitive to collect and analyze all 1,000,000 samples from the population of interest, an alternative strategy is to select some smaller number of samples, and to use a single number such as the mean concentration of U-238 from this smaller number of samples to represent the site as a whole. In a nutshell, this is sampling.

One important thing to note at this point is that sampling provides an incomplete picture of the population of interest. Since only a few of all possible samples are taken from the population of interest, the data obtained from these samples are incomplete. Because the data are incomplete, the population will not be represented *exactly*, which could therefore lead one unknowingly to making an incorrect decision about the status of the population of interest.

Regardless of the history of the site under investigation, it is extremely unlikely that every sample (every square foot section) would have exactly the same concentration of U-238. Two questions now come to mind:

- If different sample units have different concentrations of U-238, what is the true concentration of U-238 for the site as a whole?
- How well can the mean of a subset of all possible samples represent the true mean of the site as a whole?

To answer these questions, a brief discussion of some basic statistical concepts is needed. This discussion is not intended to be a course in statistics, but rather a general and intuitive discussion of some basic concepts that underlie statistical thinking.

What is the true concentration of U-238 in the surface soil at the site? Suppose every possible sample could be taken from the surface soil at the site and measured for U-238. As stated earlier, not all measured concentrations would be the same—even if there were no analytical error. Some samples would truly contain a higher concentration of U-238 than others. To get one number that represents the site as a whole, the mean of all samples could be calculated. This number would be the *true* mean concentration of U-238 for the site. While it is convenient to have one number to represent the entire site, the mean does not give a complete picture of the concentration of U-238 for this site.

If measured concentrations from all possible samples were ordered from lowest to highest, a pattern would likely be observed. For example, there may be a few very low and a few very high scores, but the great majority of the scores would likely cluster around a single point in the middle. Figure 4.12 is a graphical representation of such a pattern. (Concentration level is on the *x*-axis, and frequency of occurrence is on the *y*-axis.) This kind of graphical representation is called a distribution. It reflects the fact that the measured concentrations obtained from all possible samples vary from one another. A distribution is a more complete way of describing the concentration of U-238 for the site. It provides information that cannot be deduced from the mean alone. First, by looking at a distribution, the full range of measured concentrations can be seen. Second, it is often easy to identify a central point around which most of the measured concentrations cluster. This is roughly equivalent to the mean. Finally, a distribution provides a sense of how tightly clustered or how widely spread out the measured concentrations are. This feature of a distribution is called variability, and can be summarized by calculating the standard deviation or variance of all the measured concentrations. The concept of variance has important implications for sampling, as will be discussed below.

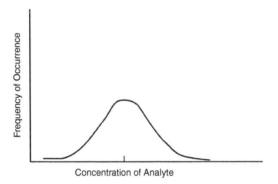

Figure 4.12 Graphical illustration of a distribution.

Three distributions commonly observed when working with environmental data sets are illustrated in Figure 4.13. These are theoretical or idealized distributions that can be thought of as hypothetical models for the site. The population distribution of real environmental data would never take the shape of one of these theoretical distributions exactly. However, the population distribution is often "close enough" to one of these theoretical distributions that it can provide a good approximation. This has tremendous benefits for data analysis.

The first distribution is often referred to as a "bell-shaped" or normal distribution. Its salient features are that it is symmetrical and unimodal—that is, it has only one "center" or mode around which a large percentage of measured concentrations clusters.

The second distribution is called a lognormal distribution. Note that it is very much like the normal, except that it contains a small number of extremely high measured concentrations. This causes the distribution to become asymmetrical, although it is still unimodal. A lognormal distribution may indicate the presence of hot spots, which create the upper tail of the distribution.

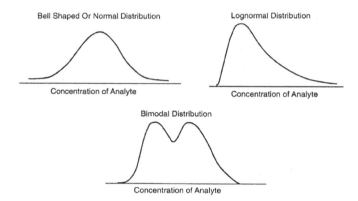

Figure 4.13 Distributions commonly observed in environmental studies.

The third distribution is called a bimodal distribution. Note that there are two "centers" or modes around which measured concentrations tend to cluster. A bimodal distribution can be either symmetrical or asymmetrical. In either case, it often indicates that the site should be subdivided in strata (see Figures 4.4 and 4.5). For example, suppose that half the site was never contaminated and the other half was heavily contaminated. If each half (or stratum) were evaluated separately, the distribution for each stratum would likely be unimodal (either normal or lognormal). However, if the site was not stratified but rather evaluated as a whole, a bimodal distribution would result. The data clustered around the lower mode represent the uncontaminated stratum, while the data clustered around the higher mode represent the contaminated stratum. If a bimodal distribution is encountered, it is a good idea to review historical information to determine whether or not the site can be subdivided into strata.

As mentioned above, distributions provide a good illustration of how tightly clustered or how widely spread out the measured concentrations in the population are. In statistical terminology, this is referred to as the "level of dispersion" or "variance" of a distribution. Numerically, the level of dispersion is summarized by calculating the standard deviation of the measured concentrations. In other words, the standard deviation provides one number that presents the dispersion of the measured concentrations in the population. The standard deviation plays an important role in calculating the number of samples that are required. Figures 4.14 and 4.15 illustrate different levels of dispersion or variability for the normal and lognormal distribution.

To recap, while it is often useful to characterize a site in terms of a single number, such as the mean or standard deviation, it is more appropriate to think of the true concentration of the site as a distribution of values.

How well can the mean of a subset of all possible samples represent the true mean of the site as a whole? Unless all possible samples are taken, the true mean concentration of U-238 for the site cannot be known. The only reasonable alternative is to take a subset of all possible samples and to calculate the mean of this subset. The mean of the subset of samples is, then, an *estimate* of the mean of the population. Should the mean of just a few samples be trusted to estimate the true mean of the site? The answer to this question depends primarily (although not solely) on how much

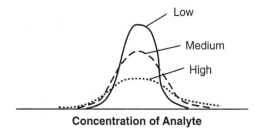

Figure 4.14 Differing levels of variability for normal distributions.

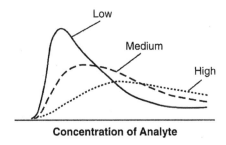

Figure 4.15 Differing levels of variability for lognormal distributions.

variance there is in the concentration of U-238 for the site as a whole. If the measured concentrations of U-238 in the population are tightly clustered (have low variance), then the mean of a few samples does a good job representing the site. However, if the measured concentrations of U-238 in the population are widely spread out, then the mean of a few samples is less likely to give an accurate representation of the site.

In either case, the mean of a few samples will never be exactly the same as the true mean of the population. So when the mean of a few samples is used to make a decision about the true condition of the site, decision errors have some real (nonzero) probability of occurring. For example, the mean of the subset of samples may lead to the conclusion that the true mean concentration of U-238 is below a specified threshold, when in fact it is actually above that threshold. The reverse may also happen where the mean of the subset of samples leads to the conclusion that the site is above the threshold, when in reality it is below the threshold. For a more-detailed discussion of decision errors, see Section 4.1.1.5.6.3.

To summarize, sampling and analyzing a subset of all possible sampling units allows one to make an educated guess or inference about the population of interest. This can be done by calculating the mean of the samples taken and using this as an estimate of the true population mean. In addition, a distribution of the measured concentrations can be used to make inferences about the shape and variance of the population. The sections below discuss several commonly used methods for selecting the subset of all possible samples. Keep in mind that any sampling approach that collects only a subset of all possible samples has some probability of leading to a decision error.

4.1.1.5.7.2.1 Simple random sampling — When little historical information about the site exists, simple random sampling is a good choice. Simple random sampling is implemented by dividing the site into possible sampling units (i.e., the area that will be represented by a single sample). A site may be divided into sampling units by overlaying a grid with spacing determined by the minimum amount of area required to collect a sample with the equipment being considered. For example, the smallest area that may be excavated by a backhoe when collecting a sample may be 10 ft^2.

Each sampling unit within the population is initially assigned a number. A subset of sampling units is then chosen by drawing the assigned numbers at random. Each of the sampling units assigned a number has an equal probability of being chosen for sampling. If some sampling units are inaccessible, they are not assigned numbers and therefore have zero probability of being selected. Because of this, they cannot be considered as part of the population and the results of any statistical hypothesis tests do not apply to these sampling units.

The biggest advantage to using a simple random sampling scheme is that it can be used when little or no historical information is available and it can provide data for almost any statistical hypothesis test, such as comparing a mean concentration to an action level. The greatest disadvantage is that the number of samples needed may be larger than that for other sampling strategies.

4.1.1.5.7.2.2 Stratified random sampling — When historical information provides sufficient detail to partition the site into relatively homogeneous, nonoverlapping areas or strata, stratified random sampling is superior to simple random sampling. Sampling units are created and numbers assigned to sampling units in the same manner as with simple random sampling. However, the selection of the sampling units varies slightly. There are two different approaches that can be used to select sampling units.

If a separate decision is to be made about each stratum, randomly choose the necessary number of samples from the specified stratum. If one decision is to be made for the site as a whole, sampling units need to be chosen from each stratum. The number of sampling units chosen from each stratum should be proportional to the size or volume of the stratum. For example, if a site has three strata such that the first covers 50% of the site, the second covers 30% of the site, and the third covers 20% of the site, then 50% of the samples should come from the first stratum, 30% from the second, and 20% from the third.

The advantage of stratified random sampling is that often fewer samples are required. This is primarily because each stratum is relatively homogeneous—or has low variance. A disadvantage may be the research required to identify relatively homogeneous, nonoverlapping strata appropriately.

4.1.1.5.7.2.3 Systematic sampling — Systematic sampling designs may be used when the objective is to search for leaks or spills, to determine the boundaries of a contaminated area, or to determine the spatial characteristics of a site. Basically, samples are collected from an evenly spaced grid where the starting point is randomly chosen.

To create a systematic sampling design, random coordinates from within the area are chosen for the first sample location. This establishes the initial location or reference point from which the grid is built. Methods for selecting systematic sampling locations are discussed in more detail in EPA (1992a). Once the grid is established, samples are systematically taken from the nodes or cross lines of the grid. The grid may be square, rectangular, or triangular in shape.

Systematic sampling may introduce a certain type of sampling bias. Because sampling occurs at the nodes, small areas of contamination may be missed if they are entirely within the grid. This could result in underestimating the contamination of the site. Conversely, if the spread of the contamination is very similar to the grid pattern, overestimation of the contamination could occur. Because of these factors, care must be taken in choosing both the size and type of sampling grid to be used. Sampling on a triangular grid pattern is often preferred because it reduces the possibility of sampling bias.

Systematic sampling can be used when the goal is to determine whether leaks or spills have occurred over a relatively large area (i.e., when the size of the potential spill or leak is small compared with the area of interest). This type of sampling is often referred to as "hot spot" sampling, where a hot spot is defined as a localized area of relatively high contamination (Gilbert, 1987). The information required for this method includes the size and shape of the site, the size and shape of a single hot spot, the type of grid that will be used, and the concentration level that defines a hot spot. From this information, the following questions may be answered:

- What grid spacing is required to find a single hot spot of a specified size with a given probability?
- What is the minimum size hot spot that can be detected for a specified grid spacing and detection probability?
- What is the probability of detecting a specified size hot spot for a given grid spacing?

This hot spot sampling procedure is designed to detect a single hot spot of a given size. It can be modified to allow for multiple hot spots; this modification is discussed in both Gilbert (1982) and EPA (1992a).

Systematic sampling can also be used when the objective is to define the boundaries of a contaminated area. The method used to employ this strategy is the same as above, but the goal is detecting a location at which the contamination drops off to a certain level, rather than to detect a hot spot of a certain concentration. Therefore, the grid spacing should be determined by the level of precision required to determine the boundaries of contamination. For example, if it is important to define precisely where the contamination levels drop to a certain concentration, a fine grid spacing is required. On the other hand, if precision is not as important, the grid size can be increased and each sample location will represent a larger area of the site.

Another instance in which a systematic design may be employed is when the spatial characteristics of the site are of interest. Systematic sampling works well to define the gradations of contamination in two and three dimensions and is called geostatistical sampling under this scenario. This type of information may be important when determining what remedial alternative is most appropriate, or when modeling contamination transport. When using geostatistical sampling, historical data are needed so that the two- or three-dimensional correlation patterns may be established. Samples are then collected to augment historical data and fine-tune the gradations in concentration contours. For additional information on spatial statistical sampling designs see Isaaks and Srivastava (1989).

4.1.1.5.7.2.4 Sequential sampling — When the site is expected to be either minimally or maximally contaminated, sequential or adaptive sampling can often dramatically reduce the number of samples required. Unlike the other sampling approaches discussed above, a sequential sampling design does not define the required sample size in advance. Instead, after a few samples are collected, a decision is made either to reject the null hypothesis (e.g., decide the site is clean), to fail to reject the null hypothesis (e.g., decide the site is contaminated), or to continue sampling.

Sequential sampling involves performing a statistical hypothesis test as results become available, rather than waiting until all the sampling results are in before running the test. The statistical hypothesis test is used to determine if the collection of additional samples is required to support the decision that the site meets or does not meet the cleanup standard. This sampling method is useful when using fast turnaround or field analytical methods in which results can be quantified very quickly. However, it should not be used in situations where the collection of an additional sample or samples requires remobilization. A more thorough discussion of sequential sampling is provided in EPA (1992a) and Bowen and Bennett (1988).

4.1.1.5.7.2.5 Factorial sampling — Factorial sampling is used primarily when sampling equipment or conducting experiments that involve holding time or analytical comparisons. When sampling equipment, it is often beneficial to categorize the equipment according to factors that might influence the level of the analyte being measured, such as the frequency with which the equipment was used. For example, dump trucks used on a daily basis over a 1-year period might reasonably be expected to be more contaminated than dump trucks used once a month over that same 1-year period. Dump trucks could be categorized in terms of "high" or "low" frequency usage and then sampled proportionately. Another example where factorial sampling is useful when various preparatory and analytical methods are used. A given measure is then categorized according to both the preparatory and analytical method that was used to obtain the measure.

In experimental settings, the goal may be to determine if one or more factors significantly contributes to the variability of the measured analyte or if any interactions between factors are significant. A general factorial design involves selecting a number of "levels" for one or more factors and collecting samples from each combination of levels. For example, if the equipment inside a storage facility needs to be characterized for disposal, the factors that may be important include the origin of the equipment, the process the equipment was used in, and the equipment size. The levels of equipment origin may be, for example, the Uranium Plant, the Reduction-Oxidation Facility, and the Plutonium-Uranium Extraction Facility. Levels of process may be material extraction, waste treatment, and product precipitation. Levels of size may be "small," "medium," and "large," with specific dimensional or volumetric definitions for each. Equipment within the facility would be categorized into one level for each factor; then a subset of each combination would be sampled to determine the mean or maximum concentration associated with all or subsets of the equipment.

4.1.1.5.7.2.6 Statistical hypothesis testing — Once the appropriate sampling design has been selected, the next step is to identify the *preferred* statistical hypothesis test to test the null hypothesis developed earlier. Running a statistical hypothesis test to determine whether a site meets the appropriate action levels is the general approach taken by EPA (1992a) and EPA (1996). After a brief discussion of statistical hypothesis testing, a number of statistical hypothesis tests will be outlined.

Although a comprehensive discussion of hypothesis testing is beyond the scope of this book, the reader should be aware that the results of a statistical hypothesis test are basically a "pass/fail" decision. "Passing" the statistical hypothesis test equates to deciding that the site is contaminated, while "failing" the statistical hypothesis test equates to deciding the site is clean. Recall that either decision may be in error, since the decision was made on the basis of collecting some small subset of all possible samples.

The interpretation of the results of any statistical hypothesis test is essentially the same, even though the details of conducting the test may vary. The goal is to reject the null hypothesis and have a high degree of confidence that the site is not contaminated. To make this determination, an "observed" statistic is calculated from the samples that were collected. This observed statistic is compared with a tabled value, often referred to as "critical" value for the statistic. If the observed statistic is significantly less than the critical statistic, the null hypothesis is rejected and the sample data support the conclusion that the site is not contaminated. If the observed statistic is significantly greater than the critical statistic, then the null hypothesis cannot be rejected and the sample data support the conclusion that the site is contaminated.

Note that in either case the true state of the site has not been determined with absolute certainty. It has not been "proved" that the population mean is above or below the Action Level. A statistical hypothesis test can only provide evidence that either supports or fails to support the null hypothesis. Nothing is ever proved with hypothesis testing because there is always some probability of making a false-positive or false-negative decision error.

Parametric and Nonparametric Statistical Hypothesis Tests. There are two basic kinds of statistical hypothesis tests: parametric and nonparametric. Both types of tests have assumptions that must be met before the results of the test can be meaningfully interpreted. However, the assumptions of a nonparametric test are often less stringent than those for the corresponding parametric test. Since the choice between a parametric and nonparametric test is based, in part, on assessing whether certain statistical assumptions have been met, it is wise to seek the advice of a statistician when making this choice. A detailed discussion of the assumptions underlying various parametric and nonparametric tests is beyond the scope of this book. However, it is important to verify any applicable assumptions before proceeding with the chosen statistical hypothesis test.

PARAMETRIC TESTS. Parametric tests are often described as having "distributional assumptions" that must be verified before the parametric test is valid. Many discussions of parametric methods state or infer that the sample data must come from a population with some known theoretical distribution, such as the normal or lognormal distribution. Discussions of this type are misleading. Parametric tests *do* make distributional assumptions—but only indirectly about the population from which the sample was drawn. Instead they make assumptions about the distribution of the sample statistic of interest. This distribution is called a sampling distribution.

The concept of a sampling distribution is best illustrated with an example. Suppose a researcher took ten samples from a population and calculated the mean of those ten samples. Now suppose the researcher repeated this process—taking a second set of ten samples and calculating a second mean. In all likelihood, the two means will be different because the two sets of ten samples were different. Now suppose the researcher repeats the same process of taking ten samples and calculating the mean an infinite number of times. The researcher now has an infinite number of means and can create a probability distribution from them. This distribution is called the sampling distribution of sample means. Sampling distributions can also be generated for other statistics such as the standard deviation.

Parametric statistical hypothesis tests, then, make assumptions about the shape of sampling distribution of the statistic of interest. A more in-depth discussion of this topic is beyond the scope of this book. However, it is important to note that the distributional assumption of the parametric test *must* be verified before the test is run. If the assumptions cannot be verified, it is inappropriate to run the parametric statistical hypothesis test since the results are unpredictable. For example, the results from a t-test cannot be meaningfully interpreted if the assumptions of the t-test are not valid. For this reason, the results from parametric tests should not be used for decision making unless the distributional assumptions of the test have been verified.

Since the preferred statistical hypothesis test is often a parametric test, this caveat is particularly worth noting. A common, but dangerous, practice is to select a parametric test as the preferred statistical hypothesis and then simply conduct that test once the data have been collected—without checking the assumptions of the test. If the assumptions happen to hold, the results from the parametric test are meaningful. However, if the assumptions do not hold, there is no way to predict whether the results over- or underestimate the true condition of the site. A researcher who has not checked his or her assumptions has no idea whether the results are meaningful or meaningless.

NONPARAMETRIC TESTS. Many discussions of nonparametric statistical hypothesis tests refer to them as "distribution-free statistical methods" or state that nonparametric tests make no assumptions about the shape of the population. This is not entirely true. Most nonparametric procedures do require that the samples be independent and some require that the population from which the samples were drawn be symmetrical. Nonetheless, the assumptions of nonparametric statistical hypothesis tests tend to be much less stringent than the assumptions of corresponding parametric tests. So in situations where the population distribution is either unknown or has some distribution other than normal, a nonparametric statistical hypothesis test may be the most appropriate choice.

COMPARISON OF PARAMETRIC AND NONPARAMETRIC STATISTICAL HYPOTHESIS TESTS. Parametric methods rely on assumptions about a sampling distribution to determine whether the null hypothesis may be rejected or not. These methods are better only if the assumptions are true. If the underlying distribution is known, a parametric test can make use of that additional information. However, if the sampling distribution is different from that assumed, the results can be unpredictable.

A primary advantage of using nonparametric methods is that they can be used for survey measurements at or near background, when some of the data are at or below instrumental detection limits. These data are sometimes reported as "less than" or "nondetects." Such data are not easily treated using parametric methods. It is recommended that the actual numerical results of measurements always be reported, even if these are negative or below calculated detection limits. However, if it is necessary to analyze data that include nonnumerical results, nonparametric procedures based on ranks can still be used in many cases.

While there are many advantages to a nonparametric test, the approach does have some drawbacks. Nonparametric tests often require more samples than parametric tests to have the same ability (power) to reject the null hypothesis. Equations for conducting nonparametric statistical hypothesis tests are available, but formulas for calculating required sample sizes can only be developed through simulation.

Many nonparametric techniques are based on ranking the measured data. The data are ordered from smallest to largest and assigned numbers or ranks accordingly. The analysis is then performed on the ranks rather than on the original data values. Nonetheless, nonparametric methods perform nearly as well as the corresponding parametric tests, even when the conditions necessary for applying the parametric tests are fulfilled. There is often relatively little to be gained in efficiency from using a specific parametric procedure, but potentially much to be lost. Thus, it may be considered prudent to use nonparametric methods in most cases.

Selecting the Preferred Statistical Hypothesis Test. The discussion below focuses on statistical hypothesis tests that are appropriate when simple random sampling has been selected as the optimal sampling design. Statistical hypothesis tests appropriate for other sampling designs (sequential, stratified random, etc.) are similar in philosophy, but more complex in implementation. If another sampling design has been chosen, consult a statistician to determine what statistical hypothesis tests are appropriate.

Note that before actually performing the preferred statistical hypothesis test, it is imperative to test whether or not the data collected meet the assumptions of that test. If they do not, an alternative statistical hypothesis test will need to be used. Given this, it is wise to collect enough data to verify the assumptions of the preferred statistical hypothesis test and ensure that the data collected will also be applicable for the alternative test, if it should become necessary to implement it.

The selection of the preferred statistical hypothesis test is determined by:

- The comparison to be made. Different statistical hypothesis tests are appropriate depending on whether one population is to be compared to an action level (e.g.,

risk-based cleanup guideline), or whether two populations are to be compared to one another (e.g., site level vs. background level).
- The parameter of interest. Different statistical hypothesis tests are appropriate for different parameters of interest.
- The type of statistical hypothesis test desired: parametric or nonparametric.

Table 4.4 presents a number of commonly used parametric and nonparametric statistical hypothesis tests. To use this table to select the preferred statistical hypothesis test, first identify the type of comparison to be made. When site closure is the issue, the appropriate comparison is often one population vs. an action level or regulatory threshold. However, the appropriate comparison may also be between two populations, such as background vs. site.

Next, recall the parameter of interest, which was identified earlier in the DQO process. Table 4.4 only lists the two most commonly used parameters of interest—the population mean and the population proportion or percentile. One should consult a statistical reference if the DQO process has identified some other parameter of interest.

Now, determine if enough historical data exist to verify the assumptions of the test. If no historical information is available, then the most reasonable choice is a nonparametric test. If historical information is available, it should be statistically evaluated at this point to determine if the assumptions of the preferred test are warranted. This is particularly important if the preferred test is a parametric test. Of course, sample data can be used to verify assumptions *after* it is collected. However, this course of action bears great risk. If only a few samples are collected, and the assumptions of the preferred test cannot be verified, then there may not be enough data to perform the corresponding nonparametric test.

Each of the statistical hypothesis tests in Table 4.4 is briefly discussed below. In keeping with other statistical discussions in this book, the goal here is not to offer a course on statistics. Rather, the goal is to give the reader a general understanding of when it is appropriate to use each test. The reader should note that all the statistical hypothesis tests discussed below proceed under the assumption that data collected are independent. This is most often assured by performing simple random sampling, which precludes any type of systematic bias from affecting the results of the sampling. Each test also makes additional assumptions that are not discussed here. Consult a statistician or statistical reference for more-detailed information about the assumptions that apply to each statistical hypothesis test discussed below. The general interpretation of the results of a statistical hypothesis test was discussed earlier and is not repeated here for each test.

One-Sample Statistical Hypothesis Tests:

One-Sample t-Test

The one-sample t-test is a parametric test that is used to determine, with some level of confidence, whether a single population mean (as estimated by a sample mean) falls at or below some prespecified limit. This is often the case when a remediated site must be declared "clean enough" to be released for public use. In this case, the population mean would be the mean concentration of the site, and the prespecified limit would be the action level.

Table 4.4 Common Parametric and Nonparametric Statistical Hypothesis Tests

Comparison	Parameter of Interest	Type of Statistical Hypothesis Test	Statistical Hypothesis Test
One population to action level	Mean	Parametric	One-sample t-test
		Nonparametric	Wilcoxon Signed Rank Test
	Proportion or percentile	Parametric	One-sample proportion test
Two populations	Mean	Parametric	Two-sample t-test (equal variances)
		Parametric	Satterthwaite's two-sample test (unequal variances)
		Nonparametric	Wilcoxon Rank Sum Test
	Proportion or percentile	Parametric	Two-sample test of proportions

When using a t-test, it is important to check the data for extremely high or low data values. When the sample size is small, the sample mean will be unduly influenced by extreme values. Therefore, it is important to check that no single value is causing the mean to be unduly high or low.

Wilcoxon Signed Rank Test

The Wilcoxon Signed Rank Test is the nonparametric version of the one-sample t-test. It can be used to determine, with some level of confidence, whether a single population mean or median falls at or below some prespecified limit. As a general rule of thumb, when an estimate of the mean is being compared against an action level, the Wilcoxon Signed Rank Test should be selected over the one-sample t-test, unless it can be demonstrated that the assumptions of the t-test have been met.

One-Sample Test of a Proportion or Percentile

A population proportion is the ratio of the number of samples that *have* some characteristic to the total number of samples in the population. A population percentile represents the percentage of samples in the population having values less than some specified threshold. An example of a one-sample test of proportions would be to determine whether, in the population, the proportion of samples having 1 pCi/g of U-238 (or less) was greater than 80%. An example of a one-sample test of percentiles would be to determine whether the population value at the 80th percentile was 1 pCi/g of U-238 (or less). These are subtly different to a statistician, but can be used interchangeably for most applications. However, note that 80% UCL on the arithmetic mean is *not* an estimate of the 80th percentile.

When testing a hypothesis that contains a proportion or percentile, it is important that the result of analyzing a sample be classified in a binary fashion—as either a "success" or "failure." In the example above, if samples with less than 1 pCi/g of U-238 are classified as a "success" and samples with more than 1 pCi/g of U-238 are classified as a "failure," then this criterion is met. However, if the concentration of U-238 was classified as "high," "medium," or "low," then the criterion would be violated and the test for proportions would be inappropriate.

TWO-SAMPLE STATISTICAL HYPOTHESIS TESTS. In concept, two-sample statistical hypothesis tests are not much more difficult to understand than one-sample tests; however, they tend to be much more difficult to calculate. Instead of comparing a population mean or population proportion to a fixed value such as an action level, two population means or proportions are compared against each other. In general, the null hypothesis is stated in terms of the difference between the two population means or proportions. In other words, claiming that Site 1 has a higher mean concentration of U-238 than Site 2 is equivalent to saying that the mean of Site 1 minus the mean of Site 2 is greater than zero. The null hypothesis can be set up to test one of the following three claims:

1. One population has a mean or proportion greater than or equal to the other.
2. One population has a mean or proportion less than or equal to the other.
3. One population has a mean or proportion different from (simply not equal to) the other.

Two-Sample t-Test
   A two-sample t-test involves comparing the means of two populations. Examples include:
   • A comparison of the site before remediation (Population 1) with the site after remediation (Population 2).
   • A comparison of a potentially contaminated site (Population 1) with a site that represents area background levels (Population 2).
   • A comparison between an upgradient (Population 1) and downgradient (Population 2) well.
   If the null hypothesis is rejected, the two populations are said to have significantly different means. Depending on how the null hypothesis is set up, the interpretation may be that one population has a mean that is significantly greater than the other, less than the other, or just different from (not equal to) the other.
   One of the assumptions of the two sample t-test is that the populations from which the two samples are drawn have the same degree of variability or have equal variances (a condition called homoscedasticity). If this assumption does not hold, then Satterthwaite's two-sample t-test, discussed below, should be used.

Satterthwaite's Two-Sample t-Test
   When the two samples come from populations that do not have equal variances, the two-sample t-test discussed above cannot be used. In this case, Satterthwaite's two-sample t-test is the appropriate alternative.

Wilcoxon Rank Sum Test
   Just as the Wilcoxon Signed Rank Test is the nonparametric alternative to a one-sample t-test, the Wilcoxon Rank Sum Test is the nonparametric alternative to a two-sample t-test. The same general rule of thumb applies. If the assumptions of the two-sample t-test cannot be verified, the Wilcoxon Rank Sum Test should be chosen in lieu of the two-sample t-test. This test does not require that the populations from which the two samples were drawn have equal variances.

Two-Sample Test of Proportions or Percentiles

A two-sample test of proportions involves comparing the proportions or percentiles of two populations. Comparisons might be similar to those listed for the two-sample t-test, but where the parameter of interest is a proportion or percentile instead of a mean. As with other two-sample tests, the null hypothesis is stated in terms of the difference between the proportions or percentiles.

Confidence Intervals: Alternative to Statistical Hypothesis Testing. An alternative to performing a statistical hypothesis test is to construct a confidence interval estimate of the population mean. When the sample mean alone is used as an estimate of the population mean, this is referred to as a "point" estimate because the sample mean is a single value. By contrast, a confidence interval is a range of values that are used to estimate the population mean. Suppose, for example, that a mean concentration of U-238 for ten soil samples is 3 pCi/g. This single value could be used to estimate the population mean. But, as discussed earlier, it is very unlikely that the sample mean will be exactly the same as the population mean. Therefore, it might be more informative to have a range of values to estimate the population mean. Suppose 1 pCi/g was added to and subtracted from the mean of 3 pCi/g yielding an interval ranging from 2 to 4 pCi/g. This range of values could also be used to estimate the population mean. This is the basic concept underlying the construction of a confidence interval.

A confidence interval can be built around a sample mean for any desired level of confidence. If, for example, a high degree of confidence is required, a 99% confidence interval can be built around the sample mean. If a lower degree of confidence will suffice, an 80% confidence interval could be built around the sample mean. Both intervals will be centered at the sample mean; however, the 99% confidence interval will be wider than the 80% confidence interval. Continuing with the example, a 99% confidence interval might range from 2.15 to 3.85 pCi/g while an 80% confidence interval might range from 2.69 to 3.31 pCi/g. Note the trade-off at work here. Requiring a higher level of confidence means accepting a larger (less precise) interval as the estimate of the population mean. Conversely, relaxing the requirement on confidence translates to having a smaller (more precise) estimate of the population mean. If both high confidence and high precision are required, the only solution is to collect a large number of samples.

Confidence intervals not only give more information about the estimate of the mean, but they can also be used to reach the same conclusion as a t-test. Suppose that the null hypothesis states that the site is assumed to be contaminated. If the action level or regulatory threshold falls within a $1-\alpha\%$ confidence interval, then the null hypothesis that the site is contaminated cannot be rejected. If the action level falls outside the $1-\alpha\%$ confidence interval, then the null hypothesis is rejected with an $\alpha\%$ chance of making a false-positive decision error.

Calculating Sample Sizes. The biggest advantage of choosing a parametric test is that formulae for calculating sample sizes, confidence intervals, and for conducting statistical hypothesis tests are well developed and readily available in a wide variety of statistical packages. For example, if the data come from a normal population, the number of samples can be calculated with the following formula:

$$n = \frac{s^2(z_{1-\alpha} + z_{1-\beta})^2}{(\mu_1 - C)^2} + (0.5)z_{1-\alpha}^2 \qquad (4.1)$$

where

n	=	number of samples required to have 1-α% confidence that upper bound of the corresponding confidence interval will be below the action level
s^2	=	variance of the sample
α	=	probability of making a false-positive error
β	=	probability of making a false-negative error
$Z_{1-\alpha}$	=	value of the standard normal distribution that cuts off 1-α of the area
$Z_{1-\beta}$	=	value of the standard normal distribution that cuts off 1-β of the area
C	=	action level
μ_1	=	lower bound of the gray region

If the data come from a lognormal distribution, appropriate formulae for calculating the number of samples required and confidence intervals can be found in Gilbert (1987). These formulae use the natural log transformed mean and standard deviation as well as terms to indicate the desired level of confidence. While the formula for calculating confidence limits from lognormal data yields results that can be interpreted like those for the normal case, the formula for calculating the number of samples is a bit different and deserves discussion.

When calculating the number of samples required in the lognormal case, the equation for n yields the number of samples required to estimate the true *median* of a lognormal distribution with some prespecified, tolerable relative error in the estimation of that median. (Since lognormal distributions are skewed, the median is preferred to the mean as the measure of central tendency.) Also specified is the amount of confidence needed that the tolerable relative error has not been exceeded. Note that n is the number of samples required to meet two criteria simultaneously: the tolerable level of relative error and the confidence level that the relative error will not be exceeded.

The equation for calculating the number of samples required in the lognormal case is

$$n = \frac{Z_{1-\alpha/2}^2 s_y^2}{[\ln(d+1)]^2 + Z_{1-\alpha/2}^2 s_y^2/N} \qquad (4.2)$$

where

n	=	number of samples required to have 1-α% confidence that the estimate of the median will contain no more than d relative error
s^2	=	variance of the sample population
α	=	probability of making a false-positive error
$Z_{1-\alpha}$	=	value of the standard normal distribution that cuts off 1-α of the area
N	=	size of the population
d	=	prespecified relative error in the estimate of the median that can be tolerated.

An example may be helpful here to understand this calculation more fully. Suppose it is important to have high confidence that a very precise estimate of the median will be obtained when sampling from a population of size 1000. The number of samples needed to achieve this goal could be calculated by setting d to a low value such as 10%, setting $Z_{1-\alpha/2}$ to a high value such as 2.575 for 99% confidence, and setting N to 1000. If the population variance were equal to 1, then 422 samples would need to be collected from a population of 1000 to have 99% confidence that the estimate of the population median would contain no more than 10% relative error. If a "looser" estimate of the median will suffice, if, for example, 20 or 30% relative error can be tolerated, then n becomes 166 or 88, respectively. Likewise, lowering the required level of confidence will also reduce the number of samples required.

When a nonparametric test is chosen, there are no "canned" equations for calculating the number of samples needed. The reason for this is that the sample size equations themselves are parametric. That is, they make assumptions about the population distribution. Since no assumption about the population distribution is made in the nonparametric case, these equations cannot be used. Therefore, sample size formulae for parametric methods are often used to approximate the number of samples required. EPA (1992a) multiples Equation 4.1 by a "fudge" factor of 1.16 to obtain the approximate number of samples required. Great caution should be taken when using this approach. The number of samples calculated in this manner should be considered a lower bound, or minimum, number of samples. The actual number of samples required may be much larger, depending on the nonparametric test used. For example, the nonparametric Wilcoxon Signed Rank Test requires five or more samples for $\alpha = 0.05$. If fewer than five samples are taken, the site will always be declared contaminated with the Wilcoxon Signed Rank Test—regardless of how small the actual data values are.

4.1.1.5.7.3 Methods of Collecting Radiological Data — At radiological sites data collection activities may involve:

- Scanning
- Direct measurements
- Sampling and analysis

"Scanning" is a measurement technique performed by moving a portable gross alpha or gross beta/gamma radiation detector at a constant speed above a surface to semiquantitatively detect areas of elevated activity. Scanning is the least expensive method of collecting radiological data, and is only able to provide results in terms of gross activity. Scanning is most often used to identify hot spot areas from which direct measurements and/or sampling and analysis activities are later performed. The results from these measurements are reported in counts per minute (cpm).

"Direct measurements" are obtained by holding a detector above the location being surveyed for a period of time. If gross alpha or gross beta/gamma measurements are being collected, the detector need only be held above the measurement location long enough for the reading to stabilize (around 5 to 10 sec). The results from these measurements are reported in cpm. On the other hand, if pCi/g measurements are to be collected for one or more isotopes using a method such as *in situ* gamma spectroscopy (Section 4.2.2.1.2.3), the length of time that the detector needs to be held above the measurement location depends upon the required detection limit. The longer the counting time, the lower the detection limit. Direct measurements are more expensive than scanning measurements, but are less expensive than sample collection and analysis.

"Sampling and analysis" is the process of collecting a sample of an environmental medium for on-site or off-site laboratory testing. Sampling and analysis is the most expensive method of collecting radiological data, but it is also the most reliable. Results from sampling and analysis activities are typically reported in units of pCi/g, or pCi/L. When designing a radiological investigation, one should utilize a balance of scanning, direct measurements, and sampling and analysis methods to develop an investigation that is both cost-effective and technically defensible.

4.1.1.5.7.4 Identifying Impacted Areas — Prior to developing a data collection design, process knowledge should be used to divide carefully the site or facility under investigation into "nonimpacted" vs. "impacted" areas. Nonimpacted areas are those areas that have no reasonable potential for residual contamination. Impacted areas are those areas that have some potential for residual contamination (EPA, 1997).

The impacted areas should then be further divided into the following three classifications:

Class 1 Areas: Areas that have, or had prior to remediation, a potential for radioactive contamination (based on site operating history) or known contamination (based on previous radiation surveys) above the derived concentration guideline level ($DCGL_w$*). Examples of Class 1 areas include:

- Site areas previously subjected to remedial actions;
- Locations where leaks or spills are known to have occurred;
- Former burial or disposal sites;
- Waste storage areas;
- Areas with contaminants in discrete solid pieces of material and high specific activity.

Class 2 Areas: Areas that have, or had prior to remediation, a potential for radioactive contamination, but are not expected to exceed the $DCGL_w$. To justify changing the classification from Class 1 to Class 2, there should be measurement data that provide a high degree of confidence that no individual measurement would exceed the $DCGL_w$. Other justifications for reclassifying an area as Class 2 are as follows:

- Locations where radioactive materials were present in an unsealed form;
- Potentially contaminated transport routes;
- Areas downwind from stack release points;

* Note that the "w" in $DCGL_w$ stands for Wilcoxon Rank Sum test, which is the statistical test recommended in EPA (1997) for demonstrating compliance when the contaminant is present in background. The sign test recommended for demonstrating compliance when the contaminant is not present in background also uses the $DCGL_w$.

- Upper walls and ceilings of buildings or rooms subjected to airborne radioactive releases;
- Areas handling low concentrations of radioactive materials;
- Areas on the perimeter of former contamination control areas.

Class 3 Areas: Any impacted areas that are not expected to contain any residual radioactivity, or are expected to contain levels of residual radioactivity at a small fraction of the $DCGL_w$, based on site operating history and previous radiation surveys. Examples of areas that might be classified as Class 3 include:

- Buffer zones around Class 1 or Class 2 areas;
- Areas with very low potential for residual contamination but insufficient information to justify a nonimpacted classification.

4.1.1.5.7.5 Developing an Integrated Survey Design — Since radiological sites very often have small areas of elevated activities that could potentially be missed using standard statistical sampling approaches, EPA (1997) recommends that one develop an approach that combines a scanning survey for detecting small areas of elevated activity with a statistical sampling design.

Table 4.5 identifies a recommended integrated survey design to confirm that remediation or decontamination and decommissioning objectives have been met.

Table 4.5   Recommended Integrated Survey Designs

Survey Unit Classification		StatisticalTest Required	Sampling and/or Surveying Measurements	Scanning
Impacted	Class 1	Yes	Systematic	100% Coverage
	Class 2	Yes	Systematic	10–100% Coverage
	Class 3	Yes	Random	Judgmental
Nonimpacted		No	No	None

Random measurement patterns are used for Class 3 survey units to ensure that the measurements are independent and meet the requirements of the statistical tests. Systematic grids are used for Class 2 survey units because there is an increased probability of small areas of elevated activity. The use of a systematic grid allows one to draw conclusions about the size of any potential areas of elevated activity based on the area between measurement locations, while the random starting point of the grid provides an unbiased method for determining measurement locations for the statistical tests. Class 1 survey units have the highest potential for small areas of elevated activity, so the systematic sampling grid needs to be relatively tight.

Scanning surveys for Class 1 areas are designed to detect small areas of elevated activity that may have been missed by the direct measurements/samples collected at systematic grid nodes. For this reason, a 100% scanning coverage is recommended.

Scanning for Class 2 areas is similarly designed to detect small areas of elevated activity that may have been missed by the direct measurements/samples collected at systematic grid nodes. However, since the probability of detecting areas of elevated activity is lower for Class 2 areas, a less rigorous scanning effort may be appropriate. The level of scanning effort should be proportional to the potential for finding areas of elevated activity. For example, if the activity levels are expected to be close to the release limit, then a larger percentage of scanning coverage is needed. On the other hand, if the activity levels are near background levels, a much lower percentage of scanning coverage is needed.

Since Class 3 areas have the lowest potential for containing small areas of elevated activity, scanning should focus on those areas with the highest potential for showing contamination based on professional judgment (e.g., corners, ditches, drains).

The flowcharts presented in Figures 4.16 and 4.17 identify the principal steps and decisions that need to be made in the site characterization and remediation, and site closeout sampling process, respectively. These flowcharts apply to both remediation and decontamination and decommissioning projects. Implementing the process outlined in these two flowcharts will ensure the defensibility of the remediation process.

4.1.1.5.7.6 Statistical Sampling Design Software — The following section provides a description, general guidance, and limitations on the use of the following statistical sampling design software: Visual Sample Plan (VSP), ELIPGRID-PC, and DEFT.

4.1.1.5.7.6.1 Visual sample plan software — The purpose of VSP is to provide the user with a simple and defensible tool that defines a technically defensible sampling scheme that can be used to support site characterization or postremediation confirmation sampling activities. VSP is applicable for any two-dimensional sampling plan including surface soil, sediment, building surfaces, water bodies, or other similar applications.

VSP development has been supported by the U.S. Department of Energy's National Analytical Management Program (NAMP), Pacific Northwest National Laboratory, and Oak Ridge National Laboratory–Grand Junction. A copy of the VSP Version 0.9E software can be downloaded from the following Internet address:

http://terrassa.pnl.gov:2080/DQO/software/vsp/

One of the advantages in using the VSP software is that it is very visual. As shown in Figure 4.18, VSP can simultaneously show the following views:

- Map View: Shows the site map, selected sampling areas, and defined sampling locations.
- Cost View: Presents statistics related to the number of samples, cost, probability, etc.
- Graph View: Presents a graph of the selected function.
- Plan View: Displays the (X,Y) coordinates of the sampling points.

Being able to see all four of these views simultaneously is particularly beneficial when negotiating error tolerance requirements with regulatory agencies. For example, by projecting the VSP software up on a wall screen, regulators can see the

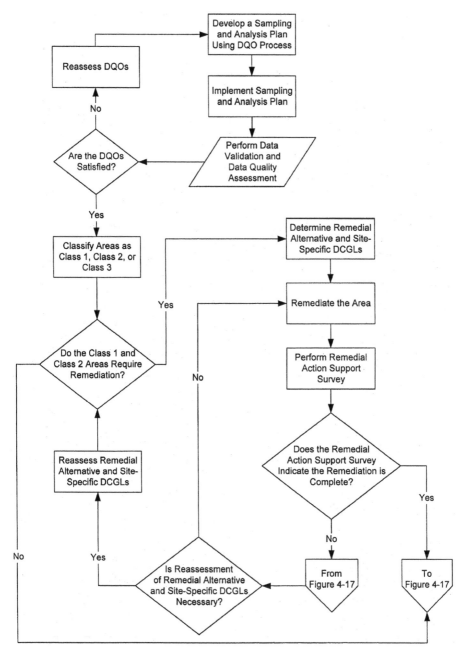

Figure 4.16 Site characterization and remediation sampling process.

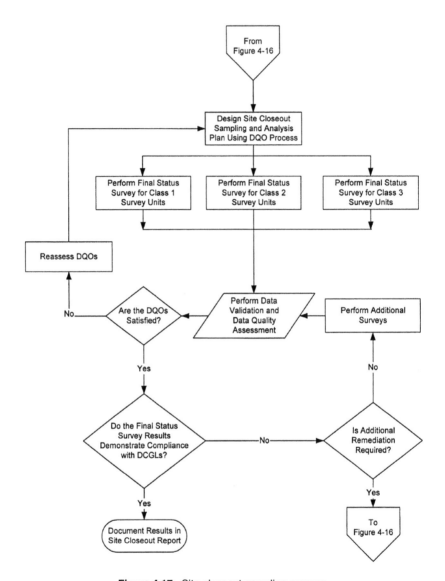

Figure 4.17 Site closeout sampling process.

impact that their stringent α error (false positive), β error (false negative), and width of the gray region requirements have on sampling costs. With VSP, error tolerances can be adjusted repeatedly until one comes up with a sample design that balances cost and uncertainty.

VSP allows the user to either sketch a simple outline of the study area in the shape of a square, rectangle, circle, ellipse, etc., or import a detailed site map in Data Exchange Format (DXF) format. Once the site map has been established, the specific area(s) where samples are to be collected is then defined with the assistance of the computer mouse cross hairs and the left mouse button. Once the sampling

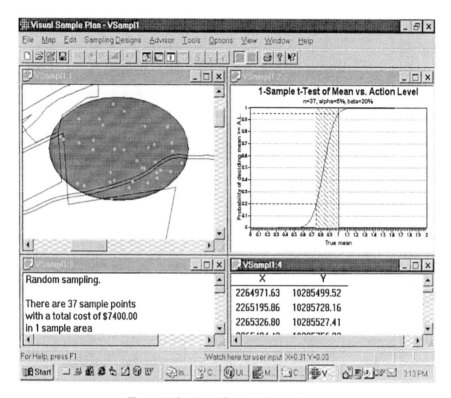

Figure 4.18   Visual Sample Plan software.

area(s) has been identified, the VSP software allows the user to select between the following two sampling designs:

- Simple random sampling;
- Systematic grid sampling (square, triangular, or rectangular grid).

Table 4.6 identifies all of the various statistical hypothesis tests that VSP is able to perform, along with the data input requirements for each test.

4.1.1.5.7.6.2 ELIPGRID-PC software — ELIPGRID-PC is a statistical sample design software that is designed to calculate the probability of locating circular or elliptical "hot spots" (targets) using a square, rectangular, and triangular grid. The original ELIPGRID algorithm was developed in 1972 as a FORTRAN IV program (Singer, 1972) which was based on a mathematical procedure for determining the probability of locating elliptical geological deposits (Singer and Wickman, 1969). In 1994, the ELIPGRID algorithm was modified to correct some problems identified in the original code and to make the algorithm available for IBM personal computers (Davidson, 1994; 1995). This modified algorithm is known as ELIPGRID-PC. A copy of the ELIPGRID-PC software can be downloaded from the following Internet address:

http://etd.pnl.gov:2080/DQO/software/elipgrid.html

Table 4.6 Visual Sample Plan Software Capabilities

Sampling Design	Type of Test	Means of Showing Compliance	Statistical Hypothesis Test	Data Input Requirements
Simple random samping	Parametric	Mean compared against action level	One-sample t-test	Define the null hypothesis (assume site is either "clean" or "contaminated") Type I error rate (α error) Type II error rate (β error) Width of the gray region (distance from lower bound of gray region to the action level) Estimated standard deviation Action level
		Mean compared against a reference mean	Two-sample t-test	Define the null hypothesis (assume site is either "clean" or "contaminated") Type I error rate (α error) Type II error rate (β error) Estimated standard deviation Allowed difference of means Difference of means at the outer bound of the gray region
		Confidence interval for a mean	NA	Select between a one-sided or a two-sided confidence interval Confidence level (%) Estimated standard deviation Largest acceptable difference between the sample mean and the true mean
	Nonparametric	Mean compared against an action level	Wilcoxon Signed Rank Test	Define the null hypothesis (assume site is either "clean" or "contaminated") Type I error rate (α error) Type II error rate (β error) Width of the gray region (distance from lower bound of gray region to the action level)

		Estimated standard deviation
		Action level
Proportion compared against a given proportion	One-sample proportion test	Define whether true proportion is ≥ or ≤ the given proportion
		Type I error rate (α error)
		Type II error rate (β error)
		Width of the gray region (delta)
		Given proportion (action level)
Comparison of two proportions	Two-sample proportion test	Define whether difference of proportions is ≥ or ≤ the specified difference
		Type I error rate (α error)
		Type II error rate (β error)
		Estimated proportion in reference area (0.0–1.0)
		Difference of proportions specified in null hypothesis (0.0–0.5)
		Width of the gray region (delta) (0.0–0.5)
Comparison of two populations	Wilcoxon Rank Sum Test	Define whether the difference of means is ≥ or ≤ the allowed difference
		Type I error rate (α error)
		Type II error rate (β error)
		Estimated standard deviation
		Allowed difference of means
		Difference of means at the outer bound of the gray region
MARSSIM Sign Test[a]	MARSSIM Sign Test[a]	Define whether the true median is ≥ or ≤ the action level
		Type I error rate (α error)
		Type II error rate (β error)
		Width of the gray region (delta)
		Estimated standard deviation
		Action level

Table 4.6 (continued) Visual Sample Plan Software Capabilities

Sampling Design	Type of Test	Means of Showing Compliance	Statistical Hypothesis Test	Data Input Requirements
Systematic grid sampling		MARSSIM Wilcoxon Rank Sum Test[a]	MARSSIM Wilcoxon Rank Sum Test[a]	Define whether the true median is $\geq$ or $\leq$ the action level
				Type I error rate (α error)
				Type II error rate (β error)
				Width of the gray region (delta)
				Estimated standard deviation
				Action level
	Parametric	Mean compared against action level	One-sample t-test	[b]
		Mean compared against a reference mean	Two-sample t-test	[c]
		Confidence interval for a mean	NA	[d]
	Nonparametric	Mean compared against an action level	Wilcoxon Signed Rank Test	[e]
		Proportion compared against a given proportion	One-sample proportion test	[f]
		Comparison of two proportions	Two-sample proportion test	[g]
		Comparison of two populations	Wilcoxon Rank Sum Test	[h]
		MARSSIM Sign Test[a]	MARSSIM Sign Test[a]	[i]
		MARSSIM Wilcoxon Rank Sum Test[a]	MARSSIM Wilcoxon Rank Sum Test[a]	[j]
	Locating hot spots by probability and hot spot size	NA	NA	Desired probability of hitting hot spot
				Grid type (square, triangular, rectangular)
				Shape of hot spot
				Area of hot spot or length of semimajor axis
				Total area to sample
				Cost of individual sample

Locating hot spots by probability and grid size	NA	NA	Grid type (square, triangular, rectangular) Length of grid side Probability of hit Shape of hot spot Total area to sample Cost of individual sample
Locating hot spots by cost and hot spot size	NA	NA	Total cost Grid type (square, triangular, rectangular) Shape of hot spot Area of hot spot or length of semimajor axis Total area to sample Cost of individual sample
Locating hot spots by predetermined grid size	NA	NA	Grid type (square, triangular, rectangular) Length of grid side Shape of hot spot Area of hot spot or length of semimajor axis Total area to sample Cost of individual sample

[a] See EPA (1997).

[b] Same data input requirements as for simple random sampling, parametric, mean compared against action level, one-sample t-test.

[c] Same data input requirements as for simple random sampling, parametric, mean compared against reference mean, two-sample t-test.

[d] Same data input requirements as for simple random sampling, parametric, confidence interval for a mean, t-test.

[e] Same data input requirements as for simple random sampling, nonparametric, mean compared against an action level, Wilcoxon Signed Rank Test.

[f] Same data input requirements as for simple random sampling, nonparametric, proportion compared against a given proportion, one-sample proportion test.

[g] Same data input requirements as for simple random sampling, nonparametric, comparison of two proportions, two-sample proportion test.

[h] Same data input requirements as for simple random sampling, nonparametric, comparison of two populations, Wilcoxon Rank Sum Test.

[i] Same data input requirements as for simple random sampling, nonparametric, MARSSIM Sign Test, MARSSIM Sign Test.

[j] Same data input requirements as for simple random sampling, nonparametric, MARSSIM Sign Test, MARSSIM Sign Test.

The ELIPGRID-PC software can be used in one of the following ways:

- To identify the probability of hitting a hot spot, assuming a specified hot spot size and shape (circular or elliptical), and assuming a specified grid size.
- To identify the required grid size, assuming a specified probability of hitting a hot spot of a specified size and shape (circular or elliptical).
- To identify the smallest hot spot likely to be hit, assuming a specified grid size.
- To identify the grid size based on a fixed sampling cost.

Figure 4.19 presents the ELIPGRID-PC main menu screen from which the user initially selects the type of investigation or program activity to be performed.

```
                    ORNL ELIPGRID-PC
                    Version 10/20/95
           PC-Based Hot Spot Probability Calculations
    ─────────────────────────────────────────────────
    Current subdirectory: C:\CLIPPER2\EDITOR\EGPC4
    ─────────────────────────────────────────────────
                       Main Menu
    ─────────────────────────────────────────────────
       P    Probability of Hitting Hot Spot
       G    Grid Size Required, Given Probability
       S    Smallest Hot Spot Hit, Given Grid
       C    Cost-Based Grid
       W    Write/Print Cost-Based Graph Data
       F    Print Graph from Graph Data File
       O    Set Program Options
       N    New Drive or Subdirectory
       D    DOS Prompt
       Q    Quit Program
    ─────────────────────────────────────────────────
    F1 key for Help                    Esc key to Exit
```

Figure 4.19 ELIPGRID-PC main menu screen.

Probability of Hitting a Single Hot Spot. The following example demonstrates how ELIPGRID-PC can be used to define the probability of hitting a single hot spot of a given size and shape when using a specified grid. The data used in the example provided below were derived from Example 10.3 of Gilbert (1987).

Example: Assuming a rectangular sampling grid with a long side to short side ratio of 2 to 1, determine the average probability of finding an elliptical hot spot that is twice as long as it is wide with a semimajor axis length of 40% of the short side of the rectangular grid.

Step 1: Choose main menu option: Probability of Hitting Hot Spot.
Step 2: Choose: Screen Input.
Step 3: Choose: Rectangular Grid.
Step 4: Enter output file name if a screen output file is desired. Otherwise, press Esc.
Step 5: Enter the following data:

Shape of the elliptical hot spot:	0.50
Length of semimajor axis:	0.40 m
Angle of orientation to grid:	99.0° (assumes random angles)
Length of short side of rectangular grid:	1.00 m
Total area to sample:	Leave as 0.0
Individual sample cost:	Leave as 0.00

As shown in Figure 4.20, the result is a probability of 12.6% of finding the elliptical hot spot, while the probability of *not* hitting the hot spot is 87.4%.

File Options Windows Help	
ELIPGRID-PC	
Determine Probability of Hitting Hot Spot for Rectangular Grid in Meters	
Shape of the elliptical hot spot: 0.50 Length of semimajor axis: 0.40 m Angle of orientation to grid: 99.0° Length of short side of rect. grid: 1.00 m Long side/short side ratio: 2.0 Total area to sample F10 = acres: 0.0 m² Individual sample cost: 0.00	Shape = short axis/long axis. F10 calculates axis from area. Angle can be 0° to 90°. Use 99.0° for average of multiple, "random" angles. Leave area and sample cost at 0, if cost information is not desired.
Probability of hitting at least once = 12.6% Probability of NOT hitting hot spot = 87.4%	
Enter = Continue Esc = Abort	

Figure 4.20 Input screen for probability of hitting hot spot.

Required Grid Size — The following example demonstrates how ELIPGRID-PC can be used to define required grid size to find a hot spot of given size and shape with specified confidence. The data used in the example provided below were derived from Example 10.1 of Gilbert (1987).

Example: Assuming a square sampling grid and a desired 90% probability of detecting a circular hot spot with a radius of 1 m, what grid size should be used?

Step 1: Choose main menu option: Grid Size Required, Given Probability.
Step 2: Choose: square grid type.
Step 3: Enter output file name if a screen output file is desired. Otherwise, press Esc.
Step 4: Enter the following data:

Shape of the elliptical hot spot:	1.00
Length of semimajor axis:	1.00 m
Angle of orientation to grid:	99.0° (assumes random angles)
Desired probability of hitting:	90.0%
Total area to sample:	Leave as 0.0
Individual sample cost:	Leave as 0.00

As shown in Figure 4.21, the required grid has a spacing of 1.79 m.

File Options Windows Help
ELIPGRID-PC
Determine Size of Square Grid in Meters

Shape of the elliptical hot spot:	1.00	Shape = short axis/long axis.
Length of semimajor axis:	1.00 m	F10 calculates axis from area.
Angle of orientation to grid:	99.0°	Angle can be 0° to 45°. Use 99.0° for "random" angles.
Desired probability of hitting:	90.0%	Use 10% to 99.9%.
Total area to sample F10 = acres:	0.0 m²	Leave area and sample cost at 0,
Individual sample cost:	0.00	if cost information is not desired.

Grid size found = 1.79m
Probability of hitting hot spot = 90.0%

Enter = Continue   Esc = Abort

Figure 4.21 Input screen for required grid size.

Size of Smallest Hot Spot Likely to Be Hit. The following example demonstrates how ELIPGRID-PC can be used to determine the smallest hot spot that will likely be hit with a specified grid size and a specified probability of hitting the target. The data used in the example below were derived from Example 10.2 of Gilbert (1987).

Example: Assuming a square grid with 2-m sides, what is the smallest hot spot that could be hit with a 90% probability?

Step 1: Choose the main menu option: Smallest Hot Spot Hit, Given Grid.
Step 2: Enter output file name if a screen output file is desired. Otherwise, press Esc.
Step 3: Enter the following data:

Shape of the elliptical hot spot:	1.00
Angle of orientation to grid:	99.0° (assumes random angles)
Length of any side of square grid:	2.00 m
Desired probability of hitting:	90%
Total area to sample:	Leave as 0.0
Individual sample cost:	Leave as 0.0

As shown in Figure 4.22, there is a 90% probability of hitting a circular hot spot with an area of 3.9 m² or more when a square grid with 2-m sides is used.

Cost-Based Grid Determination. The following example demonstrates how ELIPGRID-PC can be used to determine the size of the sampling grid based on a fixed sampling cost, and to calculate the resulting probability of hitting a hot spot of a specified size.

File Options Windows Help	
ELIPGRID-PC	
Determine Size of Smallest Hot Spot for Square Grid in Meters	
Shape of the elliptical hot spot: 1.00 Angle of orientation to grid: 99.0° Length of any side of square grid: 2.00 m Desired probability of hitting: 90.0% Total area to sample F10 = acres: 0.0 m² Individual sample cost: 0.00	Shape = short axis/long axis. Angle can be 0° to 45°. Use 99.0° for "random" angles. Use 10% to 99.9%. Leave area and sample cost at 0 if cost information is not desired.
Area of smallest hot spot hit = 3.9 m² Grid size = 2.00 m Given probability of hitting = 90.0%	Search iterations: 11 Random angle iterations: 1518
Enter = Continue Esc = Abort	

Figure 4.22 Input screen for required grid size.

Example: Assuming a square sampling grid for a 2-acre site, a maximum sampling cost of $200,000, an individual sample cost of $700, and a circular hot spot 25 m² in size, what size grid should be used and what will be the probability of hitting a hot spot at least once?

Step 1: Choose main menu option: Cost-Based Grid.
Step 2: Choose: square grid type.
Step 3: Enter output file name if a screen output file is desired. Otherwise, press Esc.
Step 4: Enter the following data:

Shape of the elliptical hot spot: 1.00
Length of semimajor axis: 2.82 m (or press F10 and enter 25 m²)
Angle of orientation to grid: 99.0° (the random angle case)
Total area to sample: 8.09 m² (or press F10 and enter 2.0 acres)
Individual sample cost: $700.00
Minimum cost of graph data: $150,000.00

As shown in Figure 4.23, for a total cost of $199,500, a square grid of 5.33 m provides an 85.2% chance of hitting the specified minimum-size hot spot at least once.

ELIPGRID-PC Limitations. The primary limitations of the ELIPGRID-PC software are as follows:

• Calculations made by the software should be treated as approximations.
• The software is not able to take site layout into consideration; rather, the total site area and grid size are used in making calculations.
• The program is based on finding two-dimensional hot spots, and is not able to take the third dimension (depth) into consideration.

File Options Windows Help	
ELIPGRID-PC	
Determine Size of Cost-Based Square Grid in Meters	
Shape of the elliptical hot spot: 1.00 Length of semimajor axis: 2.82 m Angle of orientation to grid: 99.0° Total area to sample (F10 = acres): 8,093.0 m² Individual sample cost $: 700.00 Desired cost of grid $: 200,000	Shape = short axis/long axis. F10 calculates axis from area. Angle can be 0° to 45°. Use 99.0° for "random" angles. Program will search for cost with error < ±1 sample cost.
Grid size found = 5.33 m Probability of hitting hot spot = 85.2% Required number of samples (approximate) = 285 Total cost for above number of samples = $199,500.00	
Enter = Continue Esc = Abort	

Figure 4.23 Input screen for required grid size.

4.1.1.5.7.6.3 DEFT software — The Data Quality Objectives Decision Error Feasibility Trials (DEFT) software (EPA, 1994b) was developed by Research Triangle Institute for the EPA Quality Assurance Management Staff (QAMS) to supplement DQO guidance provided in EPA QA/G-4 (EPA, 1994a). A copy of the DEFT Version 4.0 software and user manual can be downloaded from the following EPA Web address:

http://es.epa.gov/ncerqa/qa/qa_docs.html#G-4D

The DEFT software was developed as a tool to assist the user in completing DQO Steps 6 and 7 by allowing quick generation of cost information for multiple sampling designs with varying DQO constraints (e.g., error tolerances, width of gray region). Through this process, the user is able to develop a cost-effective sampling program that meets the minimum data quality requirements.

However, the DEFT Version 4.0 software has some limitations in that it assumes:

- The population has a normal distribution which is not always appropriate for a radiological site.
- The t-test is used to analyze all data sets no matter what the distribution.
- The population mean is being compared against a fixed action level. The software does not address proportions, percentiles, or comparisons between two populations.
- Sample locations can be randomly selected (e.g., soil sampling locations can be randomly selected; groundwater sampling locations are *not* typically randomly selected due to hydrogeologic considerations).
- No temporal issues apply (Section 4.1.1.5.4.3).

If any of these assumptions are inappropriate for a particular project, one should consider using an alternative sampling design software (e.g., Visual Sample Plan).

Entry Screens. The DEFT software presents multiple data entry screens where the user is prompted to enter information from DQO Steps 1 through 6. The information requested by the software includes:

- Parameter of interest (e.g., mean);
- Minimum and maximum values (range) of the parameter of interest;
- Action level;
- Null and alternative hypotheses;
- Bounds of the gray region;
- Estimate of the standard deviation;
- Sampling and analysis costs;
- Probability limits on decision errors for the bounds of the gray region;
- Additional limits on decision errors.

When requesting information, the DEFT software identifies in which DQO process step the information may be found. Default values are offered for users who wish to use the software as a learning tool. Previous entries are summarized in the lower right-hand corner of each screen.

Input Verification Screen. Once all of the DQO constraints are entered, the DEFT software displays the Input Verification Screen. This screen is used to verify the inputs from the data entry screens. Any incorrect inputs can be corrected at this time.

The Design/DQO Summary Screen. After the user has verified all of the inputs, a Design/DQO Summary Screen is displayed. This screen summarizes information on the current sampling design, DQOs, sample size, and cost. The program always starts with the assumption that a simple random sampling design will be used. Other sampling designs that the user may select include composite sampling and stratified sampling.

When a simple random sampling design is selected, the DEFT software assumes that a t-test will be used to analyze the data and thus uses the corresponding sample size formula. The performance curve calculations are also based on the t-test.

When a composite sampling design is selected, the software is used to compute the number of composite samples required to meet the DQOs based on a given number of aliquots per composite sample. To determine the number of composite samples, an estimate of the ratio of the relative standard deviation of measurement error to total standard deviation is required, along with the number of aliquots to be contained within each composite sample. The user will be prompted to enter this information the first time a composite sampling design is selected. The user may then vary the number of aliquots within a composite sample by pressing "C" and the ratio by pressing "R." The number of aliquots within a composite sample must be fewer than 100 and the ratio must be less than 1 and greater than 0.

When a stratified sampling design is selected, the study area is divided into two or more nonoverlapping subsets (strata) that cover the entire site. Strata should be defined so that physical samples within a stratum are more similar to each other than to samples from other strata. Once the strata have been defined, the DEFT software assumes each stratum will be sampled separately using a simple random design.

To estimate the sample size required for a stratified design, the DEFT software requires information regarding each individual stratum. There is a limit of four strata total in the DEFT software. For each stratum, the user will need to provide a weighing factor (weight) and an estimate of the standard deviation. The stratum weight is the proportion of the volume or area of the environmental medium contained in the stratum in relation to the total volume or area of the investigation site. The sum of the strata weights must equal 1.0, so the program automatically computes the weight of the final stratum. The default weight corresponds to an equal weighing among the remaining strata. The estimated standard deviation for each stratum must be less than two times the range of the population parameter; the default value is the current estimate for the total standard deviation.

The user may change the current sampling design or change the sample size at any time. The total cost can be modified by adjusting the analytical or field sampling costs. At this point, the DEFT software verifies if the decision error limits are satisfied. If not, the limits that are not satisfied are marked "Not Satisfied."

Several options are available on the Design/DQO Summary Screen for the user's convenience. These options include:

- Displaying the decision performance goal diagram with performance curve overlaid;
- Saving the current Design/DQO Summary Screen in a file;
- Restoring the Original DQOs.

DEFT Software Example. For the following example, assume that a radiologically contaminated building (Building 101) was recently decontaminated and is to be sampled to verify that the decontamination objectives have been met. Determine how many samples (or measurements) are needed to verify that the building is free from U-238 contamination based on the following assumptions:

Minimum concentration of U-238:	1.5 pCi/g (based on historical data)
Maximum concentration of U-238:	20.0 pCi/g (based on historical data)
Action level for U-238:	10.0 pCi/g
Gray region:	8.0 to 10.0 pCi/g
Acceptable percent of false positives:	0.05 (5%)
Acceptable percent of false negatives:	0.20 (20%)
Standard deviation:	5.00 (based on historical data)
Null hypothesis (H_0):	Mean > 8.00 pCi/g
Cost for analyzing sample in laboratory:	$300.00/sample
Cost of collecting sample:	$100.00/sample

As shown in Figure 4.24, a total of 40 samples (or measurements) are needed to meet the specified error tolerances at a cost of $16,000. Note that in the above example, if the primary COCs for Building 101 included U-238, Ra-226, Th-230,

DESIGN/DQO SUMMARY SCREEN
For the Sampling (D)esign of: Simple Random Sampling

Total Cost: $16,000.00
   (L)aboratory Cost per Sample: $300.00
   (F)ield Cost per Sample: $100.00

(N)umber of Samples: 40

   Data Quality Objectives

(A)ction Level: 10.00
Gray Region: 8.00–10.00
Null Hypothesis: mean > 10.00
(S)tandard Deviation: 5.00

Decision Error Limits:

Conc.		Prob(error)	Type
---	(1)	---	F(-)
---	(2)	---	F(-)
8.00	(B)(3)	0.2000	F(-)
10.00	(A)(4)	0.0500	F(+)
---	(5)	---	F(+)
---	(6)	---	F(+)

(G)raph   Sa(V)e File   (O)riginal DQOs   E(X)it

Figure 4.24 DEFT Design/DQO Summary Screen.

and Th-232, then the above calculation should be performed for each isotope separately. The largest required number of samples (or measurements) should be collected.

4.1.1.6 Preparing a DQO Summary Report

The results from the seven-step DQO process should be summarized in a DQO
Summary Report. Appendix A provides a template that should be used to assist in
the writing of this report. The accompanying CD-ROM provides an electronic
version of the template provided in Appendix A.

4.1.1.7 Sampling and Analysis Plan

A Sampling and Analysis Plan is prepared to provide the field sampling team
with direction on how to implement the sampling design identified in Step 7 of the
DQO process. The Sampling and Analysis Plan is composed of a summary of the
DQOs, a Quality Assurance Project Plan, and a Field Sampling Plan. Figure 4.25
provides a recommended outline for a Sampling and Analysis Plan. Appendix B
provides a template that should be used to assist in the writing of the Sampling and
Analysis Plan. The attached CD-ROM provides an electronic version of the template
provided in Appendix B.

Figure 4.25 Recommended outline for Sampling and Analysis Plan.

1.0   INTRODUCTION
     1.1   BACKGROUND
     1.2   CONTAMINANTS OF CONCERN
     1.3   DATA QUALITY OBJECTIVES
         1.3.1 Statement of the Problem
         1.3.2 Decision Rules
         1.3.3 Error Tolerance and Decision Consequences
         1.3.4 Sample Design Summary
2.0   QUALITY ASSURANCE PROJECT PLAN
     2.1   PROJECT MANAGEMENT
         2.1.1 Project/Task Organization
         2.1.2 Quality Objectives and Criteria for Measurement Data
         2.1.3 Special Training Requirements/Certification
         2.1.4 Documentation and Records
     2.2   MEASUREMENT/DATA ACQUISITION
         2.2.1 Sampling Process Design
         2.2.2 Sampling Methods Requirements
         2.2.3 Sample Handling, Shipping, and Custody Requirements
         2.2.4 Analytical Methods Requirements
         2.2.5 Quality Control Requirements
         2.2.6 Instrument/Equipment Testing, Inspection, and Maintenance Requirements
         2.2.7 Instrument Calibration and Frequency
         2.2.8 Inspection/Acceptance Requirements for Supplies and Consumables
         2.2.9 Data Acquisition Requirements (Non-Direct Measurements)
         2.2.10 Data Management
         2.2.11 Sample Preservation, Containers, and Holding Times
         2.2.12 Field Documentation
     2.3   ASSESSMENT/OVERSIGHT
         2.3.1 Assessments and Response Actions
         2.3.2 Reports to Management
     2.4   DATA VALIDATION AND USABILITY
         2.4.1 Data Review, Validation, and Verification Requirements
         2.4.2 Validation and Verification Methods
         2.4.3 Reconciling Results with DQOs
3.0   FIELD SAMPLING PLAN
     3.1   SAMPLING OBJECTIVES
     3.2   SAMPLING LOCATIONS AND FREQUENCY
     3.3   SAMPLING AND ONSITE ENVIRONMENTAL MEASUREMENT PROCEDURES
     3.4   SAMPLE MANAGEMENT
     3.5   MANAGEMENT OF INVESTIGATION-DERIVED WASTE
4.0   HEALTH AND SAFETY
5.0   REFERENCES

4.1.1.7.1   Quality Assurance Project Plan

The Quality Assurance Project Plan provides guidance on analytical performance requirements and proper handling requirements. Specifically, the Quality Assurance Project Plan provides:

- Details on the analyses to be performed;
- Analytical performance requirements (e.g., detection limits, precision, accuracy);
- Quality control requirements (e.g., blanks, duplicates, spikes);
- Sample bottle and preservation requirements;
- Sample handling, shipping, and custody requirements;
- Instrument testing, inspection, and maintenance requirements;
- Data management requirements;
- Field documentation requirements;
- Assessment and audit requirements;
- Data validation/verification requirements.

The Quality Assurance Project Plan should be prepared by a chemist, radiochemist, or data quality specialist who was involved in the preparation of the DQO Summary Report, in consultation with representatives from the radiological laboratory performing the analyses. The template provided in Appendix B and the accompanying CD-ROM should be used to guide the preparation of the Quality Assurance Project Plan.

4.1.1.7.1.1 Data Quality Indicators — Data Quality Indicators (DQIs) are qualitative and quantitative descriptors used in interpreting the degree of acceptability or utility of data (EPA, 1998). The principal DQIs are precision, bias, accuracy, representativeness, comparability, and completeness. The five principal DQIs are also referred to as the PARCC parameters. Secondary DQIs include sensitivity, recovery, memory effects, limit of quantitation, repeatability, and reproducibility. Establishing acceptance criteria for the DQIs sets quantitative goals for the quality of data generated in the analytical measurement process. Of the principal DQIs, precision, bias, and accuracy are the quantitative measures, representativeness and comparability are qualitative, and completeness is a combination of both quantitative and qualitative measures.

4.1.1.7.1.1.1 Principal data quality indicators — The following section provides definitions for each of the principal data quality indicators.

Precision. Precision is a measure of agreement among replicate measurements of the same property, under prescribed similar conditions. This agreement is calculated as either the range or as the standard deviation (s). It may also be expressed as a percentage of the mean of the measurements, such as relative range (for duplicates) or relative standard deviation (RSD).

For analytical procedures, precision may be specified as either intralaboratory (within a laboratory) or interlaboratory (between laboratories) precision. Intralaboratory precision estimates represent the agreement expected when a single laboratory uses the same method to make repeated measurements of the same sample.

Interlaboratory precision refers to the agreement expected when two or more laboratories analyze the same or identical samples with the same method. Intralaboratory precision is more commonly reported; however, where available, both intralaboratory and interlaboratory precision should be listed in the data compilation.

When possible, a sample subdivided in the field and preserved separately is used to assess the variability of sample handling, preservation, and storage along with the variability of the analysis process.

When collocated samples are collected, processed, and analyzed by the same organization, intralaboratory precision information on sample acquisition, handling, shipping, storage, preparation, and analysis is obtained. Both samples can be carried through the steps in the measurement process together to provide an estimate of short-term precision. Likewise, the two samples, if separated and processed at different times or by different people and/or analyzed using different instruments, provide an estimate of long-term precision (EPA, 1998).

Bias. Bias is the systematic or persistent distortion of a measurement process that causes errors in one direction. Bias assessments for environmental measurements are made using personnel, equipment, and spiking materials or reference materials as independent as possible from those used in the calibration of the measurement system. When possible, bias assessments should be based on analysis of spiked samples rather than reference materials so that the effect of the matrix on recovery is incorporated into the assessment. A documented spiking protocol and consistency in following that protocol are important to obtaining meaningful data quality estimates. Spikes should be added at different concentration levels to cover the range of expected sample concentrations. For some measurement systems (e.g., continuous analyzers used to measure pollutants in ambient air), spiking samples may not be practical, so assessments should be made using appropriate blind reference materials.

For certain multianalyte methods, bias assessments may be complicated by interferences among multiple analytes, which prevents all of the analytes from being spiked into a single sample. For such methods, lower spiking frequencies can be employed for analytes that are seldom or never found. The use of spiked surrogate compounds for multianalyte gas chromatography/mass spectrometry (GC/MS) procedures, while not ideal, may be the best available procedure for assessment of bias (EPA, 1998).

Accuracy. Accuracy is a measure of the closeness of an individual measurement or the average of a number of measurements to the true value. Accuracy includes a combination of random error (precision) and systematic error (bias) components that result from sampling and analytical operations. Accuracy is determined by analyzing a reference material of known concentration or by reanalyzing a sample to which a material of known concentration or amount of constituent has been added. Accuracy is usually expressed either as a percent recovery (P) or as a percent bias (P − 100). Determination of accuracy always includes the effects of variability (precision); therefore, accuracy is used as a combination of bias and precision. The combination is known statistically as mean square error.

Mean square error (MSE) is the quantitative term for overall quality of individual measurements or estimators. To be accurate, data must be both precise and unbiased. Using the analogy of archery, to be accurate, one must have one's arrows land close together and, on average, at the spot where they are aimed. MSE is the sum of the variance plus the square of the bias. (The bias is squared to eliminate concern over whether the bias is positive or negative.) Frequently, it is impossible to quantify all of the components of the mean square error—especially the biases—but it is important to attempt to quantify the magnitude of such potential biases, often by comparison with auxiliary data (EPA, 1998).

Representativeness. Representativeness is a measure of the degree to which data accurately and precisely represent a characteristic of a population parameter at a sampling point or for a process condition or environmental condition. Representativeness is a qualitative term that should be evaluated to determine whether *in situ* and other measurements are made and physical samples collected in such a manner that the resulting data appropriately reflect the media and phenomenon measured or studied (EPA, 1998).

Comparability. Comparability is the qualitative term that expresses the confidence that two data sets can contribute to a common analysis and interpolation. Comparability must be carefully evaluated to establish whether two data sets can be considered equivalent in regard to the measurement of a specific variable or groups of variables. In a laboratory analysis, the term *comparability* focuses on method type comparison, holding times, stability issues, and aspects of overall analytical quantitation.

There are a number of issues that can make two data sets comparable, and the presence of each of the following items enhances their comparability:

- Two data sets should contain the same set of variables of interest.
- Units in which these variables were measured should be convertible to a common metric.
- Similar analytic procedures and quality assurance should be used to collect data for both data sets.
- Time of measurements of certain characteristics (variables) should be similar for both data sets.
- Measuring devices used for both data sets should have approximately similar detection levels.
- Rules for excluding certain types of observations from both samples should be similar.
- Samples within data sets should be selected in a similar manner.
- Sampling frames from which the samples were selected should be similar.
- Number of observations in both data sets should be of the same order or magnitude.

These characteristics vary in importance depending on the final use of the data. The closer two data sets are with regard to these characteristics, the more appropriate it will be to compare them. Large differences between characteristics may be of only minor importance, depending on the decision that is to be made from the data.

Comparability is very important when conducting meta-analysis, which combines the results of numerous studies to identify commonalities that are then hypothesized to hold over a range of experimental conditions. Meta-analysis can be very misleading if the studies being evaluated are not truly comparable. Without proper consideration of comparability, the findings of the meta-analysis may be due to an artifact of methodological differences among the studies rather than due to differences in experimentally controlled conditions. The use of expert opinion to classify the importance of differences in characteristics among data sets is invaluable (EPA, 1998).

Completeness. Completeness is a measure of the amount of valid data obtained from a measurement system, expressed as a percentage of the number of valid measurements that should have been collected (i.e., measurements that were planned to be collected). Completeness is not intended to be a measure of representativeness; that is, it does not describe how closely the measured results reflect the actual concentration or distribution of the pollutant in the media sampled. A project could produce 100% data completeness (i.e., all samples planned were actually collected and found to be valid), but the results may not be representative of the pollutant concentration actually present.

Alternatively, there could be only 70% data completeness (30% lost or found invalid), but, due to the nature of the sample design, the results could still be representative of the target population and yield valid estimates. Lack of completeness is a vital concern with stratified sampling. Substantial incomplete sampling of one or more strata can seriously compromise the validity of conclusions from the study. In other situations (for example, simple random sampling of a relatively homogeneous medium), lack of completeness results only in a loss of statistical power. The degree to which lack of completeness affects the outcome of the study is a function of many variables ranging from deficiencies in the number of field samples acquired to failure to analyze as many replications as deemed necessary by the QAPP and DQOs. The intensity of effect due to incompleteness of data is sometimes best expressed as a qualitative measure and not just as a quantitative percentage.

Completeness can have an effect on the DQO parameters. Lack of completeness may require reconsideration of the limits for the false-negative and false-positive error rates because insufficient completeness will decrease the power of the statistical test (EPA, 1998).

4.1.1.7.1.1.2 Secondary data quality indicators — The following section provides definitions for each of the secondary data quality indicators.

Sensitivity. Sensitivity is the capability of a method or instrument to discriminate between measurement responses representing different levels of a variable of interest. Sensitivity is determined from the value of the standard deviation at the concentration level of interest. It represents the minimum difference in concentration that can be distinguished between two samples with a high degree of confidence (EPA, 1998).

Recovery. Recovery is an indicator of bias in a measurement. This is best evaluated by the measurement of reference materials or other samples of known composition. In the absence of reference materials, spikes or surrogates may be added to the sample matrix. The recovery is often stated as the percentage measured with respect to what was added. Complete recovery (100%) is the ultimate goal. At a minimum, recoveries should be constant and should not differ significantly from an acceptable value. This means that control charts or some other means should be used for verification. Significantly low recoveries should be pointed out, and any corrections made for recovery should be stated explicitly (EPA, 1998).

Memory Effects. A memory effect occurs when a relatively high concentration sample influences the measurement of a lower concentration sample of the same analyte when the higher concentration sample precedes the lower concentration sample in the same analytical instrument. This represents a fault in an analytical measurement system that reduces accuracy (EPA, 1998).

Limits of Quantitation. The limit of quantitation is the minimum concentration of an analyte or category of analytes in a specific matrix that can be identified and quantified above the method detection limit and within specified limits of precision and bias during routine analytical operating conditions (EPA, 1998).

Repeatability. Repeatability is the degree of agreement between independent test results produced by the same analyst using the same test method and equipment on random aliquots of the same sample within a short time period (EPA, 1998).

Reproducibility. Reproducibility is the precision that measures the variability among the results of measurements of the same sample at different laboratories. It is usually expressed as a variance, and low values of variance indicate a high degree of reproducibility (EPA, 1998).

4.1.1.7.2 Field Sampling Plan

The Field Sampling Plan is the component of the Sampling and Analysis Plan that provides details on the following:

- Sampling objectives
- Sampling locations and frequency
- Sampling procedures
- Sample management
- Handling of investigation-derived waste

The Field Sampling Plan should be prepared by a geologist or sampling specialist who was involved in the preparation of the DQO Summary Report. The template provided in Appendix B and the accompanying CD-ROM should be used to guide the preparation of the Field Sampling Plan.

4.2 SCANNING AND DIRECT MEASUREMENT METHODS

The following section provides the reader with guidance on the types of radiation detectors and meters that are commonly used to support soil remediation and facility decontamination and decommissioning activities. Also provided are details on commercially available systems that link these (and other detectors) to more-sophisticated computer systems, global positioning units, visual monitors, robots, pipe crawlers, etc. This is not an attempt to provide a summary of all the detectors, meters, or systems that are commercially available, but rather to provide details on multiple proven methods that should be taken into consideration when designing a radiological investigation.

While the author believes the detectors and methods presented in this book should be considered as potentially appropriate radiological investigation instruments and techniques, this book is not specifically endorsing or marketing these products. The author chose this approach over a more general discussion in hopes that it would provide the reader with more useful information.

4.2.1   Typical Radiation Instrumentation Used in Radiological Investigations

There are a number of detector systems that represent the basic set of instruments used at a radiological site. For buildings, these detectors include Geiger–Mueller (GM) detectors, alpha scintillation detectors, swipe counters, and gas proportional detectors. Except for the swipe counter, most instruments are small and inexpensive and can be operated in handheld mode or in a mounted configuration (e.g., as a floor scanner). The swipe counter is used to count alpha and beta radiation levels typically on small cotton disks or "swipes" (see Section 4.3.3.1). A swipe is used to wipe over an area (usually 100 cm^2) picking up potentially contaminated debris that is then counted in the swipe counter. The swipe counter measures removable contamination that may be compared to regulatory limits. Table 2.3 provides the regulatory limits for removable surface activity as required by 10 CFR 835 Appendix D. The GM, alpha scintillation, and proportional detectors are typically used in some combination to measure alpha radiation (scintillation and proportional detectors), beta radiation (GM and proportional detectors), and gamma radiation (GM detector) on building surfaces. Another option is to use a "PhoSwich" scintillation detector, which combines a ZnS(Ag) scintillation material adhered to plastic scintillation material for detecting both alpha and beta radiation. A more-detailed description of these detectors is found in Chapter 3.

For surveys of radionuclides in soils or sediment, the basic set of instruments includes NaI detectors, pressurized ion chambers (PICs), dose rate instruments, and sometimes *in situ* gamma spectroscopy systems. All these detectors may be used to measure gamma and X-ray radiation, but are not effective for alpha and beta radiation. Note that some detectors contain thin windows that can be uncovered to measure beta radiation levels in air. For contaminants that do not have a strong gamma or X-ray component, soil sampling is required. Handheld NaI detectors, PICs, and dose rate instruments are similar in that they measure gamma or X-ray radiation, but they have different instrument configurations, detector responses, applications, etc. The NaI detector is often used in gamma walkover surveys where the detector is swung in a serpentine motion over potentially contaminated soil (see Section 4.2.2.1.2.3). The hard metal, airtight casing protects the crystal from damage while scanning close to the ground through grassy fields or wooded areas. The detector response is typically provided in counts per minute, but can be calibrated to radionuclide soil concentration under some circumstances. A PIC and a dose rate instrument have a similar configuration in that they are essentially gas-filled boxes that measure gamma or X-ray radiation levels. However, the PIC measures the exposure rate in air (in units of roentgens per hour or R/h) while the dose rate instrument is calibrated to report a tissue-equivalent dose rate (in units of rem/h).

Both detectors are typically used to take static measurements at some distance away from a potential source. The R/h or rem/h measurements may be used in risk assessment calculations, for comparison against radiation standards or guidelines, to locate areas with elevated radiation levels, or to monitor radiation levels generally.

The *in situ* system (see Section 4.2.2.1.2.3) is a more complex and expensive detector than the others listed above, but is often used to replace some discrete samples or to conduct nonintrusive surveys when gamma-emitting radionuclides are present. *In situ* systems are also useful when there is high background radiation (e.g., from a nearby uranium mill tailings pile), where handheld instruments are less effective. They can either be used in buildings or outdoors. Handheld *in situ* gamma spectroscopy systems are commercially available with NaI crystals (around $10,000 per system). The more expensive and bulkier HPGe systems typically run between $30,000 and $50,000, but can cost twice that amount. When considering whether to use an *in situ* system, project managers must weigh the benefits of using the system (less soil sampling, instant results, large-area samples, etc.) against the detractors, which include:

- High initial cost (especially for HPGe systems);
- Liquid nitrogen temperatures involved with HPGe systems;
- QC soil sampling requirements to verify the accuracy of the measurements.

There is another set of simple detectors that is often used during radiological studies, but these detectors contain no electronics (also called passive detectors; see Section 4.2.2.4). They are thermoluminescent detectors (TLDs), alpha or etch track detectors, and electrets. These instruments are often distributed around the boundary of the study area to monitor radiation levels and radon releases. Passive detectors are inexpensive forms of insurance used to document that there were no releases during site activities. If results from analysis of these detectors indicate a release in excess of site limits, modifications can then be made to site activities based on the results.

A brief description of the basic instruments described above is provided in Table 4.7. This table lists the basic instruments, their common use, common models from various manufacturers, and comparable detectors that can be used as replacements. Note that many of these detectors are used in combination to investigate both buildings and outdoor areas. Additionally, these detectors may be used as parts of more complex detection systems. For example, the NaI detector is often connected to a global positioning system (GPS) unit for performing walkover surveys (Section 4.2.2.1.2.2). When used in this way, the NaI detector provides radiation level measurements while the GPS unit provides coordinates for each measurement. Maps of large areas can then be generated to pinpoint the location of elevated gamma or X-ray radiation levels.

Tables 4.8 through 4.10 present examples of the types of detailed information that can be downloaded off the Internet for various handheld meters, detectors, and swipe counters. The information provided in Tables 4.8 and 4.9 was derived from the Ludlum Measurement, Inc., Web page, which can be visited at www.ludlums.com. Details on Eberline and Duratek instrumentation can be downloaded from the following Web addresses: www.eberlineinst.com and www.gtsduratek.com.

Table 4.7 Basic Instruments Used in Radiological Studies

Instrument	Common Use	Common Models with Manufacturer	Comparable Detectors
GM detector	Scanning surfaces of building material, equipment, and personnel for beta/gamma radiation	Ludlum Model 44-9 Eberline Model HP-210	Gas proportional detector or alpha/beta PhoSwich scintillation detector
Alpha scintillation detector	Scanning surfaces of building material, equipment, and personnel for alpha radiation	Ludlum Model 43-5 Eberline Model AC3-7	Gas proportional detector or alpha/beta PhoSwich scintillation detector
Swipe counter	Scanning swipes of building material, equipment, or personnel for loose alpha/beta-emitting contaminants	Ludlum Model 43-10-1	GM detector, alpha scintillation detector, and gas proportional detector which can also be used to count swipes
Gas proportional detector	Scanning surfaces of building material, equipment, and personnel for alpha/beta radiation	Ludlum Model 43-20 Eberline Model HP-100C	GM detector, alpha scintillation detector, or alpha/beta PhoSwich scintillation detector
NaI detector	Measuring gamma or X-ray radiation levels; often used to perform walkover surveys	Ludlum Model 44-10 Eberline Model SPA-3	PIC and dose rate instrument
PIC	Measuring gamma or X-ray radiation levels in R/h	Eberline Model RO-20	NaI detector which can be calibrated to produce R/h readings; dose rate instrument
Dose rate instrument	Measuring gamma or X-ray radiation levels in rem/h	Bicron Micro-rem Dose Rate Meter	PIC
In situ gamma	Identifying and quantifying gamma- and X-ray-emitting radionuclides in surface soil or on surface materials in buildings	Vendor Only: Canberra EG&G Oxford	NaI detector, PIC and dose rate instrument, although these do not quantify
TLD	Measuring alpha, beta, gamma, X-ray, or neutron radiation levels	Vendor Only: SAIC ICN	PIC and dose rate instrument, although the TLD will measure over large periods of times (weeks or months) averaging out short-term fluctuations
Etch track detectors	Measuring radon concentrations	Landauer's Radtrak®	Electret or air sampler (with filter)
Electret	Measuring radon concentrations	Rad Elec's E-PERM®	Etch track detectors or air sampler (with filter)

Table 4.8 Examples of Ludlum Handheld Meters

Handheld Meter	Meter Specifications	Compatible Detectors	Use
Ludlum Model 3-90 Pancake GM Survey Meter	*Sensitivity:* Typically 3300 cpm/mR/h (Cs-137 gamma) *Meter Dial:* 0–5000 cpm *Linearity:* Reading within ±10% of true value with detector connected *High Voltage:* Adjustable from 200–1500 V *Threshold:* 30 ± 10 mV *Area:* Active—15 cm², Open—12 cm²	Internal pancake style GM detector with 1.7 ± 0.3 mg/cm² window	Alpha, beta/gamma surveys
Ludlum Model 3 Survey Meter	*Meter Dial:* 0–2 mR/h or 0–5000 cpm *Linearity:* Reading within ±10% of true value with detector connected *High Voltage:* Adjustable from 200–1500 V *Threshold:* 30 ± 10 mV	Geiger–Mueller (GM), scintillation Model 44-2 1 × 1 in. Gamma Scintillator Model 44-3 Low Energy Gamma Scintillator Model 44-6 GM Detector Model 44-7 End Window GM Detector Model 44-9 Pancake GM Detector Model 44-38 Energy Compensated GM Detector	Alpha, beta/gamma surveys
Ludlum Model 3-97 Survey Meter	*Internal:* 1 × 1 in. NaI(Tl) scintillator; sensitivity typically 175 cpm/μR/h (Cs-137 gamma) *External:* Model 44-38 energy-compensated GM detector; sensitivity typically 1200 cpm/mR/h (Cs-137) *Meter Dial:* 0–2 mR/h or 0–2400 cpm *Linearity:* Reading within ±10% of true value with detector connected *High Voltage:* Adjustable from 200–1500 V *Threshold:* 30 ± 10 mV	Model 44-2 1 × 1 in. NaI(Tl) Scintillator Model 44-38 Energy Compensated GM Detector	Beta/gamma surveys
Ludlum Model 5 Geiger Counter	*Energy Response:* Within ±15% of true value from 60 keV to 3 MeV *Saturation:* In excess of 1000 R/h *Meter Dial:* 0–2 mR/h *Linearity:* Reading within ±10% of true value with detector connected *High Voltage:* Adjustable from 200–1500 V *Threshold:* 30 ± 10 mV	Two internal energy compensated GM tubes	Gamma surveys

Table 4.8 (continued)   Examples of Ludlum Handheld Meters

Handheld Meter	Meter Specifications	Compatible Detectors	Use
Ludlum Model 12SA MicroR Meter	*Sensitivity:* Typically 175 cpm/µR/h (Cs-137 gamma) *Meter Dial:* 0–3 µR/h *Linearity:* Reading within ±10% of true value *High Voltage:* Adjustable from 200–1500 V *Threshold:* 30 ± 10 mV *Area:* Active—15 cm², open—12 cm²	Model 44-2 1 × 1 in. NaI(Tl) Scintillator	Low-level (µR) gamma surveys

Data courtesy of Ludlum Measurement, Inc.

4.2.2   Radiological Detection Systems

This section presents a number of more complex radiological detection systems that are designed to support soil remediation; facility decontamination and decommissioning; and surveying of tanks, drums, canisters, crates, and remote areas. These systems are designed for large and/or complex radiological sites, and sites that have very high levels of radiation. Since some of the methods presented may be used to support multiple types of projects, they are in some cases addressed in more than one of the following sections. For example, *in situ* gamma spectroscopy is an effective method for supporting both soil remediation and facility decontamination and decommissioning projects.

4.2.2.1   Soil Characterization and Remediation

The following section provides details on a number of aboveground and downhole radiological scanning and direct measurement methods that should be considered to support site characterization and soil remediation activities. Table 4.11 summarizes the data provided and primary use of each of the soil remediation scanning and direct measurement methods presented in this section.

4.2.2.1.1   Airborne Gamma Spectrometry

Airborne gamma spectrometry surveys are typically performed as an initial screening method to support the radiological characterization of large areas impacted by fallout resulting from nuclear accidents (e.g., Chernobyl, Three Mile Island), nuclear weapons testing, or very large environmental sites (e.g., Hanford Nuclear Reservation). Airborne gamma spectrometry surveys may also be performed periodically during remedial activities to track progress, or following remedial activities to ensure no hot spots have been overlooked. These surveys are performed using either a helicopter or winged aircraft equipped with a multichannel spectrometer system and gamma ray sensors for measuring both terrestrial and atmospheric radiation.

Three naturally occurring radioactive elements that emit gamma radiation of sufficient intensity to be measured by airborne gamma ray spectrometry include K-

Table 4.9 Examples of Ludlum Handheld Detectors

Detector	Specifications
Ludlum Model 43-68 100 cm² Gas Proportional Detector	*Indicated Use:* Alpha/beta survey *Recommended Counting Gas:* P-10 (10% methane, 90% argon) *Entry Window:* Typically 0.8 mg/cm² aluminized mylar Window Area: 126 cm² active, 100 cm² open *Efficiency:* Typically 15% (C-14), 30% (Tc-99), 20% (Pu-239), <1% gamma *Background:* Alpha < 5 cpm (when operating on the alpha only plateau region), beta < 400 cpm *Compatible Instruments:* Ludlum Model 12, 16, 18, 2000, 2200, 2221, 2224, 2225, 2241, 2350-1 *Operating Voltage:* Alpha typically 1000–1500 V; Beta/gamma, typically 1600–1800 V *Counter Threshold Setting:* Typically 2–5 mV
Ludlum Model 43-89 100 cm² Alpha-Beta Scintillator	*Indicated Use:* Alpha/beta surveys *Scintillator:* ZnS(Ag) adhered to 0.010-in.-thick plastic scintillation material *Entry Window:* Typically 1.2 mg/cm² aluminized mylar *Window Area:* 125 cm² active, 100 cm² open *Efficiency:* Typically 16% (Pu-239), 5% (Tc-99), 16% (Sr-90/Y-90) *Background:* Alpha < 3 cpm; beta typically < 300 cpm *Compatible Instruments:* Ludlum Model 2224, 2225, 2929 *Tube:* 1.5-in.-diameter magnetically shielded photomultiplier *Operating Voltage:* Typically 500–1200 V
Ludlum Model 43-90 100 cm² Alpha Scintillator	*Indicated Use:* Alpha surveys *Scintillator:* ZnS(Ag) *Entry Window:* Typically 0.8 mg/cm² aluminized mylar *Window Area:* 125 cm² active, 100 cm² open *Efficiency:* Typically 20% (Pu-239) *Compatible Instruments:* General-purpose survey meters, rate meters, and scalers *Tube:* 1.5-in.-diameter magnetically shielded photomultiplier *Operating Voltage:* Typically 500–1200 V
Ludlum Model 44-3 Low Energy Gamma Scintillator	*Indicated Use:* I-125 and X-ray survey *Scintillator:* 1 in. diameter × 1 mm thick NaI(Tl) scintillator *Entry Window:* 15 mg/cm² *Window Area:* 5 cm² active and open *Recommended Energy Range:* Approximately 10–60 keV *Background:* Typically 40 cpm/µR/h *Sensitivity:* Typically 675 cpm/µR/h (I-125) *Efficiency:* Typically 19% for I-125 *Compatible Instruments:* General-purpose survey meters, rate meters, and scalers *Tube:* 1.5-in.-diameter magnetically shielded photomultiplier *Operating Voltage:* Typically 500–1200 V
Ludlum Model 44-6 GM Detector	*Indicated Use:* Beta gamma survey Detector: 30 mg/cm² stainless steel wall halogen quenched GM *Sensitivity:* Typically 1200 cpm/mR/h (Cs-137) *Beta Cutoff:* Approximately 200 keV *Dead Time:* Typically 95 µsec

Table 4.9 (continued)   Examples of Ludlum Handheld Detectors

Detector	Specifications
Ludlum Model 44-9 Pancake GM Detector	*Compatible Instruments:* General-purpose survey meters, rate meters, and scalers *Operating Voltage:* 900 V *Indicated Use:* Alpha beta gamma surveying, frisking *Detector:* Pancake-type halogen quenched GM *Entry Window:* 1.7 ± 0.3 mg/cm² mica *Window Area:* 15 cm² active, 12 cm² open *Efficiency:* Typically 5% (C-14), 22% (Sr-90/Y-90), 19% (Tc-99), 15% (Pu-239) *Sensitivity:* Typically 3300 cpm/mR/h (Cs-137) *Dead Time:* Typically 80 μsec
Ludlum Model 44-10 2 in. × 2 in. NaI Gamma Scintillator	*Compatible Instruments:* General-purpose survey meters, rate meters, and scalers *Operating Voltage:* 900 V *Indicated Use:* High-energy gamma detection *Scintillator:* 2 in. diameter × 2 in. thick NaI(Tl) scintillator *Sensitivity:* Typically 900 cpm/μR/h (Cs-137) *Compatible Instruments:* General-purpose survey meters, rate meters, and scalers *Tube:* 2-in.-diameter magnetically shielded photomultiplier *Operating Voltage:* Typically 500–1200 V

Data courtesy of Ludlum Measurement, Inc.

40, U-238, and Th-232. Cs-137 and several other anthropogenic isotopes also emit gamma radiation of sufficient intensity to be detected by the airborne gamma spectrometry. Cs-137, K-40, U-238, and Th-232 emit gamma rays at an energy level of 662, 1460, 1765, and 2614 keV, respectively.

The variations in the types of ground cover at the site should be carefully evaluated prior to interpreting the results from the airborne gamma spectrometry surveys since concrete, asphalt, building structure, and other types of ground cover will shield the activity of the underlying soil. The attenuation of gamma rays in most materials is proportional to the electron density of the material. For this reason, the absorption of gamma rays both in the ground and by the mass of the air between the surface and the aircraft must be taken into account (IAEA, 1991).

4.2.2.1.1.1 Instrumentation — Sodium iodide (NaI) and some other crystalline substances give off scintillations (flashes of light) when struck by alpha or beta particles, or gamma rays. These scintillations can act as a gamma-ray sensor and are counted by a photomultiplier, thus giving an indication of the intensity of radiation in the vicinity. Gamma-ray sensors are designed primarily for use in multichannel spectrometer systems. Each gamma-ray sensor contains multiple detectors each of which is composed of a rectangular, thallium-activated NaI crystal (up to 4 L in size). A photomultiplier tube is attached to the end of each detector. The detectors are mounted on a shock-absorbing chassis, and enclosed in a well-insulated protective aluminum container (Figures 4.26 and 4.27).

Each NaI crystal is typically connected to a module that contains a microprocessor, pulse detection circuit, coincident pulse recorder, pulse converter, tempera-

Table 4.10 Examples of Ludlum Swipe Counters

Counter	Specifications
Ludlum Model 43-10 Alpha Sample Counter	*Indicated Use:* Alpha sample counting Scintillator: ZnS(Ag) *Window:* Windowless *Sample Holder:* Anodized aluminum tray with a 1-in.-diameter sample ring to allow for 1- or 2-in.-diameter samples *Maximum Sample Size:* 2-in.-diameter × 0.4-in. thick *Efficiency:* Typically 42% (Pu-239) *Compatible Instruments:* Typically used with Ludlum Model 1000, 2000, 2200, 2221 *Tube:* 2-in.-diameter magnetically shielded photomultiplier *Operating Voltage:* 1350–1600 V
Ludlum Model 43-78-1 Large Sample Beta Counter	*Indicated Use:* Large-sample beta counting *Scintillator:* Plastic scintillation material *Window:* Windowless *Sample Holder:* Anodized aluminum tray *Maximum Sample Size:* 5-in. diameter × 0.18-in. thick *Efficiency:* Typically 40% (Sr-90/Y-90), 12% (C-14), 35% (Tc-99) *Compatible Instruments:* Typically used with Ludlum Model 1000, 2000, 2200, 2221 *Tube:* 5-in.-diameter magnetically shielded photomultiplier *Operating Voltage:* 500–1200 V
Ludlum Model 120 Gas Proportional Sample Counter	*Indicated Use:* Alpha/beta sample counting *Scintillator:* P-10 (10% methane, 90% argon) *Window:* Typically 0.4 mg/cm² aluminized mylar *Sample Tray:* Adjustable height from 0.0–0.35 in. *Maximum Sample Size:* 2-in. diameter × 0.3-in. thick *Efficiency:* Typically 35% (Th-230), 42% (Sr-90/Y-90), 10% (C-14) *Compatible Instruments:* Ludlum Model 12, 16, 18, 2000, 2221, 2241, 2350-1 *Operating Voltage:* Alpha (900–1300 V), Beta-gamma (1300–1700 V) *Counter Threshold Setting:* Typically 2–5 mV

Data courtesy of Ludlum Measurement, Inc.

ture sensor, high voltage supply, program memory, and nonvolatile crystal parameter memory. Each module is then linked together and connected to the spectrometer processing module. Each second, the pulse results are examined and output as digital values of counts per second. As a general rule, one should use the largest volume of sodium iodide crystals as practical when performing a survey: 17 or 33 L for helicopter surveys and 33 or 50 L for fixed-wing surveys.

The natural gamma ray spectrum (0 to 3000 keV) can be resolved into as many as 256 channels, with each channel ranging from 10 to 12.5 keV in width. Some spectrometer software allows for the acquisition and storage of both upward- and downward-looking detectors simultaneously, which allows corrections for both background and Compton scattering to be made.

4.2.2.1.1.2 Flight Line Direction and Speed — For study areas where little or no information is available regarding distribution patterns of radiological contami-

Table 4.11 Soil Remediation Scanning/Direct Measurement Methods

Scanning/Direct Measurement Method	Data Provided	Primary Use
Airborne gamma spectrometry	Strong gamma-emitting isotopes	NaI detectors mounted on a helicopter or fixed-wing aircraft are used to survey very large study areas to define where gamma-emitting radionuclides are concentrated
Mobile Surface Contamination Monitor-II (MSCM-II)	Gross beta/gamma	System composed of three large-area radiation detectors mounted on a tractor; ties gross beta/gamma data to global positioning coordinates; used to support site characterization and postremediation verification surveys
Global Positioning Environmental Radiological Surveyor (GPERS-II)	Gross beta/gamma	Backpack unit that ties gross beta/gamma data to global positioning coordinates; used to support site characterization and postremediation verification surveys
In-situ gamma spectroscopy	Gamma-emitting isotopes	HPGe detector mounted on a portable cryostat (tripod); used to collect isotopic measurements from surface soil without collecting samples
Downhole gross gamma logging	Gross gamma	NaI detector lowered down borehole to assist in identifying depth intervals showing elevated gamma activity; short count times of 1–2 min typically used
Downhole HPGe measurements	Gamma-emitting isotopes	HPGe detector lowered down borehole and used to collect isotopic measurements from selected depth intervals; count time dependent on detection limit requirements
Cone penetrometry	Gamma-emitting isotopes, metals, soil moisture/resistivity, soil stratigraphy, video imagery, pH, temperature, permeability	Steel cone is hydraulically pushed into the ground while *in situ* measurements are continuously collected and transported to the ground surface for data interpretation; method able to collect samples of soil, water, and soil vapor for laboratory analysis
Large-area survey monitor	Gross alpha	Used to collect *in situ* measurements of plutonium in soil, debris, and buried containers for criticality control

nants, airborne gamma ray spectroscopy surveys are typically flown in either a north–south or east–west direction. On the other hand, if the results from historical investigations show radioactive contaminants to be distributed in an elongated pattern (i.e., fallout elongated along dominant wind direction at the time of release), consideration should be given to flying the airborne surveys parallel to the long axis of the elongation. This approach will minimize the total number of flight line passes needed to complete the survey, and will minimize the chances of missing hot spots.

While helicopter surveys are typically flown at speeds ranging from 30 to 60 mph, fixed-wing aircraft surveys are typically flown at speeds ranging from 80 to

Figure 4.26 NaI detectors mounted in protective aluminum containers.

Figure 4.27 NaI detectors mounted for helicopter survey.

120 mph. Since instrument readings are collected at a standard time interval (e.g., one reading per second), slower airspeeds are preferred since they reduce the horizontal distance between measurement positions.

When designing a sampling program that includes airborne radiological surveys, the preferred flight direction, flying speed, flying height, and horizontal distance between flying lines should be clearly defined in the Sampling and Analysis Plan and should be based on the radiological target size, shape, and orientation.

4.2.2.1.1.3 Flying Heights and Spacing — In order for an airborne gamma-ray survey to provide reliable results, it is essential that the airborne survey be performed at a constant height. Since gamma rays are attenuated by air in an exponential fashion, lower flying heights provide more reliable survey results since anomalies can more easily be distinguished from background. For terrain that has very little topographic relief, flying heights as low as 100 to 150 ft are used. Lower flying heights and slower speeds are particularly desirable when mapping sites where contamination is suspected to be only slightly elevated above background levels. However, it is important to recognize that lower flying heights have significantly greater safety risks, particularly in areas that have significant topographic relief. Generally, flying heights of 300 to 400 ft are used for mapping nuclear fallout (IAEA, 1991).

It is important to select a spacing between flight lines that is consistent with the project-specific objectives defined by the DQO process (Section 4.1.1.5). For example, the selected spacing should take into consideration the size and shape (i.e., circular, elliptical) of the target one is looking for, the flying height, and the potential consequences (e.g., low, moderate, severe) of missing the target. If the potential consequences of missing the target are severe, the selected spacing between flight lines should be tight. On the other hand, if the potential consequences for missing the target are only minor, a wider spacing may be more appropriate. When mapping a very large area that is known to be contaminated (e.g., nuclear fallout), one should consider initially using a wide line survey to identify the general areas where higher-level activity is concentrated. Then, a tight line survey would follow to define the boundaries of the high-activity areas more clearly.

For additional details on implementing an airborne gamma spectrometry survey, see IAEA (1991) or visit the Scintrix Earth Science Instrumentation Web site at www.idsdetection.com/products.html/pd_sesi.html.

4.2.2.1.2   Land-Based Surveys

Land-based surveys are often performed as an initial screening method to identify the boundaries of surface radiological contamination at smaller sites (e.g., <50 acres), or are performed to define more clearly the boundaries of surface contamination identified by airborne gamma spectrometry surveys (see Section 4.2.2.1.1).

Some of the land-based survey methods that should be considered to support soil remediation projects include:

- Mobile Surface Contamination Monitor-II
- Global Positioning Environmental Radiological Surveyor
- *In situ* gamma spectroscopy
- Downhole gamma logging
- Downhole HPGe measurements
- Cone penetrometry
- Large area survey monitor

4.2.2.1.2.1 Mobile Surface Contamination Monitor-II — The Mobile Surface Contamination Monitor-II (MSCM-II) monitoring system was developed by Westinghouse Hanford Company for the U.S. Department of Energy, and represents a first-generation integration of Global Positioning System (GPS) technology with radiation detection equipment. This system has been used at the Hanford Nuclear Reservation to support the characterization of many radiological sites.

The MSCM-II utilizes a GPS receiver integrated with three large-area plastic radiation detectors mounted on a large tractor. The unit has a titanium beta window to allow detection of high-energy beta radiation as well as gamma radiation. An unshielded gamma detector is positioned on top of the detector unit and is used to measure cosmic and terrestrial radiation outside the columnated area. The cosmic and terrestrial radiation measurements are then subtracted from the columnated measurements.

The positioning information is provided via the U.S. Government Navstar satellites in orbit around the Earth. The GPS requires a minimum of two satellites and as many as eight satellites to determine position. The position data are then joined with the corresponding radiation measurements. A Geographical Information System (GIS) can then be used to plot the resulting data on a base map.

For further details on the capabilities of this system, contact ThermoRetec, Inc., at www.thermohanford.com or (509) 371-1506.

4.2.2.1.2.2 Global Positioning Environmental Radiological Surveyor — The Global Positioning Environmental Radiological Surveyor (GPERS-II) system was developed by ThermoRetec, Inc., to add additional mobility to radiation surveys, particularly for those areas that are difficult to access by vehicles. GPERS-II was the ThermoRetec fourth-generation integration of GPS technology with radiation detection equipment.

GPERS-II is a smaller, more compact version of the MSCM II unit (see Section 4.2.2.1.2.1). GPERS-II is a backpack unit that integrates all the capability of the MSCM-II in a unit that a single person can carry (Figures 4.28 and 4.29). The GPERS-II has been developed in a modular form that allows more versatility and reliability. The radiation detection system utilized on the GPERS-II is an off-the-shelf commercial unit. In the event of failure of the system, the broken unit can be removed, a new unit plugged in, and the survey restarted very quickly. For highly radioactive environments the GPERS-II unit can be mounted on a robotic or remotely operated vehicle (RADCART) such as that shown in Figure 4.30.

The GPERS-II unit utilizes the latest advances in GPS receivers that have the ability to do on-the-fly positional differentiation. This process takes the offset that is programmed into the Navstar satellites and allows correction of the position to less than 1 m anywhere in the world. This offset that is broadcast by the satellites is controlled by the U.S. military and changes regularly. The differential correction ability is necessary to ensure correct positions are obtained. Without this ability, the position could be off as much as 1 mile.

One of the primary advantages of the GPERS-II system is that it is very versatile, in that it can be mounted on a motorized vehicle, cart, or wagon, in addition to being carried as a backpack unit. It also has the advantage of being able to receive input from two different detectors simultaneously. This eliminates the need to perform two separate surveys with two separate detectors. One of the limitations of the GPERS-II unit is that the performance of the system is limited by the 1-meter accuracy of the positioning system. A second limitation is that the radio signals are unable to penetrate building walls, and are impacted by heavily forested areas, or steep canyon walls.

For further details on the capabilities of this system, contact ThermoRetec, Inc., at www.thermohanford.com or (509) 371-1506.

Figure 4.28 Field survey using GPERS-II backpack unit. (Courtesy of ThermoRetec, Inc.)

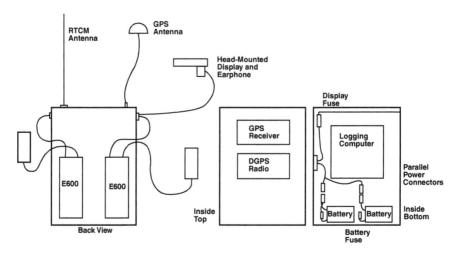

Figure 4.29 Components to GPERS-II backpack unit. (Courtesy of ThermoRetec, Inc.)

4.2.2.1.2.3 In Situ *Gamma Spectroscopy* — Recent advances in computer technology have made it possible to develop laboratory-grade portable gamma spectroscopy systems that can be effectively utilized in the field. An *in situ* gamma spectroscopy system consists of a high-purity germanium (HPGe) detector mounted on a portable cryostat, a tripod for variable-height field measurements, a notebook computer with a multichannel analyzer board, gamma spectra analysis software, and calibrations for the likely sample geometries. These systems are lightweight and portable, but sufficiently rugged for most field data collection efforts. The advantage of using *in situ* gamma spectroscopy is that it can provide quantitative soil concentration results, and therefore for some sites it can be used to reduce the need for soil

Figure 4.30 GPERS-II unit attached to remotely operated vehicle. (Courtesy of ThermoRetec, Inc.)

sampling and fixed laboratory analyses. The Canberra *In Situ* Object Counting System discussed in Section 4.2.2.2.3 is a type of *in situ* gamma spectroscopy instrument that can be used to support either remediation or facility decontamination and decommissioning operations (see Figures 4.36 and 4.37, later). An NaI detector may also be used to collect *in situ* gamma spectroscopy measurements if there is only one isotope of concern (e.g., Ra-226) at a site, or if there are multiple isotopes present in relatively high concentrations. This should be taken into consideration since an HPGe detector is much more expensive than an NaI detector.

The basic principles behind *in situ* gamma spectroscopy measurements are the same as those for gamma spectroscopy in a fixed laboratory. The *in situ* spectrometer measures gamma interactions within the HPGe detector and converts these data to concentrations of each gamma-emitting radionuclide per unit mass or volume of sample. The major difference is that for *in situ* measurements, the spectrometer is viewing a much larger sample volume than is possible through laboratory analysis. For an unshielded, 3-ft-high *in situ* measurement, the detector is "viewing" or collecting data from the area on the ground surface represented by a circle around the detector with a radius of approximately 33 ft. The depth (as well as the exact area) of detector view is a function of gamma energy, but for most radionuclides this depth is approximately 6 to 12 in. A conceptual picture of the soil volume (sample size) viewed by the *in situ* spectrometer is a bowl with a radius of approximately 33 ft, and a depth at the center of the bowl of approximately 1 ft, tapering to the surface at a distance of approximately 33 ft (DOE, 1996b).

The sample size or volume of soil viewed by the *in situ* spectrometer can be adjusted through collimation (partial shielding of the detector), and by adjustment of the height of the detector above the ground surface. Because unshielded *in situ* measurements look at a large surface area (and thus a large soil volume), the concentration results reported from such measurements represent average concentrations over the field of view of the detector (approximately 3200 ft^2).

Unlike sampling, where only a small portion of the soil at a site can be measured, *in situ* gamma measurements can be used to cover a site more comprehensively. Due to the built-in averaging characteristics of *in situ* gamma spectroscopy measurements, it is desirable to use this method in combination with gamma walkover surveys to define clearly the boundaries of the areas showing elevated activity. If contaminants are distributed relatively uniformly across the study area and through-out the depth viewed by the *in situ* detector, then the average results reported by the *in situ* measurements should closely match results from samples collected and composited from the same general area. However, if localized areas of elevated contamination (hot spots) are present, the *in situ* measurements will average these areas across the entire volume (area and depth) viewed by the detector. Conse-quently, the results from an uncollimated *in situ* measurement of an area with hot spots would not be expected to match the results from samples taken from the hot spots.

In situ gamma spectroscopy measurements should be considered to support characterization surveys prior to and during remedial action, as well as to perform verification surveys following remediation. This technology provides waste minimi-zation potential by reducing the total number of samples collected for laboratory analysis. However, it is important to keep in mind that for smaller projects it may be less expensive to send soil samples to the laboratory for analysis than either to rent or to purchase an *in situ* gamma spectroscopy system.

A typical *in situ* gamma spectroscopy system can perform measurements at the <1 pCi/g level for most common contaminants from reactor facilities (e.g., Cs-137, Co-60) and from ore-processing facilities (e.g., Ra-226, Th-232), and measurements at the levels of a few pCi/g for fuel processing and reprocessing facilities (Am-241, U-235, U-238). The results are available immediately for use to determine the need and strategy for further measurements, or to guide the excavation effort. A GPS can be easily integrated to provide automatic sample location (latitude, longitude, and elevation) to within several feet, and to record this with the sample information. These data can be used in combination with a GIS to present this information to the user rapidly in a variety of cartographic schemes.

The field gamma spectroscopy results can be transmitted back to a base station via telemetry link and directly into the GIS for an immediate view of the survey effort, and to guide the selection of the next measurement point. *In situ* gamma spectroscopy can be effectively used to support initial site characterization, excava-tion activities, or to verify that a site is clean.

Further information on *in situ* gamma spectrometry is presented in Section 4.2.2.2.3.

4.2.2.1.2.4 Downhole Gross Gamma Logging — Downhole gross gamma log-ging is an effective method for quickly identifying borehole depth intervals showing elevated gamma activity. This method is most often used to support site character-ization activities by assisting in the selection of the depth intervals to collect down-hole HPGe measurements (see Section 4.2.2.1.2.5), or to assist in the selection of the sampling intervals to send to the laboratory for analysis.

Once the maximum depth of a borehole has been reached, an NaI detector is lowered down through the inside of the drilling augers (or open borehole) to the bottom of the hole. Gross gamma activity counts are then collected at systematic time and depth intervals. For most site characterization studies gross gamma activity measurements are collected at 1-ft depth intervals through the entire length of the borehole, using a 1-min count time at each interval.

Once the entire depth of a borehole has been counted, the cpm readings are then used to identify the depth intervals showing the highest gross gamma activity, as well as the first depth interval to show background activity. The advantage to using the downhole gross gamma logging method is that it is inexpensive to implement and can help minimize the number of samples that need to be sent to the laboratory for analysis. For a site characterization study, typically only three soil sampling intervals need to be selected from a borehole for analytical testing. These intervals include the surface sampling interval (for human health and risk assessment purposes), the interval showing the highest gross gamma activity, and the first interval below this that shows background gross gamma activity. The remaining sampling intervals should be archived in case they are needed at a later time.

4.2.2.1.2.5 Downhole HPGe Measurements — Downhole HPGe measurements are collected in the same manner as downhole gross gamma logging (see Section 4.2.2.1.2.4), except an HPGe detector is lowered down through the inside of the drilling augers (or open borehole) as opposed to an NaI detector. This method is capable of providing laboratory-quality gamma isotopic data in the field. The length of the count time at each of the sampling intervals is dependent upon the detection limit requirements. Typically, count times range from 20 to 60 min in length.

4.2.2.1.2.6 Cone Penetrometry — A cone penetrometer consists of a steel cone that is hydraulically pushed into the ground while *in situ* measurements are continuously collected and transported to the ground surface for data interpretation and visualization. The hydraulics are housed in a vehicle (see Figure 4.84 later) or a crane transportable unit (Figure 4.31) that can provide ballast of up to 80,000 lb. Penetration rates for most soil types are approximately 40 to 50 ft/h.

A cone penetrometer is used frequently to support site characterization studies since it is generally more cost-effective than drilling and sampling, and there are no drill cuttings to clean up after the push rods are removed from the hole. This method provides continuous subsurface screening-quality radiological, chemical, physical, and/or electrical data. Since this method provides real-time data, the results from one push location can be used to assist in selection of the next location.

The cone penetrometer can be fitted with a number of different sensors that can be used to support site characterization activities. These sensors can provide the following types of data:

- Gamma spectroscopy
- X-ray fluorescence
- Soil moisture and resistivity
- Soil stratigraphy
- Video imagery
- pH, temperature
- Permeability

Figure 4.31 Crane transportable cone penetrometer unit.

The cone penetrometer is able to collect samples of soil, water, and soil vapor for laboratory analysis, and can install piezometers for soil vapor and groundwater measurements. Refer to Section 4.3.3.7.1 for additional information on how the cone penetrometer (direct push method) can be used to support groundwater investigations (DOE, 1996a).

4.2.2.1.2.7 Large Area Survey Monitor — The Large Area Survey Monitor is manufactured by BNFL Instruments Ltd. and is used to collect *in situ* measurements of plutonium in soil, debris, and buried containers for criticality control. This system utilizes polyethylene-wrapped He-3 detectors to measure the plutonium from an area that measures 8×8 ft with a depth of 3 ft. The system also has an optional gamma measurement capability.

The Large Area Survey Monitor is composed of a detector module and a control module, which can be separated by a distance of up to 300 ft, which in turn allows for remote operation in highly radioactive environments (Figure 4.32). The calibration of the system is checked automatically between measurement points. An installed Cf-252 source is used to provide quality control checks. The system is able to achieve a detection limit of <10 g of plutonium at a depth of 3 ft, when the moisture content of the soil is <15 wt% water. The spatial resolution is <8 in. in all dimensions.

The detector module is typically moved between sampling points using either a crane or forklift, and is sealed to allow it to be decontaminated whenever it becomes contaminated.

For further details on the capabilities of this monitor, refer to the BNFL Instruments Web page at www.bnfl-instruments.com or call (505) 662-4192.

4.2.2.2 Building Decontamination and Decommissioning

The following section provides details on a number of scanning and direct measurement methods that should be considered to support facility decontamination and decommissioning activities. Table 4.12 summarizes the data provided and primary use of each of the decontamination and decommissioning scanning and direct measurement methods presented in this section.

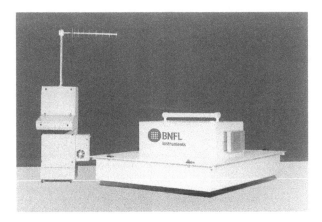

Figure 4.32 Large area survey monitor. (Courtesy of BNFL Instruments, Ltd.)

4.2.2.2.1   GammaCam™

The GammaCam radiation imaging system is manufactured by AIL Systems, Inc., and is a portable system that provides two-dimensional images of gamma ray-emitting objects (Figure 4.33). This system is particularly useful in supporting building decontamination and decommissioning activities and work in nuclear reactors since it can quickly identify those building surfaces and interior areas with elevated activity levels. This system also has the ability to provide images through various types of coverings (e.g., paint, metal drum surfaces). Figure 4.34 shows an example of how a radiation leak appears on a photograph taken with the GammaCam.

This system provides a real-time pseudo-color image of gamma ray sources, superimposed on a black-and-white video picture. The superimposed radiation and visual images are displayed on a standard portable PC screen. The radiation image appears as rings of color, with each color relating to the relative strength of the gamma radiation source. Since the GammaCam system is not able to differentiate between isotopes on its own, it can only provide dose or exposure rate estimates for a specific isotope that is known to be present.

The GammaCam can help minimize worker exposure since it can be operated remotely or at large stand-off distances. It can also help to minimize cost by focusing building decontamination activities on those areas known to be contaminated. The collection of radiation images routinely throughout the decontamination process can help to identify any contaminated areas that may have been missed, and provide management with visual evidence of progress to date. Once building decontamination and decommissioning activities are complete, copies of the images collected throughout the effort could be released to the public to give people confidence the decontamination and decommissioning activities were successful. This system can also be used to help minimize cost by assisting in the separation of waste material that needs to be disposed of in a radiological landfill from that which can be disposed of in a municipal landfill.

Table 4.12 Building Decontamination and Decommissioning Scanning/Direct Measurement Methods

Scanning/Direct Measurement Method	Data Provided	Primary Use
GammaCam™	Images of gamma ray and X-ray emitting objects	Used as a method to identify quickly where gamma-emitting hot spots are located on building surfaces; provides two-dimensional images of gamma ray-emitting objects superimposed on a black-and-white video picture; data can also be used in conjunction with software to produce three-dimensional models of contamination; method designed primarily for high radiation areas
RadScan™ 700	Images of gamma ray emitting objects	Used as a method to identify quickly where gamma-emitting hot spots are located on building surfaces; pinpoints the origin of measured radiation by overlaying radiometric data overtop a video image; instrument head pans 340° in horizontal direction, tilts 40° below horizontal and 90° above horizontal; method designed primarily for high radiation areas
In Situ Object Counting System (ISOCS)	Gamma-emitting isotopes	Uses a germanium detector to collect *in situ* gamma spectroscopy isotopic data from most any building surface (floors, wall, ceiling, objects), soil surface (aboveground and downhole), waste drums, etc.
Laser Assisted Ranging and Data System (LARADS)	Gross alpha, gross beta, gross gama	Collects real-time gross alpha/gamma data from building interior or exterior surfaces; detector is attached to a tracking system; radio or wired data communication system transmits measurement data to a portable computer that can be operated remotely; activity measurements are attached to x, y, and z positional coordinates
3-D Imaging Decommissioning In-Situ Plutonium Inventory Monitor	Gross alpha	Monitor used to support alpha plant decommissioning operations by providing three-dimensional mapping of residual plutonium on plant vessels, drums, gloves, boxes, etc.
TRU Piece Monitor	Gross alpha	Monitor used to support decommissioning of plutonium facilities by measuring total plutonium content of single items of waste (e.g., folded piping, metal plating) using neutron coincidence counting
IonSens Alpha Pipe Monitor	Gross alpha	Monitor is used to measure the internal and external surfaces of cut pipe and scaffolding poles for alpha activity thus permitting the categorization of the material as either low-level waste or free release
IonSens 208 Large Item Monitor	Gross alpha	Used to monitor exposed surfaces of complex items, such as process equipment, folded piping, etc., for the purpose of categorizing the material as either low-level waste or suitable for free release; detects the ions produced by alpha particle interactions in the air
Pipe Surveying	Gross alpha, gross gamma, organic vapor, combustible gas, oxygen, pH/temperature, video	Often performed to determine if contaminated sludge or other material is present in a pipe, to determine if any cracks are present, and to determine where a pipe ultimately discharges

Figure 4.33 GammaCam.

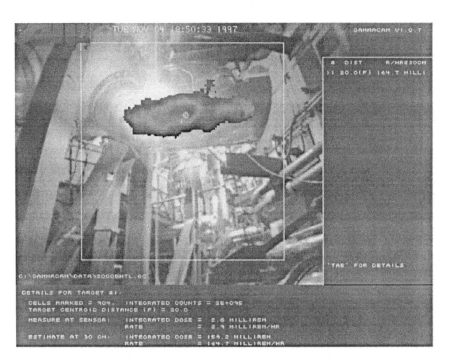

Figure 4.34 Radiation leak detected by GammaCam.

The GammaCam is most often attached to a tripod that holds the system in place. However, for highly radioactive environments, the system may be attached to a remote-operated vehicle. Figure 4.34 illustrates how the system was used to identify the primary source of radiation in one section of a reactor building.

The GammaCam utilizes a high-density terbium-activated scintillating glass detector that operates in the spectral range of <80 keV to >1.3 MeV. The system is sensitive to a 1 µR integrated dose for a Cs-137 point source. The exposure time is selected by the user and may range from 10 msec to 10 min. The field of view for the imaging system can be adjusted from as narrow as 25° to as wide as 50°. The instrument resolution varies from 1.3° to 2.6°, and is dependent upon the width of the field of view. The narrower the field of view, the better the instrument resolution. The instrument can be used under temperature conditions ranging from 5 to 40°C, and has the following system power requirements: 215 W, 110 to 240 VAC 50/60 Hz. The instrument utilizes Windows 95 software, and sets up for use in just a few minutes.

The GammaCam has been commercially available since 1996 and has been used to support radiological studies at the following facilities:

- Argonne National Laboratory, Illinois
- Los Alamos National Laboratory, New Mexico
- Hanford Nuclear Reservation, Washington
- Peach Bottom Atomic Power Station, Pennsylvania
- Arkansas Nuclear One Site, Arkansas
- Wolf Creek Nuclear Operations, Kansas

For additional information about the GammaCam, visit the AIL Systems, Inc., Web page at www.ail.com.

4.2.2.2.2   RadScan™ 700

The RadScan 700 Gamma Scanner made by BNFL Instruments is similar to the GammaCam in that it is designed to pinpoint the origin of measured radiation by overlaying radiometric data on a video image. The head of the RadScan 700 can pan as much as 340° in a horizontal direction and tilt as much as 40° below horizontal and as much as 90° above horizontal (Figure 4.35). This instrument is particularly useful in supporting building decontamination and decommissioning activities and work in nuclear reactors since it can quickly scan large building surface areas to identify hot spots. This instrument provides real-time data and utilizes gamma spectroscopy to provide a spectrum of colors to identify either specific count rates or dose rates for individual isotopes (e.g., Cs-137, Pu-240, U-235) associated with each hot spot.

For building decontamination, the RadScan 700 is similar to the GammaCam in that it can help minimize cost by focusing decontamination activities on those areas known to be contaminated. The collection of radiation images routinely throughout the decontamination process can help to identify any contaminated areas that may have been missed, and provide management with visual evidence of progress to date. Once building decontamination and decommissioning activities are complete, copies of the images collected throughout the effort could be released to the public to give them confidence the decontamination and decommissioning activities were successful. This system can also be used to help minimize cost by assisting in the separation of waste material that needs to be disposed of in a radiological landfill from that which can be disposed of in a municipal landfill.

The RadScan 700 can be deployed in many ways including the use of a fixed standard mobile crane, tripod, or remotely operated vehicle. This system uses a

Figure 4.35 RadScan 700 Gamma Scanner.
(Courtesy of BNFL Instruments,
Ltd.)

detachable detector head which is optimized for both plutonium and fission product gamma energies. This system performs all self-checking functions automatically, and can be operated from a remote location using a dedicated PC-based power and control module.

This system has three modes of operation which include the count mode, range corrected mode, and normalized mode. The count mode displays the raw count rate(s) at the detector from objects within the current viewing zone with no correction for distance. The range corrected mode allows the operator to correct the real-time count rate(s) to what it would be if the object were at a range specified by the user. The normalized mode allows the operator to normalize the real-time count rate(s) to the count rate from an object specified by the user, and to correct for range. This enables the operator to estimate whether the zone contains more or less activity relative to the normalization zone.

A viewing zone circle is displayed on the instrument screen. This viewing zone circle is color coded to indicate the reliability of the displayed count rate information. When the count rate at the detector is relatively high, the circle defining the field of view will become green within a few seconds of the scanning head coming to rest. This indicates that enough counts have been received to enable the count rate to be determined within operator pre-set uncertainty limits. On the other hand, when the count rate at the detector is low, the circle will remain red until enough counts have been received. The system also informs the operator if the count rate exceeds the limit of the detector system.

The RadScan 700 can be set to scan an area of interest automatically within operator-defined vertical and horizontal limits. Threshold limits can also be set on the count rate information to define gamma hot spots. The count rate and scan information for all measurement positions is time-recorded by the PC. Once initiated, the auto-survey feature displays an estimate for the duration of the complete scan. If accepted, the entire scan is performed automatically. Once completed, the video of the auto-survey scan can be replayed at the operator's convenience.

All scan data collected during auto survey mode are date- and time-stamped and stored as individual files on the system hard disk drive or floppy disk. These files are directly compatible with Excel for Windows which allows two-dimensional contour maps to be generated.

The RadScan 700 is commercially available and uses a CsI (TI) scintillator photodiode. The detection range for the instrument is from 1 to 50 m. At a 1-m distance, the system is able to detect Cs-137 with activity levels as low as 1 μCi or as high as 0.2 Ci. The PC can be positioned at a distance of 40 m or greater from the inspection head.

For further details on the capabilities of this monitor, refer to the BNFL Instruments Web page at www.bnfl-instruments.com or call (505) 662-4192.

4.2.2.2.3   In Situ *Object Counting System*

The In Situ Object Counting System (ISOCS) developed by Canberra is a type of *in situ* gamma spectroscopy instrument (also see Section 4.2.2.1.2.3) that is able to provide laboratory-quality gamma spectroscopy results at any location and for essentially any object of interest. The ISOCS can be used to support soil characterization, remediation, and facility decontamination and decommissioning activities. The ISOCS system includes a germanium detector with a multiattitude cryostat, an InSpector-2000 multichannel analyzer, a laptop computer, Genie-2000 gamma spectroscopy software, ISOCS mathematical efficiency calibration software, a set of adjustable shield/collimator components, and a special equipment cart to facilitate aiming of the detector and movement of all system components. Figures 4.36 and 4.37 show how the ISOCS can be used to support facility decontamination and decommissioning activities as well as soil remediation and drum and container monitoring.

The ISOCS mathematical calibration software allows accurate efficiency calibrations to be performed for a wide variety of sample shapes, sizes, densities, and distances between the sample and the detector. The objects can be very small or very large (<3000 ft), or the objects can be very close or very far away (<1500 ft). Efficiency values can be calculated for photon energies in the range of 45 to 7000 keV. Photon attenuation effects due to collimators and shielding components (if present) can be included in the efficiency calibration process. Attenuation effects due to the sample material itself, the container walls (if any), and the air between the sample and the detector are also included in the calculations. Any material at any density can be created by the user and incorporated into the samples and/or containers.

The ISOCS system can be a valuable tool in a wide variety of measurement applications. Some of the possible measurement applications involve quantitative analyses of objects or structures for which traditional sample collection is unsafe and/or impractical.

The ISOCS system can be used for determinations of radioactivity content in near-surface soil. A common protocol for *in situ* measurements of soil activity is to place the detector in a downward-looking position at a height of 1 m above the soil surface. The soil matrix can be modeled as a cylindrical plane or rectangular plane with area large enough (~400 m²) to be considered "infinite" for photon energies of interest. The activity distribution within the soil can be modeled as a

Figure 4.36 ISOCS supporting facility decontamination and decommissioning.

Figure 4.37 ISOCS supporting remediation and drum/container monitoring.

thin plane, uniform, or exponentially decreasing with depth (each with or without a
layer of background-level overburden at the surface). Subsurface soil monitoring
(well logging) can also be performed using an ISOCS system with a detector mounted
with a suitable small-diameter cryostat. The soil matrix surrounding the borehole
can be modeled using one of the standard ISOCS geometry templates. The versatile
shield set can be reassembled to allow counting of samples in the field. For example,
the ISOCS system can also be used to assess standing water or water samples
collected in the field, as well as air particulate filters and iodine cartridges collected
to evaluate natural background activity levels and facility effluents.

The ISOCS system can be used to perform direct measurements of all common
waste containers, including drums, boxes, and bags. Different source activity distributions within the container (e.g., hot spots at the center or outer edge vs. uniform
concentration) can be modeled and quickly compared to evaluate the effect on
measured activity values. These measurements are potentially faster, more accurate,
and less hazardous than opening the waste containers to collect samples. Analysis
results can be used for inventory and shipping purposes.

The ISOCS system can be used to determine nuclide-specific contamination
levels on floors, walls, and ceilings. *In situ* measurements of these surfaces are faster

and safer than smear surveys or other sample collection/analysis methods, and they are more accurate for nonuniform activity distributions. ISOCS measurements of activity inside pipes, valves, and ducts can also be performed. All of these components can be easily modeled with one of the standard ISOCS geometry templates. In the event of a spill or leak of radioactive material, ISOCS measurements can be performed to evaluate the activity of standing pools or piles of material, as well as residual surface contamination after the loose material has been removed.

For additional information about the ISOCS, visit the Canberra Web page at www.canberra.com or call (800) 243-4422.

4.2.2.2.4 Laser-Assisted Ranging and Data System

The Laser-Assisted Ranging and Data System (LARADS) was developed by ThermoRetec, Inc. The LARADS provides real-time data on the location and activity of radiological contamination on interior and exterior building surfaces.

The LARADS is composed of a tracking system, radio or wired data communication device, a portable computer that can be operated on site or from a remote location, and customized software. The LARADS is capable of interfacing with many different kinds of radiological detectors for assisting in defining contaminant distribution, dose rates, etc. The LARADS, like the GPERS-II (see Section 4.2.2.1.2.2), is capable of logging two channels of radiological instrumentation simultaneously.

The activity measurements, and x, y, and z positional coordinates, collected using the LARADS are electronically recorded. These data can be superimposed on computer-aided design drawings or digital photographs of the walls and floor. The location of each reading is recorded using a laser-assisted mapping system (similar to that utilized by civil land surveyors for mapping geographical locations). Figures 4.38 and 4.39 show how the LARADS can be used to support both facility decontamination and decommissioning operations as well as soil remediation activities.

Figure 4.38 LARADS used to support building floor scanning surveys. (Courtesy of ThermoRetec, Inc.)

Figure 4.39 LARADS used to support soil remediation surveys. (Courtesy of ThermoRetec, Inc.)

The LARADS is characterized by the following features:

- Compatible with readily available portable or handheld radiological survey;
- Applicable to a variety of building surfaces including floors, walls, ceilings, and interior/exterior structures;
- Capable of providing well-documented release surveys or characterization surveys involving low and/or high levels of contamination;
- Adaptable to mobile platform configurations for large-area surveys;
- Capable of electronic storage of survey and radiation detection data;
- Capable of generating reports that are easily integrated with GIS for use in generating computer-aided design drawings.

When surveying large surface areas, the LARADS can be configured with large detectors on a mobile platform. The system may also be configured with high-level radiation detectors for use in characterizing more extreme radiological conditions. The LARADS setup time to level tripod and connect system cabling and is approximately 20 min per room, regardless of size.

A few of the advantages of the LARADS system are that it can produce automated survey reports, and is able to remove a technician's subjective observations from the radiological data collection and report generation processes. The LARADS survey reports provide a clear and concise representation of exactly where and how much contamination is present on building surfaces. Because of the accuracy of the positional subsystem used on the LARADS, detector velocity is captured along with the radiological and positional data. The surveyor is given real-time feedback in the field to maintain detector velocities below predetermined values. This greatly aids in achieving survey required minimum detectable activity, minimum detectable count rate, and minimum scan sensitivity levels.

For further details on the capabilities of this system, contact Thermo Hanford, Inc., at www.thermohanford.com or at (509) 371-1506.

4.2.2.2.5   3-D Imaging Decommissioning In-Situ Plutonium Inventory Monitor

The 3-D Imaging Decommissioning In-Situ Plutonium Inventory Monitor (DISPIM) is manufactured by BNFL Instruments Ltd. and is used to support alpha plant decommissioning operations by providing three-dimensional mapping of residual plutonium in plant vessels, drums, glove boxes, etc., prior to removing them from process lines.

The 3-D Imaging DISPIM system is composed of a detector assembly, mobile processing area, and a mobile operator data input area (Figure 4.40). The detector assembly is composed of 24 moderated He-3 detector modules with built-in charge amplifiers. The system is able to reach a detection limit of 1 g of plutonium (assuming a typical isotopic composition of 25% Pu-240) with an 8-h count time. The system is able to monitor a Pu-240 isotopic composition ranging from 2 to 30%, with an accuracy of ±30%. The system is able to target the x, y, z coordinates of a hot spot with an accuracy of ±15 cm. The system can be supplemented with high-resolution gamma spectrometer for gamma measurement of plutonium isotopic composition.

Figure 4.40   3-D Imaging DISPIM. (Courtesy of BNFL Instruments, Ltd.)

The 3-D Imaging DISPIM system is able to monitor a wide range of object shapes up to 180 ft³ in size. The system is precalibrated at the factory. The calibration is then checked before collecting each *in situ* measurement by placing a calibration check source Cf-252 within (or on the outer surface) of the item being surveyed. After the calibration check, the measurement is carried out. Three-dimensional representations of the located hot spots are displayed on the monitor along with the plutonium content and measurement uncertainty. A color printer provides a hard copy of the three-dimensional image and data output.

For further details on the capabilities of this monitor, refer to the BNFL Instruments Web page at www.bnfl-instruments.com or call (505) 662-4192.

4.2.2.2.6 TRU Piece Monitor

The TRU Piece Monitor is manufactured by BNFL Instruments Ltd. and is used to support decommissioning of plutonium facilities by measuring total plutonium content of single items of waste (e.g., folded piping, metal plating) using neutron coincidence counting. After the measurement is collected, the waste item is transferred into a waste drum. The monitor is able to keep a running total of the amount of plutonium added to each waste drum to ensure the total does not exceed criticality or waste disposal limits.

The TRU Piece Monitor contains 12 He-3 neutron detectors, and a 20% efficiency high-purity germanium detector. The monitor can accept a maximum package size of $12 \times 12 \times 12$ in. (Figure 4.41). The monitor has a detection limit capability of <50 mg plutonium, with a precision of $\pm 2\%$. The monitor is designed to measure plutonium content within the range of 50 mg to 850 g.

Figure 4.41 TRU Piece Monitor. (Courtesy of BNFL Instruments, Ltd.)

The system outputs the results from the measurement to the console along with a nuclear safety (worst-case) assessment. The system uses sealed sources to verify the calibration of the germanium detector between measurements at the same time the neutron background is confirmed to be within acceptable limits. The neutron coincidence count is collected in a series of short time periods, which are analyzed for statistical consistency.

For further details on the capabilities of this monitor, refer to the BNFL Instruments Web page at www.bnfl-instruments.com or call (505) 662-4192.

4.2.2.2.7   IonSens Alpha Pipe Monitor

The IonSens Alpha Pipe Monitor is manufactured by BNFL Instruments Ltd. and is used to measure the internal and external surfaces of cut pipe and scaffolding poles for alpha activity, thus permitting the categorization of the material as either low-level waste or free release. This monitor is able to classify metallic piping with dimensions that range from 3 to 20 ft in length and 2 to 6 in. in width (Figure 4.42).

Figure 4.42 IonSens Alpha Pipe Monitor. (Courtesy of BNFL Instruments, Ltd.)

This monitor is unique in that it is designed to detect the ions produced by alpha particle interactions in the air surrounding the pipe being surveyed, which is more reliable than attempting to measure the alpha activity directly. The system is composed of three separate modules:

1. Air inlet module
2. Measurement module
3. Detection head module

The air inlet module contains an air filter that removes particulates, dust, and ions from the air drawn into the system from the outside environment. The measurement module is where the piping to be surveyed is held. A cradle system is used in the measurement module to ensure that all piping is held centrally in the chamber. The detection head module contains the ion detector, HEPA filter, fan unit, data-processing electronics, and computer. The monitor automatically performs background and calibration check measurements to maintain instrument accuracy.

For further details on the capabilities of this monitor, refer to the BNFL Instruments Web page at www.bnfl-instruments.com or call (505) 662-4192.

4.2.2.2.8   IonSens 208 Large Item Monitor

The IonSens 208 Large Item Monitor is manufactured by BNFL Instruments and was developed for the purpose of monitoring exposed surfaces of complex items, such as process equipment, folded piping, etc. (Figure 4.43). It was also developed for the purpose of categorizing the material as either low-level waste or suitable for

Figure 4.43 IonSens 208 Large Item Monitor. (Courtesy of BNFL Instruments, Ltd.)

free release. This monitor is typically used to support building decommissioning operations.

The IonSens 208 detects the ions produced by alpha particle interactions in the air surrounding the item being surveyed. Thus, a single measurement is sensitive to contamination located on all exposed surfaces of an item. The system is composed of the following three modules:

1. Air inlet module
2. Measurement module
3. Detection head module

The air inlet module filters air that is drawn into the measurement module and removes particulates, dust, and ions. The measurement module is composed of a front-loading chamber where the item to be surveyed is loaded into the system. This module can accept items with dimensions up to $3 \times 3 \times 2.5$ ft. The detection head module contains the ion detector, HEPA filter, fan unit, data-processing electronics, and computer.

The IonSens 208 is automated to perform a variety of diagnostic routines to verify the system is operating properly. Once a measurement sequence is complete, the system displays the total alpha activity measured against the free-release criteria. These results are stored in the system computer hard drive. At regular intervals the operator is prompted to calibrate the instrument and perform a background measurement. The software guides the operator through the calibration and background measurement process.

This system is capable of rotating the item in the sample chamber to improve the measurement accuracy, and can be supplemented with scintillation detectors for beta/gamma measurements.

For further details on the capabilities of this monitor, refer to the BNFL Instruments Web page at www.bnfl-instruments.com or call (505) 662-4192.

4.2.2.2.9   Pipe Surveying

The following section provides the reader with information regarding pipe surveying tools, which are available to assist building decontamination and decommissioning and soil remediation activities. The primary objective behind performing a pipe survey is to determine if contaminated material is present in the pipe, determine if there are any cracks in the pipe where the contamination could leak into the surrounding formation, and determine where the pipe ultimately discharges. If contaminated material is found within a pipe, special tools are available that can clear it from the pipe. Currently, pipe surveying technology is undergoing rapid advancement due to its practical applicability, and its relative cost-effectiveness.

Surveying tools are advanced through a pipe using either a self-propelled crawler unit or fiberglass push cable. The crawler unit is currently commercially available in sizes small enough to investigate 6-in.-diameter pipes. Smaller-diameter pipes can be investigated by advancing tools using a fiberglass push cable. Some of the investigative tools that are currently available for pipe surveying include:

- High-resolution video cameras
- NaI radiation detectors
- Organic vapor analyzers
- Combustible gas indicators
- Temperature and pH meters

Pipe crawler units are commonly designed to meet site-specific requirements and conditions, such as:

- Pipe diameters
- Pipe bend angles
- Pipe surface texture
- Surveying distances
- Analytical instrumentation
- Site radiation or chemical contamination levels

Depending on the pipe geometry, it is often necessary to build the crawler unit in modules, which are linked together in a trainlike configuration. When a modular system is used, a separate module can be used for the motor, scanning instruments, and video camera(s) (Figure 4.44).

Wheels or tracks are used to propel the crawler units. Wheeled units use four or six wheels, which are narrow, and chamfered to the approximate radius of the pipe. These wheels combined with a direct current–powered motor provide traction to negotiate most sludge and mudlike environments (Figure 4.45).

Tracked systems can be advantageous for short-distance applications (<100 ft) in large-diameter pipes or ducts, due to their ability to turn on their own center. This feature can be very useful when negotiating complex runs, particularly when there is a significant amount of debris. With this one exception, wheeled systems outperform tracked systems in most other applications.

Figure 4.44 Modular pipe surveying instruments.

There are some very sophisticated devices that have been developed for characterizing pipes that utilize what is known as inchworm or spider technology. These types of devices can negotiate small-diameter piping or tubing systems. However, these devices are very slow, and are more prone to breakdown than are the simpler wheel or track units.

The distance a system can travel is primarily dependent on the weight of the tractor package, and the size and weight of the telemetry cable. As a general rule, a tractor can only pull its own weight in cable and associated drag. Onboard multiplexing or using fiber-optic telemetry transmission cable can help to reduce the drag. However, fiber-optic cable is very fragile and therefore is not recommended since piping interrogation activities are commonly performed in rugged environments. Some wireless telemetry-controlled systems have also been developed; however, they tend to be unreliable. When using wireless systems in metal pipes, there is the potential for confusing the transmission signal receiver, and thus there is the risk of losing a system in the pipe.

Another approach that can be used for long-distance application (>100 ft) is powering the crawler and the sensor instruments with onboard batteries. With this approach, the cables are only needed to provide trickle charge to the crawler package and to transmit information. This approach can achieve horizontal distances of 1500 ft or more.

Some accessories that can be designed modularly and that can be used to expand the capability of the crawler unit include:

- Pan-and-tilt for vision and/or directional sensors
- Laser distance and scaling instrumentation
- Pipe-clearing tooling
- Inclinometry for measuring settling or slope

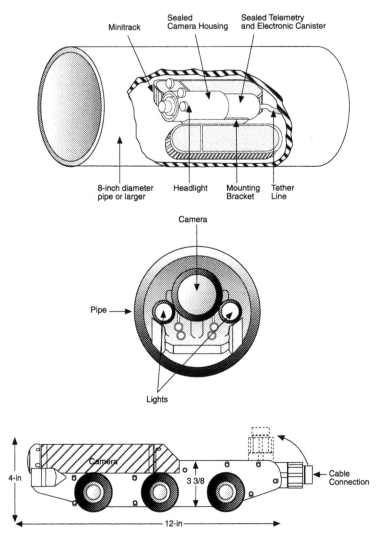

Figure 4.45 Example of wheel and track crawling unit. (From Byrnes, M.E., *Field Sampling Methods for Remedial Investigations*, Lewis Publishers, Boca Raton, FL, 1994. With permission.)

A new technology that is being investigated uses chemically sensitive microchips to identify and measure specific elements or chemical species within the pipe system. Two other technologies in development at this time include a sensor package to detect contaminants that have leaked from a breached piping system and an *in situ* pipe repair system.

If a specific detector requires the pipe to be free of debris prior to performing the survey, a squeegee-like tool can be attached to a crawler unit (Figures 4.46 and 4.47). Tethers are attached to all devices to allow retrievability in the event the drive unit fails.

Figure 4.46 Squeegee-like pipe-cleaning tool.

Figure 4.47 Squeegee-like pipe-cleaning tool removing sediment from pipe.

The manual push method is generally required for pipes smaller than 6-in. in diameter, and utilizes $^1/_8$ to $^5/_8$-in. solid flexible polyethylene rods to push and retrieve sampling tools (Figure 4.48). At this time, as far as the author is aware, the longest push accomplished in mockup is approximately 700 ft, and approximately 350 ft in actual field application. The same investigative tools available for the pipe crawler are also available for the push method. To keep the sampling instrument or camera properly positioned in the pipe, instrument-bearing skids are used. For larger-diameter pipes (3- to 6-in.), wheeled "dogbone"-shaped skids are most effective, while skids with runners are used in smaller-diameter pipes. In an effort to maximize distance, small three-wheeled devices can be attached to the push-rod every several feet to reduce the friction of the cable rubbing against the pipe.

The manual push approach is very cost-effective at approximately $2.00/ft for the push cable. If short distances (<350 ft) are the intent, then the push method can be effectively utilized in pipes up to 10-in. in diameter. This approach is less effective for larger-diameter piping since the push rod can bunch up more easily and thus reduce the acquired distances.

Information regarding the location or position of a sampling tool in a pipe can be acquired by means of a footage encoder, pressure wheels and transducers, or

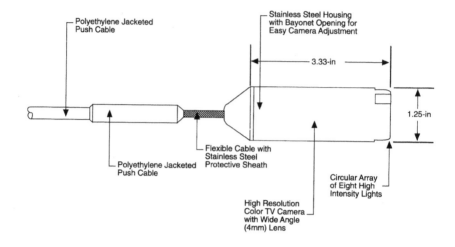

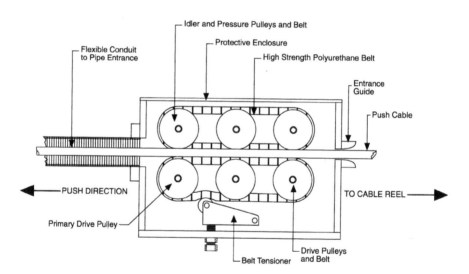

Figure 4.48 Small-diameter television camera and push cable mechanism. (From Byrnes, M.E., *Field Sampling Methods for Remedial Investigation*, Lewis Publishers, Boca Raton, FL, 1994. With permission.)

inclinometers, which are all commercially available. Fiber-optic encoders have also been used with some success, although the setup of such devices is more complex.

Software systems are currently being developed that can provide precise system positioning information from a photogrammetry-based program. The only requirement for this method is a physical reference point within the area of interest. This method can also be used to navigate equipment within an area remotely, and may replace more complex methods. Information can be displayed and recorded on a cathode ray tube or digital display indicating distance, pipe identification number, date, and time.

There is instrumentation now available that accepts real-time voltage or current inputs, translates them into reportable values (i.e., rad/h), and displays the results in real time on the video screen. This device has a high/low-level alarming capability and will accept analog, log, nonlinear, or quadrature inputs. This instrument is particularly valuable for providing a single-source documentation package of raw data and acquired information.

Imaging systems are available in inexpensive models, which can easily be damaged in radioactive environments, or radiation-resistant models which are built to withstand higher levels of radiation. The two basic types of cameras available include the tube and the charged coupled device (CCD). The tube design cameras are limited by high lighting requirements. The CCD systems are typically much less radiation resistant but have the advantage of low lighting requirements and can be utilized in the Black-White/Chroma format. Radiation resistance testing performed on some CCDs has shown a significant difference in radiation resistance from one model to another. Overall, they are all able to last to approximately the low to mid-10^4 rad range for accumulated dose. For short-term exposure, the limits vary from 20 to 200 rad/h. As the exposure levels increase, the video screen begins "sparkling." High-level exposures can completely white-out the image. This effect goes away once the camera is removed from the radiation field.

Another relatively new technology utilizes a charge injection device, which manufacturers claim has significant radiation resistance. The accumulated dose threshold for this camera is not significantly higher than the CCD at 10^5 rad.

Tube cameras have significant radiation resistance capability. Short-term dose effects are not seen until approximately 10^5 rad, and the long-term accumulation threshold is in the 10^8 rad range. The trade-off occurs in the high light requirements and long length of the color system. The black-and-white system has the problems associated with black-and-white images. The latest in radiation-resistant color CCD cameras is a lead-shielded, optics-line-of-sight-removed system that reportedly provides even longer life and functionality in radiation environments than the tubed cameras for even less cost. These systems have high resolution and zoom capability and can employ CCD front ends that can provide clear imaging in the <1 Lux illumination range. The disadvantage of these systems is the bulky package and weight. Diameters of the camera are approximately 3.5 in. and the lead-shielded unit weighs approximately 30 lb.

Another example of a pipe surveying setup is shown in Figure 4.49. For further information on pipe and remote surveying instrumentation, contact Pacific Northwest National Laboratory at (509) 375-2243.

4.2.2.3 Tank, Drum, Canister, Crate, and Remote Surveying

The following section provides details on a number of scanning and direct measurement methods that should be considered to support the characterization of tanks, drums, canisters, and crates that contain highly radioactive materials. This section also provides details on remote surveys instruments. Table 4.13 summarizes the data provided and primary use of each of the scanning and direct measurement methods presented in this section.

Figure 4.49 Example of another pipe surveying setup.

4.2.2.3.1   Tank Surveying

This section provides details on methods and tools that are available to support the characterization of large underground radioactive waste storage tanks, such as those present at the Hanford Nuclear Reservation.

4.2.2.3.1.1 Light-Duty Utility Arm — Light-Duty Utility Arm is a robotic arm on the end of a telescoping mast that can be lowered into large (1 million gal) waste tanks to deploy a video camera, physical property sensors, and sampling tools (Figure 4.50). Previously, waste samples could only be obtained directly under a tank riser using auger tools or core sampling devices.

The Light-Duty Utility Arm was developed as a joint effort among the following companies:

- Idaho National Engineering and Environmental Laboratory
- Oak Ridge National Laboratory
- Sandia National Laboratories
- Spar Aerospace Ltd.
- Westinghouse Savannah River Company
- Los Alamos Technical Associates
- Pacific Northwest National Laboratory
- Southwest Research Institute
- Westinghouse Hanford Company

The Light-Duty Utility Arm system provides a mobile, multiaxis positioning system that can access large waste storage tanks through existing openings (e.g., risers) in the tank that are as small as 12 in. in diameter. This flexible, adaptive system provides a robotic platform capable of deploying *in situ* surveillance, confined sluicing, inspection, and waste sampling. Analytical tools such as sensors can be attached to the system to measure physical properties of tank waste. The arm is equipped with a gripper for handling various sampling tools, and has a stereo video camera attached to it to document tank waste contents.

Table 4.13 Container Scanning and Direct Measurement Methods

Project Type	Scanning/Direct Measurement Method	Data Provided	Primary Use
Tank Surveying	Light-Duty Utility Arm	High-resolution video, physical property sensors, sampling tools	Robotic arm on the end of a telescoping mast that can be lowered into large waste tanks to deploy a video camera, physical property sensors, and sampling tools; can enter tanks through openings as small as 12 in. diameter
	Raman probe	Organic and inorganic chemicals using Raman spectroscopy	Used to support the surveying of high-level waste tanks for concentrations of organic compounds
	Near-infrared spectroscopy fiber-optic probe	Water content of high-level radioactive waste	Remote fiber-optic probe developed as an alternative to thermogravimetric analysis for measuring the water content of high-level radioactive waste using near-infrared spectroscopy
	Corrosion probe	Electrochemical noise generated by electrochemical reactions	Used to monitor the corrosivity rate within high level waste underground storage tanks
	Pipeline Slurry Monitor	Percentage of suspended solids, density, viscosity, mass flow rate, particle size	Designed to monitor transport properties of slurry in waste transfer lines to help prevent blockage and ensure safe transportation of waste to treatment facility; real-time data enable operators to maintain acceptable control limits and thereby prevent pipeline blockage
	Fluidic Sampler	Homogeneous tank waste samples for laboratory analysis	Allows tank waste samples to be collected while tank contents are being agitated; provides more representative samples for analysis
Drum, canister, and crate surveying	TRU Drum Monitor	Total plutonium and plutonium isotopic composition	Used to assess the plutonium isotopic composition of waste contained in drums; used to ensure criticality limits, transportation limits, and/or that disposal facility limits are not exceeded

Table 4.13 (continued) Container Scanning and Direct Measurement Methods

Project Type	Scanning/Direct Measurement Method	Data Provided	Primary Use
	TRU Crate/Box Monitor	Total plutonium and isotopic composition	Monitor uses passive neutron coincidence counting combined with high-resolution gamma spectrometry to measure the total plutonium content and isotopic composition of waste contained within a crate/box
	Plutonium Can Contents Monitor	Total plutonium and isotopic composition	Used to measure the total plutonium content and isotopic composition of filled PuO$_2$ product cans
	RTR-3 Radiography System	Radiographic image	Portable real-time imaging system used to examine physical properties of both special nuclear material and its storage container; the radiographic images see "inside" the storage container without the use of intrusive techniques
Remote Surveying	Andros Mark VI	High-resolution video, gross gamma, smear sample collection	Designed for deployment in high radiation areas where personnel are prohibited access; used to collect characterization data such as gross gamma readings, video, and smear samples remotely

Figure 4.50 Light-Duty Utility Arm.

The robotic tank inspection end effectors presented in Figures 4.51 through 4.54 are tools that can be mounted on the Light-Duty Utility Arm. These tools are specifically designed to:

• Collect radiation and chemical measurements from within the tank;
• Identify corrosion, cracks, pits, and weld defects in tank interiors and permanently record their locations;
• Collect liquid and solid waste samples;
• Assist in the collection of tank heel samples for chemical and radiological analysis;
• Collect stereoscopic images from within the tank;
• Scarify the surfaces of the tank.

Figure 4.51 Remote tank inspection end effector.

Figure 4.52 High-resolution video system.

Figure 4.53 Light-weight scarifier.

Figure 4.54 Stereoscopic camera system.

The Light-Duty Utility Arm has a radial reach of at least 13.5 ft from the tank riser centerline. The deployment mast extends to at least 62.5 ft. This system is operated remotely, thus reducing harmful exposures to personnel.

In September 1996, the Light-Duty Utility Arm was deployed into Tank 106-T at the Hanford Nuclear Reservation. Tank 106-T is a 530,000-gal-capacity single-shell tank, which leaked radioactive waste to the surrounding soil. The arm deployed a high-resolution stereo video system to inspect the tank dome, risers, and walls. Valuable inspection data recorded by the Light-Duty Utility Arm are being used to better understand what caused the tank to leak.

The Light-Duty Utility Arm is currently being used at the Hanford Nuclear Reservation Tank AX-104 to retrieve waste samples for chemical and radiological analysis and leach testing. The analytical results from samples retrieved from the floor of the tank, and from the tank liner ribs and dome, are to be used to calculate the residual waste inventory in support of a risk assessment.

Currently, plans are in place for Oak Ridge Reservation and Idaho National Engineering and Environmental Laboratory to use modified versions of the Light-Duty Utility Arm to gather information needed to support a record of decision, and tank closure study.

For additional information about the Light-Duty Utility Arm, contact the Pacific Northwest National Laboratory at (509) 372-4331 or visit their Web site at www.pnl.gov/tfa.

4.2.2.3.1.2 Raman Probe — The U.S. Department of Energy Tank Focus Area and Lawrence Livermore National Laboratory developed the Raman probe for use in supporting the surveying of high-level waste tanks to assess the species and concentrations of organic compounds (Figure 4.55). The Raman probe is deployed by using a cone penetrometer to lower the probe through a riser in the tank. The cone penetrometer is then used to advance the probe through the tank waste. A sealed sapphire window was added to the cone penetrometer to adapt the probe for in-tank use. Laser light passes through the sapphire window and interrogates the sample or the environment, and the resulting data signal is then transmitted back to the optical collection system for subsequent data analysis. The sapphire window is brazed into the stainless steel cone penetrometer.

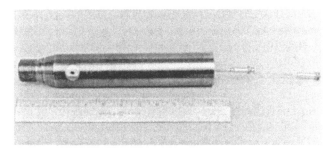

Figure 4.55 Raman probe.

The Raman probe utilizes Raman spectroscopy, which is an optical method that detects many organic and inorganic chemicals in the media surrounding the probe. When laser light is beamed through a sapphire window, the light hits the sample and causes molecules to vibrate in a distinctive way. This creates a vibrational "fingerprint." The vibrational signal is then captured and transmitted via fiber-optic cables to an analyzer where it is compared to known vibrational signals. The final result is an analysis of the material surrounding the probe (DOE, 1999a).

In 1998, the Raman probe was tested on *in situ* trichloroethene and tetrachloroethene soil contamination at the Savannah River Site, and on 135 archived waste samples from 14 tanks from the Hanford Nuclear Reservation. The Raman probe demonstration on the Hanford waste resulted in more than 99% accuracy for compound identification, and greater than 93% accuracy in identifying both the compounds and the concentration. For additional details on the Raman probe, visit the U.S. Department of Energy Office of Science and Technology Web page at http://ost.em.doe.gov.

4.2.2.3.1.3 Near-Infrared Spectroscopy Fiber-Optic Probe — The U.S. Department of Energy Tank Focus Area and Hanford Tank Waste Remediation System Program supported the development of the near-infrared spectroscopy system which utilizes a remote fiber-optic probe to provide real-time measurements of the water content in high-level radioactive wastes. This technology was developed as a cost-effective and safer alternative to the thermogravimetric analysis techniques.

The near-infrared spectroscopy system uses a low-power light source to measure optical molecular absorption within a sample that is contained within a hot cell. This light beam is passed through a fiber-optic cable from the analyzer system to the probe which has been inserted through penetrations in the hot cell. The system takes advantage of the near-infrared spectroscopy optical reflection from a sample to determine total water concentration. Water concentrations are determined from the near-infrared spectra using calibration models that are built from simulant and real waste standards with known moisture contents (DOE, 1999b).

The remote fiber-optic probe is designed to survive harsh caustic and radiation environments that characterize tank waste, and is compatible with a cone penetrometer (Section 4.3.3.7.1), Light-Duty Utility Arm (Section 4.2.2.3.1.1), or other mechanical deployment systems. The near-infrared spectroscopy system has been successfully deployed in hot cells at the Hanford Nuclear Reservation.

For additional details on the near-infrared spectroscopy system visit the U.S. Department of Energy Office of Science and Technology Web page at http://ost.em.doe.gov.

4.2.2.3.1.4 Corrosion Probe — The electrochemical noise-based corrosion probe was developed under the direction of Lockheed Martin Hanford Corporation and the U.S. Department of Energy Tank Focus Area for the purpose of monitoring the corrosivity rate within high-level waste underground storage tanks. This system is designed to measure corrosion rates and detect changes in waste chemistry that are responsible for triggering the onset of pitting and cracking. Corrosion probes are used to signal when it is time to add additional corrosion inhibitor to a tank (DOE, 1999c). Figures 4.56 and 4.57 illustrate the corrosion probe and electrode array.

Figure 4.56 Corrosion probe.

Figure 4.57 Corrosion probe electrode array.

Electrochemical noise consists of low-frequency and small-amplitude signals that are spontaneously generated by electrochemical reactions occurring on corroding surfaces. The electrochemical noise-based corrosion probe uses three nominally identical electrodes (working, counter, and pseudoreference) immersed in the tank being surveyed. Electrochemical noise is measured as time-dependent fluctuations in corrosion current between the working and counter electrodes. Electrochemical noise potential is measured as the time-dependent fluctuation of the difference in the corrosion potential between the working/counter electrode assembly and the pseudoreference electrode. A zero-resistance ammeter electrically joins the working and counter electrodes. The zero-resistance ammeter measures the potential differences between the working/counter electrode assembly and the pseudoreference electrode.

The corrosion probe is deployed through a tank riser with a minimum diameter of 10 cm. The probe has three electrode arrays distributed along the length of the probe body. The two upper electrode arrays are positioned to monitor vapor-phase corrosion, while the lowermost array is immersed approximately 55 cm into the liquid waste. The corrosion data collection system consists of a data logger for each probe, process control software, and a commercially available data acquisition system.

The corrosion probe has successfully been deployed in Tanks 241-AZ-101, 241-AN-107, and 241-AN-102 at the Hanford Nuclear Reservation. For additional details on the corrosion probe, visit the U.S. Department of Energy Office of Science and Technology Web page at http://ost.em.doe.gov.

4.2.2.3.1.5 Pipeline Slurry Monitors — When transferring radioactive waste from storage tanks to treatment facilities, sludge wastes are typically mobilized and mixed with liquid wastes to create a slurry of liquid and suspended solids. This slurry is then transferred by pipeline to the treatment facility. Since slurries created from tank waste often have a high viscosity and solids content, they are difficult to pump and generate large backpressures. Pipelines can also become blocked.

Pipeline slurry monitors are designed to monitor the transport properties (e.g., percentage of suspended solids, density, viscosity, mass flow rate, particle size) of the slurry in the transfer lines since this can help prevent blockage and ensure safe transportation of the waste. These in-line monitors provide real-time measurements of slurry properties which enables operators to maintain acceptable control limits and thereby prevent pipeline blockage (DOE, 1999d).

In May 1999, the Endress + Hause, Inc., Promass 63M meter was successfully deployed at an Oak Ridge Reservation demonstration where the meter was installed in the Slurry Monitoring Test Loop at the Gunite and Associated Tank Slurry Transport System. In this demonstration waste was transported from the Gunite and Associated Tank W-9 to the Melton Valley Storage Tanks.

The Promass 63M Coriolis meter is commercially available from Endress + Hauser, Inc. Further information is available at http://www.us.endress.com/ or http://ost.em.doe.gov.

4.2.2.3.1.6 Fluidic Sampler — Historically, waste tank mixing pumps needed to be turned off to allow the collection of grab or core samples for laboratory analysis. Since the mixing pumps are used to homogenize the tank contents, turning these pumps off to sample often led to the collection of nonrepresentative samples.

The Fluidic Sampler, which is manufactured by AEA Technology Engineering Services, Inc., allows sampling to be performed remotely while the tank contents are being agitated (Figure 4.58). This approach to tank sampling minimizes the risk of personnel exposure and leads to the collection of samples that are more representative of the tank contents. The only exposure that personnel receive during the sampling process is when transferring the sample bottle from the sampler to a cask. Using the Fluidic Sampler also reduces secondary waste since there is nothing to dispose of after a sample is taken. Figure 4.59 shows the above-tank portion of the Fluidic Sampler (DOE, 1999e).

Figure 4.58 Fluidic Sampler reverse-flow diverter pump.

Figure 4.59 Fluidic Sampler shielded sampling station.

Over the past 20 years, approximately 400 Fluidic Samplers have been installed and used at nuclear installations in the United Kingdom with no reported failures. In 1998, the Fluidic Sampler was installed and used to sample the contents of Tank 48 at the Savannah River Site. At the Hanford Nuclear Reservation, the Fluidic Sampler is being adapted to allow the collection of samples from multiple tank depths.

For additional details on the Fluidic Sampler, contact AEA Technology Engineering Services, Huntersville, NC, or visit the U.S. Department of Energy Office of Science and Technology Web page at http://ost.em.doe.gov.

4.2.2.3.2 Drum, Canister, and Crate Surveying

The following sections present a number of surveying or monitoring systems that can be used to determine the radiological composition of drums, canisters, and crates without having to open the containers for sampling.

4.2.2.3.2.1 TRU Drum Monitor — The TRU Drum Monitor is manufactured by BNFL Instruments and was developed for the purpose of assessing the plutonium isotopic composition of waste contained in drums (Figure 4.60). The assessment of plutonium isotopic composition of drum contents may be required to address criticality concerns, or to determine if drum contents exceed transportation or disposal facility limits.

Figure 4.60 TRU Drum Monitor.

The TRU Drum Monitor is designed to accommodate 55 gal and 500 L drums. The monitor contains a minimum of 36 (and as many as 60) He-3 neutron detectors and one high-purity germanium detector. Once a drum is loaded into the monitor, the drum is rotated during the measurement period to allow a better all-around view of the drum.

The monitor can reach a detection limit of 20 mg of total plutonium with a 1000 sec count. The dynamic range for the monitor is from 20 mg to 850 g of total plutonium. The monitor is generally able to provide results with an accuracy of ±15%. The monitor performs its own self-check to ensure it is in proper operating order. At preset intervals, it prompts the operator for a calibration check. Once the plutonium calibration standard is loaded into the measurement chamber, the calibration routine is automatically carried out.

The TRU Drum Monitor can come equipped with a bar code reader that ties the identity of the loaded drum to the results from the survey. On-screen outputs include drum identification, weight, plutonium mass, overall error, and plutonium isotopic composition. The full assay information, including the gamma spectrum and neutron count information, is stored within the system hard drive and can be copied to a floppy disk or an optical disk drive.

For further details on the capabilities of this monitor, refer to the BNFL Instruments Web page at www.bnfl-instruments.com, or call (505) 662-4192.

4.2.2.3.2.2 TRU Crate/Box Monitor — The TRU Crate/Box Monitor is manufactured by BNFL Instruments Ltd. and is used to identify the total plutonium content of miscellaneous decommissioning waste contained in crates or boxes measuring up to $8.2 \times 9.75 \times 11.3$ ft in size (Figure 4.61). The monitor uses passive neutron coincidence counting combined with high-resolution gamma spectrometry to measure the plutonium of wastes contained within a crate/box.

Figure 4.61 TRU Crate/Box Monitor.

Prior to loading into the main neutron counting chamber, the crate/box passes between two high-resolution gamma spectrometry detectors. These data are automatically combined with the neutron coincidence counting data to enable matrix and isotopic corrections to provide an accurate total plutonium mass assay. The two high-resolution gamma spectrometry detectors allow the TRU Crate/Box Monitor to identify and quantify many gamma-emitting fission products, activation products, and actinides. The system is protected against electrical interferences common to an industrial plant environment.

The TRU Crate/Box Monitor contains 100 (or more) polyethylene moderated He-3 detectors distributed in 26 independent detection modules, and two high-resolution gamma spectrometry detectors. Under typical operating conditions, the system has a detection limit of <1 g of plutonium with a 4-h count time. The dynamic range of the system is 1 to 850 g of plutonium, and has an assay accuracy of ±25%.

The performance of the system is automatically checked by exposing a radioactive source to gamma detectors at routine intervals. At preset intervals, the monitor prompts the user for a calibration check. The calibration check is performed by loading a plutonium calibration standard into the measurement chamber. The calibration routine is then automatically carried out.

For further details on the capabilities of this monitor, refer to the BNFL Instruments Web page at www.bnfl-instruments.com, or call (505) 662-4192.

4.2.2.3.2.3 Plutonium Can Contents Monitor — The Plutonium Can Contents Monitor is manufactured by BNFL Instruments Ltd. and is used to measure the total plutonium content and isotopic composition of filled PuO_2 product cans (Figure 4.62). This monitor can be incorporated into plant process operations where cans automatically pass from the filling station through the monitor en-route to the PuO_2 storage area. As cans are retrieved from the storage area for repackaging or export from the facility, they can be passed through the monitor a second time to verify that the correct can has been selected. This system has been used successfully at the U.K. Sellafield Site.

Figure 4.62 Plutonium Can Contents Monitor.

The Plutonium Can Contents Monitor contains 10 He-3 gas-filled neutron detectors and one high-resolution gamma spectrometry detector. The monitor has a dynamic range of <1 to 8 kg of PuO_2 with a measurement time of 30 min. The monitor has an accuracy of less than ±5% for plutonium mass and less than ±3% for isotopic composition.

The initial calibration of the monitor is performed by the manufacturer using a sealed plutonium standard of known mass and isotopic composition. The calibration is then verified and refined at the plant using actual process material to ensure the instrument is optimized for actual process material configurations. The monitor includes a number of built-in system test and measurement control procedures to ensure the integrity of the measurement data.

For further details on the capabilities of this monitor, refer to the BNFL Instruments Web page at www.bnfl-instruments.com, or call (505) 662-4192.

4.2.2.3.2.4 RTR-3 Radiography System — Radiography is a proven technique that can be used for nondestructive evaluation. For example, radiography can be used to examine physical properties of both the special nuclear material and its storage container. The radiographic images see "inside" the storage container without the use of intrusive techniques.

The Science Applications International Corporation (SAIC) RTR-3 system is a portable real-time imaging system that can be operated by a single individual. The system consists of a compact imager, an integrated control unit, and a portable X-ray source. The imager has an 8 × 10 in. X-ray-sensitive screen. The integrated control unit incorporates a portable computer with a 9.5-in. flat panel display and internal modem. The X-ray source is of a modular design, producing short-duration pulses having a maximum energy of 150 kV. All controls use the familiar Microsoft Windows interface. System setup/takedown time is approximately 10 min. Stretch, zoom, pan, scroll, sharpening, and smoothing adjustments are all found on the control unit. The images are in standard Tag Image File format, which is compatible with a wide variety of industry image enhancement software. The system can be operated using AC batteries as a power source.

The SAIC RTR-3 Real-Time Radiography System was recently used to perform a baseline radiographic inspection of special nuclear material storage containers at the Hanford Nuclear Reservation Plutonium Finishing Plant. As part of this investigation, a total of 44 different containers were examined. The items examined included a variety of material forms (oxide, metal, scraps) and container configurations. The RTR-3 system was successful at distinguishing can seams, can types, material position, material shape, and plastic bags. Images were successfully acquired through as many as four layers of metal containers.

For additional information about the RTR-3, contact SAIC at www.saic.com.

4.2.2.3.3   Remote Surveying

Remote surveying should be performed in highly radioactive environments or environments where there are confined spaces. Using remote surveying instruments to collect chemical and radiological data and samples for laboratory analysis reduces the risk of personnel exposure.

4.2.2.3.3.1 Andros Mark VI — The ANDROS Mark VI remote hazardous duty system was developed by REMOTEC and is designed for deployment in areas where personnel are prohibited access, such as high radiation areas (Figure 4.63).

The ANDROS Mark VI was initially deployed for remote collection of characterization data such as gross gamma readings, video, and smear samples. The system is capable of operating off of 110 VAC, or battery power, and can drive on and off of the custom-designed, crane-deployable lifting fixture. Communication is via standard communication, fiber-optic umbilical lines, or radio frequency.

In August 1998, the Pacific Northwest National Laboratory deployed the ANDROS Mark VI robot into the 221-U Facility Railroad Tunnel. Prior to this deployment, the radiation levels within the railroad tunnel were unknown; consequently personnel access was prohibited. The robot traversed to the outer rollup door (approximately 190 ft) collecting gross radiation data and videotaping the condition

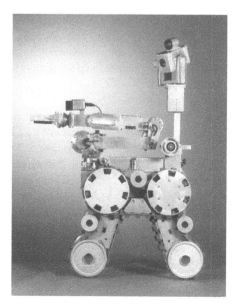

Figure 4.63 ANDROS Mark VI Remote Hazardous Duty System.

of the tunnel. Nine smear samples were taken from selected locations along the tunnel. For this deployment, the robot was configured with:

- Two camera and lighting systems (one color and one black and white);
- A real-time gross gamma detector;
- Smear sample pads;
- A lifting fixture (with camera and lights) for the robot, with a motorized cable playout and retrieval system;
- An operator control station with video recording equipment.

The deployment of the ANDROS system in the railroad tunnel was a success, and the system has since been deployed into the 750-ft ventilation tunnel.

The ANDROS Mark VI is commercially available through REMOTEC at (423) 483-0228. Additional information about the ANDROS Mark VI can be obtained from www.bhi-erc.com/canyon/canyon.htm.

4.2.2.4 Exposure Monitoring

This section provides details on two methods that should be considered to support the monitoring of radiological environments for exposure levels. These include the Alpha Track Detector and the Electret Ion Chamber. Table 4.14 summarizes the data provided and primary use of each of two monitoring methods.

4.2.2.4.1 Alpha Track Detector

The Alpha Track Detector is a 1-mm-thick strip of plastic that is sensitive to alpha particles that penetrate it (Figure 4.64). This detector is distributed by

Landauer, Inc. When an alpha particle penetrates the plastic, a submicroscopic damage center is created as a result of ionization. The damage centers resulting from contact with contaminated soil are enlarged by chemical etching in an alkaline solution at 60°C. These damage centers are then counted using a computerized microscope counting system. The number of damaged centers can then be used to estimate the dose that workers or the public are receiving from the site.

Commercially, this device has been used primarily to measure radon levels indoors, but because it is sensitive to all alpha radiation, it could potentially have an application to soil characterization as well. For soil characterization, the number of etched pits per unit on the detector is proportional to the concentration of alpha-emitting radionuclides in the soil.

Table 4.14 Soil Remediation Scanning/Direct Measurement Methods for Monitoring

Scanning/Direct Measurement Method	Data Provided	Primary Use
Alpha Track Detector	Gross alpha	Strip of plastic that is sensitive to alpha particles that penetrate it; damage centers are counted using a computerized microscope counting system
Electret Ion Chamber	Gross alpha	Alpha particles enter the chamber and ionize the air inside the chamber; the ionized charge deposits its charge on the electret; the alpha activity is inferred from the rate of charge depletion of the electret

Figure 4.64 Alpha Track Detector.

The primary limitation to this technology is that it cannot identify which specific isotopes are emitting the alpha activity. For further details on the capabilities of this system, refer to the Landauer, Inc., Web page at www.landauerinc.com.

4.2.2.4.2   Electret Ion Chamber

The Electret Ion Chamber is a short cylinder opened at one end with a charged Teflon disk (electret) attached to the opposite end. Alpha particles entering the chamber through the opened end ionize the air inside the chamber. The ionized air deposits its charge on the electret depleting the total charge. This charge depletion can be measured with a handheld voltmeter, and alpha activity in the soil is inferred from the rate of charge depletion of the electret. The chamber is designed to maximize alpha particle response and minimize beta and gamma response. Like the alpha track detectors, this technology is currently in commercial use for measuring indoor levels of radon gas.

The role of this technology in characterization is similar to that of alpha track detectors (Section 4.2.2.4.1). It may be used for verification that a remedial action has been effective or it can be used to reduce the number of samples that must be taken and analyzed during characterization activities. The advantage of the ion chamber over the alpha track detector is that measurements can be taken at the site with the portable electrometer, while the alpha track detectors must be sent to an off-site laboratory for analysis (typically 1 day to several days).

For addition details on the Electret Ion Chamber, see DOE (1996b).

4.3   MEDIA SAMPLING

When collecting samples of soil, sediment, concrete, paint, dust, or any other type of media one must carefully select the type of sample to collect. The term *sample* in this case refers to the physical collection of representative material from a media for the purpose of radiological or chemical analysis. Selecting the appropriate type of sample is important since it will influence the resulting analytical data. The four types of samples that are most frequently collected for a radiological investigation include grab, composite, swipe, and integrated. Scanning and direct measurement addressed in this section do not involve the physical collection of a sample for analysis.

4.3.1   Sample Types

The four primary sample types when collecting samples from radiological environments include:

- Grab samples
- Composite samples
- Swipe samples
- Integrated samples

The following sections briefly describe each of these sample types and provide guidance on when they should be used.

4.3.1.1 Grab Samples

A grab sample refers to the physical collection of a media sample from a single location for analysis. When collecting a grab, the sample is transferred directly from the sampling tool into the sample jar or bottle. No mixing or compositing is performed when collecting this type of sample. Care should be taken when selecting the sampling tool to be certain that the method of sample collection does not in itself composite the sample. For example, when collecting shallow soil grab samples, the scoop or slide hammer coring tool method is preferred over the hand auger method since the rotation of the hand auger blades composite the soil during the sample collection process. Generally, a grab sample is not collected over sampling intervals greater than 0.5 to 1.0 ft in depth, or outside the perimeter of a 1-ft^2 area. Grab samples should be collected when one would like to know the range of activity levels (minimum and maximum) present at a site or facility. Grab sampling is effective in collecting samples for site characterization, waste characterization, risk assessment, feasibility study, remedial design, and postremediation confirmation sampling.

4.3.1.2 Composite Samples

A composite sample is collected by either taking multiple grab samples from different locations and homogenizing them together, or by collecting samples at depth from the same sampling location and homogenizing multiple depth intervals together for analysis. Composite sampling is most frequently used to help reduce the cost of site characterization and in some cases waste characterization activities. However, it is important to recognize that the results only provide the mean activity for the composited intervals, and will not provide a reliable estimate of the range of activity levels present at a site or facility. Generally, composite sampling is not used to collect data for human health or environmental risk assessment since it is important to know the full range of activity levels that a receptor may be exposed to. Composite sampling is generally not recommended for site closeout sampling since it tends to "dilute" the analytical results. When collecting composite samples, it is important that the composited interval not be so large that the resulting data are diluted beyond the point of providing meaningful information.

4.3.1.3 Swipe Samples

Swipe samples are distinct from grab, composite, or integrated samples in that they are collected from the surfaces of walls, floors, ceiling, piping, ductwork, etc. Swipe samples are collected for the purpose of determining the amount of removable activity from a surface. Once a swipe sample has been collected, it is typically analyzed for gross alpha and/or gross beta activity using a swipe counter. Swipe samples are most often collected to support building characterization, risk assessment, and building closeout. Section 4.3.3.1 identifies the procedure used for collecting a swipe sample.

4.3.1.4 Integrated Samples

Integrated sampling involves the collection of a sample from one location over an extended period of time (e.g., weeks, months). Integrated sampling is most commonly performed when assessing surface water, groundwater, or air quality over time. It is different from composite sampling in that samples are collected from the same location over an extended time period. For example, one integrated surface water sample may be collected from a location downstream from a nuclear power plant, where a small volume of water is collected every week throughout the 3-month wet season. At the end of the 3-month period, the integrated sample is analyzed to identify the mean activity level of various isotopes in the surface water over that time period.

4.3.2 Sampling Designs

For details on statistical and nonstatistical sampling designs that should be considered to support environmental studies, refer to Section 4.1.1.5.7.

4.3.3 Media Sampling Methods

This section provides standard operating procedures for various media sampling methods that should be considered for use in supporting soil remediation and building decontamination and decommissioning activities. Sampling methods have been provided for the following:

- Swipe sampling
- Concrete sampling
- Paint sampling
- Soil sampling
- Sediment sampling
- Surface water and liquid waste sampling
- Groundwater sampling
- Air sampling
- Drum and waste container sampling

4.3.3.1 Swipe Sampling

Swipe samples are collected from material surfaces for the purpose of determining the amount of removable activity. It is important to define the amount of removable activity from building surfaces, equipment, etc., since, if not controlled, this activity can easily be spread outside the radiologically controlled area and has the potential for exposing workers or other receptors to contaminants through the inhalation and ingestion pathways.

To prevent the spread of contamination from a radiological controlled area, 10 CFR 835.1101 requires that material and equipment not be released if removable surface contamination levels on accessible surfaces of total activity exceed the values specified in 10 CFR 835 Appendix D (see Table 2.3). The only exceptions to this requirement are as follows:

- Materials and equipment exceeding the removable surface contamination values may be conditionally released for movement on-site from one radiological area immediately to another radiological area if appropriate controls are exercised.
- Materials and equipment with fixed contamination levels that exceed the total (fixed + removable) contamination values specified in 10 CFR 835 Appendix D may be released for use in controlled areas outside of radiological areas only if (1) removable surface contamination levels are below the removable surface contamination levels specified in 10 CFR 835 Appendix D; and (2) the material or equipment is routinely monitored and clearly marked or labeled to alert personnel of the contaminated status.

Once a swipe sample has been collected, it is most frequently analyzed for gross alpha or gross beta activity using a swipe counter. Swipe counters can measure gross gamma activity but are not usually used for that purpose because the measurement is highly inefficient. A swipe counter is a gas-filled detector or scintillator that is best suited for alpha and beta detection. A swipe counter is used in part because handheld detectors are not always effective at measuring alpha and beta contamination in the field. The swipe counter has a better geometry and a better counting efficiency than field instruments. If the gamma component is pronounced, one could use handheld instruments for counting.

The following procedure identifies how a swipe sample should be collected to define the amount of removable activity. For most sampling programs, a total of three people are sufficient for implementing this sampling procedure. Two are needed for sample collection, labeling, recording sampling location, and documentation, and one is needed for health and safety.

The following procedure should be used to collect swipe samples for determining the amount of removable activity.

Equipment
1. Swipe counter dry filter papers
2. 100 cm² (10 × 10 cm) plastic template
3. Plastic Ziploc bags
4. Sample labels
5. Sample logbook
6. Chain-of-custody forms
7. Chain-of-custody seals
8. Permanent ink marker
9. Health and safety and radiological scanning instruments
10. Health and safety clothing
11. Sampling table
12. Waste container

13. Plastic waste bags

Sampling Procedure
1. In preparation for sampling, confirm that all the necessary preparatory work has been completed, including obtaining property access agreements; meeting health and safety, decontamination, and waste disposal requirements; and checking the calibration of all radiological and health and safety scanning instruments.
2. Place the 100 cm² plastic template over the location where the swipe sample is to be collected.
3. Lay a piece of swipe counter dry filter paper within the 100 cm² area to be sampled. Place index finger over top of filter paper and apply moderate pressure. Using a circular motion, cover the entire sampling area (Figure 4.65).
4. Transfer the filter paper into a plastic Ziploc bag.
5. After sealing the Ziploc bag, attach a sample label and custody seal.
6. See Chapter 5 for details on preparing samples and sample shipment.
7. Survey the coordinates of the sampling point to preserve the sampling location.

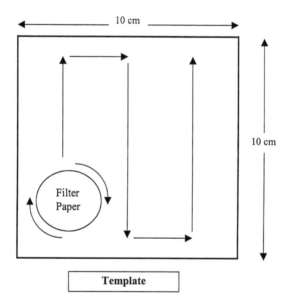

Figure 4.65 Swipe sampling method.

4.3.3.2 Concrete Sampling

Concrete samples are generally collected from building floors, walls, and/or ceilings for the purpose of confirming the results from scanning measurements or to estimate the activity level of the concrete for waste disposal purposes. The concrete drilling method and concrete coring methods presented below should be used to collect concrete samples for these purposes.

4.3.3.2.1   Concrete Drilling Method

Concrete samples are commonly collected from building walls, floors, ceilings, and/or other surfaces to verify the results from building scanning surveys. The gross alpha, beta, and/or gamma results from scanning surveys are typically used to estimate the pCi/g concentration of various isotopes in the concrete based on a set of isotope distributional assumptions. These assumptions (and calculated pCi/g concentrations) need to be verified through the collection and spectroscopy analysis of representative concrete samples. These concrete samples should be collected using a hand drill and a 0.5-in. diameter (or larger) drill bit, where multiple shallow holes (approximately 0.25 in. in depth) are drilled in the immediate vicinity of a scanning measurement hot spot. The drill cuttings are collected and sent to an on-site or fixed laboratory for gamma and/or alpha spectroscopy analysis. The number of shallow holes that are required to be drilled is dependent upon the minimum sample volume requirements of the laboratory for the analyses to be performed.

The following procedure identifies how a concrete sample should be collected to verify the results from building scanning surveys. For most sampling programs, a total of four people are sufficient for implementing this sampling procedure. Two are needed for sample collection, labeling, recording sampling location, and documentation; a third is needed for health and safety; and a fourth is needed for miscellaneous tasks such as waste management and equipment decontamination.

The following procedure should be used to collect concrete samples for radiological analysis.

Equipment
1. Hand drill and drill bit (0.5-in. or larger)
2. Stainless steel spoon
3. Ruler
4. Duct tape
5. Plastic Ziploc bags
6. Sample jars
7. Sample labels
8. Sample logbook
9. Chain-of-custody forms
10. Chain-of-custody seals
11. Permanent ink marker
12. Health and safety and radiological scanning instruments
13. Health and safety clothing
14. Sampling table
15. Waste container
16. Plastic waste bags

Sampling Procedure
1. In preparation for sampling, confirm that all the necessary preparatory work has been completed, including obtaining property access agreements; meeting health and safety, decontamination, and waste disposal requirements; and checking the calibration of all radiological and health and safety scanning instruments.
2. Use radiological scanning instruments and permanent ink marker to identify clearly the boundaries of the hot spot to be sampled.

3. Use a ruler and a small piece of duct tape to mark the maximum drilling depth of 0.25 in. on the drill bit.
4. If the hot spot to be sampled is located on a vertical wall, use a piece of duct tape to secure an open plastic Ziploc bag just beneath the hot spot to catch the drill cuttings.
5. Drill as many 0.25-in.-deep holes within the boundaries of the hot spot as needed to provide the minimum sample volume required by the laboratory.
6. Use stainless steel spoon to transfer drill cuttings into sample jar.
7. After the jar is capped, attach a sample label and custody seal to the jar and immediately place it into a sample cooler or radiologically shielded container.
8. See Chapter 5 for details on preparing sample jars and coolers for sample shipment.
9. Transfer any drill cuttings left over from the sampling into a waste container. Prior to leaving the site, all waste containers should be sealed, labeled, and handled appropriately.
10. Survey the coordinates of the sampling point to preserve the sampling location.

4.3.3.2.2   Concrete Coring Method

Concrete core samples are commonly collected from building walls, floors, ceilings, and/or other surfaces to estimate the concentration of various isotopes in the concrete if the building were to be demolished. When evaluating waste disposal options for demolished building materials, it is essential that concrete core samples be collected throughout the entire thickness of building walls, floors, ceilings, and/or other surfaces for laboratory analysis, since the results from building scanning surveys only reflect surface contamination. Core samples collected for this purpose are typically collected from the locations showing the highest activity based on scanning surveys.

For most sampling programs, a total of four people are sufficient for implementing this sampling procedure. Two are needed for sample collection, labeling, recording sampling location, and documentation; a third is needed for health and safety; and a fourth is needed for miscellaneous tasks such as waste management and equipment decontamination.

The following procedure should be used to collect concrete core samples for radiological analysis.

Equipment
1. Coring drill and coring barrel (length and diameter based on floor/wall thickness and required sample volume)
2. Hose and water supply for coring drill
3. Plastic Ziploc bags
4. Sample labels
5. Sample logbook
6. Chain-of-custody forms
7. Chain-of-custody seals
8. Permanent ink marker
9. Health and safety and radiological scanning instruments
10. Health and safety clothing
11. Sampling table

    12. Waste container

    13. Plastic waste bags

Sampling Procedure

1. In preparation for sampling, confirm that all the necessary preparatory work has been completed, including obtaining property access agreements; meeting health and safety, decontamination, and waste disposal requirements; and checking the calibration of all radiological and health and safety scanning instruments.

2. Use radiological scanning instruments and permanent ink marker to clearly identify the boundaries of the hot spot from which the core sample will be drilled.

3. Connect water supply line to core drill.

4. Begin drilling. Keep steady downward pressure on the coring barrel. Raise the core barrel every few minutes to facilitate the removal of drill cuttings.

5. Once the cut has been completed, remove the coring barrel from the hole. Typically, the concrete core will remain in the core barrel. Tap the outside of the core barrel gently with a hammer until the core drops out.

6. Place the core in a large plastic Ziploc bag.

7. After sealing the Ziploc bag, attach a sample label and custody seal and immediately place it into a sample cooler or radiologically shielded container.

8. See Chapter 5 for details on preparing coolers for sample shipment.

9. Transfer any drill cuttings left over from the coring into a waste container. Prior to leaving the site, all waste containers should be sealed, labeled, and handled appropriately.

10. Survey the coordinates of the sampling point to preserve the sampling location.

4.3.3.3 Paint Sampling

Paint samples are commonly collected from a variety of building surfaces including walls, piping, support beams, etc., in support of building decontamination and decommissioning activities. Paint is most frequently sampled in older buildings due to concerns over lead and PCB content (which has the potential for creating an RCRA or TSCA waste). Oftentimes, paint (and various types of resins) is used to fix radiological contaminants on building surfaces. Paint also provides the advantage of shielding alpha activity on these surfaces.

The following procedure identifies how a paint sample should be collected from a building surface for analytical testing. For most sampling programs, a total of four people are sufficient for implementing this sampling procedure. Two are needed for sample collection, labeling, recording sampling location, and documentation; a third is needed for health and safety; and a fourth is needed for miscellaneous tasks such as waste management and equipment decontamination.

The following procedure should be used to collect paint samples for chemical and/or radiological analysis.

Equipment

1. Stainless steel spatula

2. Stainless steel bowl

3. Plastic Ziploc bags

  4. Sample jars
  5. Sample labels
  6. Sample logbook
  7. Chain-of-custody forms
  8. Chain-of-custody seals
  9. Permanent ink marker
 10. Health and safety and radiological scanning instruments
 11. Health and safety clothing
 12. Sampling table
 13. Waste container
 14. Plastic waste bags

Sampling Procedure
  1. In preparation for sampling, confirm that all the necessary preparatory work has been completed, including obtaining property access agreements; meeting health and safety, decontamination, and waste disposal requirements; and checking the calibration of all radiological and health and safety scanning instruments.
  2 When sampling paint for chemical and/or radiological composition, use gross gamma radiological scanning instruments and permanent ink marker to clearly identify the boundaries of a painted surface showing elevated activity. Preferentially select locations where the paint is either bubbled or is peeling since this will facilitate sample collection. When samples are only being collected for chemical analysis, there is no need to perform the gross gamma scan.
  3. If the paint to be sampled is from a vertical wall, use a piece of duct tape to secure an open plastic Ziploc bag just beneath the hot spot to catch the paint scrapings. Otherwise, use a stainless steel bowl (or other container) to help catch the paint scrapings.
  4. Use a stainless steel spatula to scrape the painted surface, directing the paint chips into a Ziploc bag or stainless steel bowl. Continue this effort until the minimum sample volume required by the laboratory has been collected.
  5. Use a stainless steel spoon to transfer paint chips into a sample jar.
  6. After the jar is capped, attach a sample label and custody seal to the jar and immediately place it into a sample cooler or radiologically shielded container.
  7. See Chapter 5 for details on preparing sample jars and coolers for sample shipment.
  8. Any paint scrapings left over from the sampling should be transferred into a waste container. Prior to leaving the site, all waste containers should be sealed, labeled, and handled appropriately.
  9. Survey the coordinates of the sampling point to preserve the sampling location.

4.3.3.4 Soil Sampling

The following section provides the reader with shallow and deep soil sampling methods that should be considered to support soil remediation studies. The criteria used in selecting the most appropriate method include the analyses to be performed on the sample, the type of sample being collected (grab or composite), and the sampling depth. SOPs have been provided for each of the methods to facilitate implementation.

At radiological sites, soil sampling is most often performed for the following purposes:

- Defining the nature and extent of contamination;
- Assessing the risk that a site poses to human health and the surrounding environment;
- Defining soil distribution coefficients and geotechnical properties and supporting modeling studies and the evaluation of potential remedial alternatives;
- Performing treatability testing and other engineering evaluations;
- Verifying that remedial action objectives have been met.

When defining the nature and extent of contamination, it is typically most cost-effective to define the sources and horizontal and vertical extent of contamination using scanning and/or direct measurements (Section 4.2) combined with confirmation soil sampling. This is referred to as a judgmental sampling approach since soil samples are collected from locations that have the highest likelihood of showing contamination.

Soil sampling is often required to provide the data needed to assess the risk that a site poses to human health and the surrounding environment. Samples collected for this purpose are often taken from random or systematic locations since collecting samples from judgmental locations may grossly overestimate the risk. Soil sampling may be required to provide the data needed to support various types of engineering or modeling studies. These studies may require samples to be collected from judgmental, random, systematic, or one of the other approaches described in Section 4.1.1.5.7.2, depending upon the focus of the study.

After a soil remediation effort is complete, soil samples are collected from random or systematic locations so that the results can be evaluated statistically to determine whether or not the null hypothesis can be rejected and the site declared clean (see Chapter 8). The seven-step DQO process (Section 4.1.1.5) should be used to define the sampling approach, required number of samples, analyses to be performed, and analytical performance requirements.

The following sections present the most effective shallow and deep soil sampling methods, along with detailed procedures on how to use them.

4.3.3.4.1 Shallow Soil Sampling

Soil samples collected from a depth of 5 ft or less are generally referred to as "shallow." The most effective shallow soil sampling methods include the scoop, hand auger, slide-hammer, open-tube, split-tube or solid-tube, and thin-walled tube methods. When preparing a sampling program, considerable thought should go into selecting appropriate sampling methods, since the selected method can influence the analytical results. For example, if the contaminants of concern at a site include volatile organics, it would not be good practice to collect samples using the hand auger method, since this method churns up the soil, which facilitates volatilization. The slide-hammer, or split-tube or solid-tube sampler would be a more appropriate selection, since both of these tools can be used to remove a compacted, but undisturbed core of soil.

The hand auger is most effectively used to collect composite soil samples from sites where the contaminants of concern do not include volatile organics. When using this tool, samples for radiological and/or chemical analysis generally are not composited over intervals greater than 1 ft. since compositing larger intervals tends to dilute the composition of the sample beyond the point of providing useful data.

The slide-hammer and split-tube or solid-tube samplers can be used to collect either grab or composite samples. When collecting a grab sample, these tools are commonly lined with sample sleeves that can be quickly capped after sample collection. When composite samples are collected, the soil from the intervals, or locations to be composited, is transferred into a stainless steel bowl and homogenized with a stainless steel spoon prior to filling sample jars.

The scoop sampler can be used to collect grab or composite samples of surface soil. To collect a grab sample, the surface soil is scooped directly into a sample jar. A composite sample can be collected by scooping surface soil from the locations to be composited into a stainless steel bowl, and homogenized prior to filling sample jars.

When shallow soil samples are needed for lithology description only, the open-tube sampler or the split-tube sampler is an effective tool. The open-tube sampler has a sampling tube of small enough diameter that it can easily be advanced several feet into the ground to provide a small-diameter soil core. Since the sampling tube is open on one side, the soil lithology can be described without removing the sample from the tube. The split-tube sampler is beaten into the ground using a slide hammer or drill rig hammer. A sample sleeve should not be used in split-spoon samples when collecting samples for lithology description.. When soil samples are needed for geotechnical analysis, the thin-walled tube sampler hydraulically pushed into the ground is the preferred sampling method.

Table 4.15 summarizes the effectiveness of each of the six recommended sampling methods. A number "1" in the table indicates that a particular procedure is most effective in collecting samples for a particular laboratory analysis, sample type, or sampling depth. A number "2" indicates that the procedure is acceptable, but less effective, whereas an empty cell indicates that the procedure is not recommended. For example, Table 4.15 indicates that the slide-hammer and split/solid-tube methods are most effective in collecting soil samples for volatile organic analysis. The scoop method is considered acceptable when collecting samples for this analysis, while the hand auger, open-tube, and thin-walled tube methods are not recommended. The following sections provide further details and standard operating procedures for each of the recommended soil sampling methods.

4.3.3.4.1.1 Scoop Method. The scoop is a handheld sampling tool that is effective in collecting samples of the top 0.5-ft of soil. Figure 4.66 presents a variety of AMS scoop samplers. This method is commonly used to collect samples of discolored soil observed at the ground surface, or to collect samples from areas where, for some reason, deeper sampling is not possible. Grab or composite samples can be collected with this method by either spooning soil from one location directly into a sample jar or by compositing soil from more than one location in a stainless steel bowl prior to filling a sample jar (Figure 4.67). For additional information on scoop samplers, see www.ams-samplers.com.

Table 4.15 Rating Table for Shallow Soil Sampling Methods

	Laboratory Analyses								Sample Type			Depth		
	Radionuclides	Volatiles	Semivolatiles	Metals	Pesticides	PCBs	TPH	Geotechnical	Grab	Composite (Vertical)	Composite (Areal)	Surface (0.0–0.5 ft)	Shallow (0.0–5.0 ft)	Lithology Description
Scoop	1	2	1	1	1	1	1		1		1	1		1
Hand auger	1		1	1	1	1	1			1	2	1	1	1
Slide-hammer	1	1	1	1	1	1	1		1	1	2	1	1	2
Open-tube									1			1	1	1
Split tube/ solid tube	1/1	1/1	1/1	1/1	1/1	1/1	1/1		1/1	1/2	2/2		1	1/2
Thin-walled tube								1	1				1	

Note: 1 = preferred method; 2 = acceptable method; empty cell = method not recommended.

For most sampling programs, four people are sufficient for this procedure. Two are needed for sample collection, lithology description, labeling, and documentation; a third is needed for health and safety; and a fourth is needed for miscellaneous tasks such as waste management and equipment decontamination.

The following equipment and procedure can be used to collect shallow soil samples using the scoop method for chemical and/or radiological analysis.

Equipment
   1. Stainless steel scoop
   2. Stainless steel bowl
   3. Stainless steel spoon
   4. Sample jars
   5. Sample labels
   6. Cooler packed with Blue Ice® (Blue Ice is only required for nonradiological analyses)
   7. Trip blank and coolant blank (only required for volatile organic and nonradiological analyses, respectively)
   8. Sample logbook
   9. Chain-of-custody forms
   10. Chain-of-custody seals
   11. Permanent ink marker
   12. Health and safety and radiological scanning instruments
   13. Health and safety clothing
   14. Waste container
   15. Sampling table
   16. Plastic sheeting
   17. Plastic waste bags

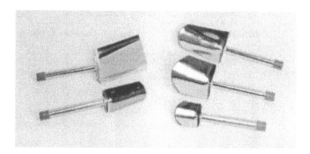

Figure 4.66 AMS scoop samplers.

Figure 4.67 Compositing soil sample.

Sampling Procedure

1. In preparation for sampling, confirm that all necessary preparatory work has been completed, including obtaining property access agreements; meeting health and safety, decontamination, and waste disposal requirements; and checking the calibration of all radiological and health and safety scanning instruments.

2. Cut a 1-ft-diameter hole in the center of the plastic sheeting, and center the hole over the sampling point. The purpose of this sheeting is to help prevent the spread of contamination.

3. Begin collecting the sample by applying a downward pressure on the scoop until the desired sampling depth is reached, then lift. Scan the sample using chemical and/or radiological scanning instruments, and record results in a bound logbook. If a grab sample is being collected, transfer the soil from the scoop directly into a sample jar. If a composite sample is being collected, transfer the soil from each location to be composited into a stainless steel compositing bowl and homogenize with a stainless steel spoon prior to filling a sample jar.

4. After the jar is capped. attach a sample label and custody seal to the jar and immediately place it into a sample cooler or radiologically shielded container. Samples for chemical analysis should be packed in Blue Ice.

5. See Chapter 5 for details on preparing sample jars and coolers for sample shipment.

6. Transfer any soil left over from the sampling into a waste container. Prior to leaving the site, all waste containers should be sealed, labeled, and handled appropriately.

7. Have a professional surveyor survey the coordinates of the sampling point to preserve the exact sampling location.

4.3.3.4.1.2 Hand Auger Method — The hand auger is an effective shallow soil sampling tool when the contaminants of concern do not include volatile organics, since the augering motion facilitates volatilization. This tool is composed of a bucket auger, which comes in various shapes and sizes, a shaft, and a T-bar handle (Figure 4.68). Extensions for the shaft are available to allow sampling at deeper intervals. However, in most soils this tool is only effective in collecting samples to a depth of 5 ft, since the sample hole typically begins to collapse at this depth. Since the auger rotation automatically homogenizes the sampling interval, this method is used to collect composite samples.

Figure 4.68   Hand auger being used to collect shallow soil sample.

For most sampling programs, four people are sufficient for this sampling procedure. Two are needed for sample collection, lithology description, labeling, and documentation; a third is needed for health and safety; and a fourth is needed for miscellaneous tasks such as waste management and equipment decontamination.

The following equipment and procedure can be used to collect shallow soil samples for chemical and/or radiological analysis.

Equipment
    1. Stainless steel hand auger
    2. Stainless steel bowl
    3. Stainless steel spoon
    4. Sample jars
    5. Sample labels
    6. Cooler packed with Blue Ice (Blue Ice is only required for nonradiological analyses)

7. Trip blank and coolant blank (only required for volatile organic and nonradio-logical analyses, respectively)
8. Sample logbook
9. Chain-of-custody forms
10. Chain-of-custody seals
11. Permanent ink marker
12. Health and safety and radiological scanning instruments
13. Health and safety clothing
14. Waste container
15. Sampling table
16. Plastic sheeting
17. Plastic waste bags

Sampling Procedure
1. In preparation for sampling, confirm that all necessary preparatory work has been completed, including obtaining property access agreements; meeting health and safety, decontamination, and waste disposal requirements; and check-ing the calibration of all radiological and health and safety scanning instruments.
2. Cut a 1-ft-diameter hole in the center of the plastic sheeting, and center the hole over the sampling point. The purpose of this sheeting is to help prevent the spread of contamination.
3. Begin collecting the soil sample by applying a downward pressure while rotat-ing the auger clockwise. When the auger is full of soil, remove it from the hole, and transfer the soil into a stainless steel bowl using a stainless steel spoon. Continue sampling in this manner until the bottom of the sampling interval is reached.
4. Composite the soil in the sampling bowl by using the stainless steel spoon to break apart any large chunks of soil; then mix and stir the soil enough to homogenize the sample thoroughly.
5. Scan sample using chemical and/or radiological scanning instruments, and record results in a bound logbook.
6. Transfer soil into a sample jar using the stainless steel spoon.
7. After the jar is capped, attach a sample label and custody seal to the jar and immediately place it into a sample cooler or radiologically shielded container. Samples for chemical analysis should be packed in Blue Ice.
8. See Chapter 5 for details on preparing sample jars and coolers for sample shipment.
9. Transfer any soil left over from the sampling into a waste container. Prior to leaving the site, all waste containers should be sealed, labeled, and handled appropriately.
10. Have a professional surveyor survey the coordinates of the sampling point to preserve the exact sampling location.

4.3.3.4.1.3 Slide Hammer Method — For collecting shallow soil core samples for chemical and/or radiological analysis, the slide hammer coring tool is recom-mended. This tool consists of a stainless steel core barrel, an extension rod, and a slide hammer (Figure 4.69). Most core barrels have an inside diameter of 2 or 2.5 in., and are 1 or 2 ft in length, but can be special ordered in other sizes. The top of the barrel is threaded so that it can be screwed into an extension rod. The barrel is also constructed so that it can accept stainless steel (or other material) sample liners which are commonly used to facilitate the removal of soil from the barrel without disturbing the sample (Figure 4.70). Without sample liners, soil must be extracted from the barrel using a spoon or knife, then transferred into a sample jar. This is not a problem when samples are being analyzed for radionuclides, metals, and nonvolatile compounds. However, it is a problem if samples are to be analyzed for volatile organics since a good portion of the volatiles can be lost into the ambient air. In contrast, sample liners can be quickly removed from the barrel and sealed with airtight Teflon caps. After labeling the liners, they can be shipped directly to the laboratory for analysis. If samples are being collected for lithology description only, clear plastic liners are available.

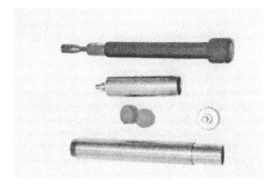

Figure 4.69 AMS slide hammer soil core sampler.

Figure 4.70 AMS core barrel sample liners.

Extension rods are available in various lengths to allow sampling at depths greater than the length of the core barrel. These rods are screwed into the core barrel at one end, and into the slide hammer at the other end. A slide hammer is used to beat the core barrel into the ground. The hammer is available in different shapes and weights to accommodate the needs of the sampler. For additional information on soil core samplers, see www.ams-samplers.com.

For most sampling programs, four people are sufficient for this sampling procedure. Two are needed for sample collection, lithology description, labeling, and documentation; a third is needed for health and safety; and a fourth is needed for miscellaneous tasks such as waste management and equipment decontamination.

The following equipment and procedure can be used to collect shallow soil samples for chemical and/or radiological analysis.

Equipment
1. Slide hammer and extension rods
2. Core barrel
3. Stainless steel sample sleeves
4. Teflon end-caps for stainless steel sleeves
5. Stainless steel bowl
6. Stainless steel spoon
7. Stainless steel knife
8. Sample jars
9. Sample labels
10. Cooler packed with Blue Ice (Blue Ice is only required for nonradiological analyses)
11. Trip blank and coolant blank (only required for volatile organic and nonradiological analyses, respectively)
12. Sample logbook
13. Chain-of-custody forms
14. Chain-of-custody seals
15. Permanent ink marker
16. Health and safety and radiological scanning instruments
17. Health and safety clothing
18. Waste container
19. Sampling table
20. Plastic sheeting
21. Aluminum foil
22. Plastic waste bags

Sampling Procedure
1. In preparation for sampling, confirm that all necessary preparatory work has been completed, including obtaining property access agreements; meeting health and safety, decontamination, and waste disposal requirements; and checking calibration of all radiological and health and safety scanning instruments.
2. Cut a 1-ft-diameter hole in the center of the plastic sheeting, and center the hole over the sampling point. The purpose of this sheeting is to help prevent the spread of contamination.
3. If sample sleeves are to be used, unscrew the core barrel from the slide hammer and load it with decontaminated stainless steel sleeves of the desired length. Avoid touching the inside surface of the core barrel and sleeves, for this will contaminate the sampler. Screw the core barrel back onto the slide hammer.
4. Use the slide hammer to beat the core barrel to the desired depth, and record the blow count in a sample logbook.
5. Remove the core-barrel from the hole by rocking it from side to side several times before lifting or reverse beating the sampler from the hole.

6. To collect a grab sample, unscrew the core barrel from the sampler and slide the sample sleeves out onto a piece of aluminum foil. Using a stainless steel knife, separate the sample sleeves. Scan the soil exposed at the ends of the sample sleeves using chemical and/or radiological scanning instruments. Record results in a bound logbook. Place Teflon caps over the ends of the sleeves to be sent to the laboratory. If sample sleeves are not being used, spoon soil from the core barrel directly into a sample jar. Scan the soil in the jar using chemical and/or radiological scanning instruments. Record results in a bound logbook. To collect a composite sample, sample sleeves are not needed; rather, soil from each of the intervals to be composited should be transferred into a stainless steel bowl and homogenized prior to filling a sample jar.
7. After the jar is capped, attach a sample label and custody seal to the jar and immediately place it into a sample cooler or radiologically shielded container. Samples for chemical analysis should be packed in Blue Ice.
8. See Chapter 5 for details on preparing samples and coolers for sample shipment.
9. Any soil left over from the sampling should be transferred into a waste container. Prior to leaving the site, all waste containers should be sealed, labeled, and handled appropriately.
10. Have a professional surveyor survey the coordinates of the sampling point to preserve the exact sampling location.

4.3.3.4.1.4 Open-Tube Sampler Method — For collecting shallow soil samples for lithology description, the open-tube sampler is recommended. This tool consists of an open-core barrel, extension rod, and T-bar handle. The AMS Soil Recovery Probe presented in Figure 4.71 is a type of open-tube sampler. Since the sampling tube is open on one side, the soil lithology can be described without removing the sample from the tube. This tool works most effectively in moist nongravelly soils. Clear plastic sleeve liners can be inserted into the open-tube sampler. When the sleeve liner is removed from the sampler, it can be capped to preserve the sample. For additional information on open tube samplers, see www.ams-samplers.com.

Figure 4.71 AMS Soil Recovery Probe.

For a large sampling program, three people are sufficient for this procedure. One is needed for sample collection and description; a second is needed for health and safety; and a third is needed for miscellaneous tasks such as waste management and equipment decontamination.

The following equipment and procedure can be used to collect shallow soil samples for lithology description.

Equipment
1. Open-tube sampler and extension rods
2. Stainless steel knife
3. Sample labels
4. Sample logbook
5. Permanent ink marker
6. Health and safety and radiological scanning instruments
7. Health and safety clothing
8. Waste container
9. Sample table
10. Plastic sheeting
11. Aluminum foil
12. Plastic waste bags

Sampling Procedure
1. In preparation for sampling, confirm that all necessary preparatory work has been completed, including obtaining property access agreements; meeting health and safety, decontamination, and waste disposal requirements; and checking calibration of all radiological and health and safety scanning instruments.
2. Cut a 1-ft-diameter hole in the center of the plastic sheeting, and center the hole over the sampling point. The purpose of this sheeting is to help prevent the spread of contamination.
3. Use the T-bar handle to rotate the sampler while pushing it into the ground.
4. Remove the sampler from the ground by pulling upward on the T-bar. To avoid injury, be certain to keep your back straight and lift with your legs.
5. Lay the sampler on the sampling table underlain by a piece of aluminum foil. Since the sampling tube is open sided, the soil can be described without removing it from the tube. When describing the lithology, it is recommended that a knife be used to slice open the sample, to reveal the sample texture.
6. If the sample is to be archived, remove the soil core through the open side of the sample tube. Wrap the core in aluminum foil; then place it in a core box. Mark the box with the name of the sampler, sampling time, date, location, and depth, and seal it shut with custody tape.
7. To collect a deeper sample from the same hole, attach an extension rod to a clean sampling tube and repeat the above procedure.
8. Transfer any soil left over from the sampling into a waste container. Prior to leaving the site, all waste containers should be sealed, labeled, and handled appropriately.
9. Have a professional surveyor survey the coordinates of the sampling point to preserve the exact sampling location.

4.3.3.4.1.5 Split-Tube or Solid-Tube Method — The split-tube or solid-tube method is very similar to the slide hammer coring method, with the exception being that a drill rig is used to beat the sampler into the ground. These samplers are composed of a split or solid sample tube, hardened shoe, soil catcher, and ball check (Figure 4.72). These samplers are available in two standard sizes, where the tubes are either 18- or 24-in. in length and have an outside diameter (O.D.) of 2 or 3 in.

If the sampler is being used to collect samples for laboratory analysis, it should be made of stainless steel. Using stainless steel sample liners to line the sample tube

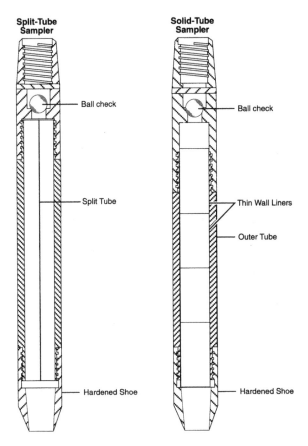

Figure 4.72 Split-tube and solid-tube sampler. (From Byrnes, M.E., *Field Sampling Methods for Remedial Investigations*, Lewis Publishers, Boca Raton, FL, 1994. With permission.)

is not a necessity; however, they are recommended when analyzing samples for volatile organics. Without sample liners, soil must be extracted from the tube and transferred into a sample jar using a stainless steel spoon. In this procedure, volatile organics can be lost into the ambient air. In contrast, sample liners can be quickly removed from the barrel, sealed with airtight Teflon caps, labeled, custody-sealed, then shipped to the laboratory for analysis. When samples are being analyzed for radionuclides, metals, nonvolatile compounds, or lithology description, sample liners are not needed.

For most sampling programs, three people are sufficient for this sampling procedure in addition to the drill rig operators. One is needed for sample collection, lithology description, labeling, and documentation; a second is needed for health and safety; and a third is needed for miscellaneous tasks such as waste management and equipment decontamination.

The following equipment and procedure can be used to collect shallow soil samples for chemical and/or radiological analysis.

Equipment
1. Stainless steel split-tube or solid-tube sampler
2. Stainless steel sample sleeves
3. Teflon end-caps for stainless steel sleeves
4. Auger drill rig with slide hammer
5. Stainless steel bowl
6. Stainless steel spoon
7. Stainless steel knife
8. Soil sample jars
9. Sample labels
10. Cooler packed with Blue Ice (Blue Ice is only required for nonradiological analyses)
11. Trip blank and coolant blank (only required for volatile organic and nonradiological analyses, respectively)
12. Sample logbook
13. Chain-of-custody forms
14. Chain-of-custody seals
15. Permanent ink marker
16. Health and safety and radiological scanning instruments
17. Health and safety clothing
18. Waste container
19. Sampling table
20. Plastic sheeting
21. Plastic waste bags

Sampling Procedure
1. In preparation for sampling, confirm that all necessary preparatory work has been completed, including obtaining property access agreements; meeting health and safety, decontamination, and waste disposal requirements; and calibrating all health and safety and sampling equipment.
2. Cut a 1-ft-diameter hole in the center of the plastic sheeting, and center the hole over the sampling point. The purpose of this sheeting is to help prevent the spread of contamination.
3. Have the drillers back the drill rig up to the sampling location, carefully raise the mast, then auger down to the top of the desired sampling interval.
4. Attach the sampler to a length of A-rod and lower it down the inside of the augers. Using the drill rig hammer, beat the sampler into the ground. Record the blow count in a sample logbook.
5. After removing the A-rod from the hole, detach the split tube or solid tube from the sampler.
6. To collect a grab sample using a split tube, break the tube open to reveal the sample sleeves. Using a stainless steel knife, separate the individual sleeves. Scan the soil exposed at the ends of the sample sleeves using chemical and/or radiological scanning instruments. Record results in a bound logbook. Place Teflon caps over the ends of those to be sent to the laboratory for analysis. If sample sleeves are not being used, spoon soil from the split tube directly into a sample jar. Scan the soil in the jar using chemical and/or radiological scanning instruments. Record results in a bound logbook.

To collect a grab sample using a solid tube, slide the sample sleeves out of one end of the solid tube. Using a stainless steel knife, separate the individual sleeves. Scan the soil exposed at the ends of the sample sleeves using chemical and/or radiological scanning instruments. Record results in a bound logbook. Place Teflon caps over the ends of those to be sent to the laboratory for analysis. If sample sleeves are not being used, spoon soil from the solid tube directly into a sample jar. Scan the soil in the jar using chemical and/or radiological scanning instruments. Record results in a bound logbook.

To collect a composite sample, there is no need to use sample sleeves. Rather, soil from each of the intervals to be composited should be transferred into a stainless steel bowl and homogenized prior to filling a sample jar. Scan the soil in the jar using chemical and/or radiological scanning instruments. Record results in a bound logbook.

7. After the jar is capped, attach a sample label and custody seal to the jar and immediately place it into a sample cooler or radiologically shielded container. Samples for chemical analysis should be packed in Blue Ice.

8. See Chapter 5 for details on preparing samples and coolers for sample shipment.

9. Transfer any soil left over from the sampling into a waste container. Prior to leaving the site, all waste containers should be sealed, labeled, and handled appropriately.

10. Have a professional surveyor survey the coordinates of the sampling point to preserve the exact sampling location.

4.3.3.4.1.6 Thin-Walled Tube Method — What is unique about this method is that the sample tube is hydraulically pushed into the ground using a drill rig, as opposed to being driven into the ground with a hammer. The advantage of pushing the sampler into the ground is that the soil is not artificially compacted in the sampling process. Consequently, the thin-walled tube is the preferred method for collecting samples for geotechnical analysis. Some of the more common geotechnical tests run on soil samples include porosity, hydraulic conductivity, specific gravity, grain size distribution, Atterberg limits, compaction, consolidation, compression, and shear.

The thin-walled tube method utilizes a thin-walled sampling tube which has a standard O.D. of 3-in., and length that allows the collection of a 30-in. sample (Figure 4.73). There are four holes at the top of the tube, which are used to connect the sampler to a sampling rod. Thin-walled tubes are available in either low carbon steel, or stainless steel. If only geotechnical analyses are to be performed on the sample, low carbon steel is acceptable. However, if the soil sample is to be tested for chemical or radiological composition, the sampler should be made of stainless steel.

For most sampling programs, three people are sufficient for this sampling procedure in addition to the drill rig operators. One is needed for sample collection, lithology description, labeling, and documentation; a second is needed for health and safety; and a third is needed for miscellaneous tasks such as waste management and equipment decontamination.

The following equipment and procedure can be used to collect shallow soil samples for geotechnical testing.

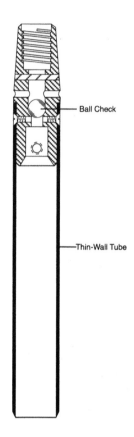

Figure 4.73 Thin-walled tube sampler. (From Byrnes, M.E., *Field Sampling Methods for Remedial Investigations*, Lewis Publishers, Boca Raton, FL, 1994. With permission.)

Equipment
1. Thin-walled tube sampler
2. Sample tube end-caps
3. Auger drill rig
4. Sampling knife
5. Sample labels
6. Sample logbook
7. Chain-of-custody forms
8. Chain-of-custody seals
9. Permanent ink marker
10. Health and safety and radiological scanning instruments
11. Health and safety clothing
12. Waste container
13. Sampling table
14. Plastic sheeting
15. Paraffin wax
16. Plastic waste bags

Sampling Procedure
1. In preparation for sampling, confirm that all necessary preparatory work has been completed, including obtaining property access agreements; meeting

health and safety, decontamination, and waste disposal requirements; and checking the calibration of all radiological and health and safety scanning instruments.

2. Cut a 1-ft-diameter hole in the center of the plastic sheeting, and center the
hole over the sampling point. The purpose of this sheeting is to help prevent
the spread of contamination.

3. Have the drillers back the drill rig up to the sampling location, carefully raise
the mast, then auger down to the top of the desired sampling interval.

4. Attach the sampler to a length of A-rod and lower it down the inside of the
augers. Using the drill rig, hydraulically push the sampler into the ground.

5. After removing the A-rod from the hole, detach the sample tube. Scan the soil
exposed at the ends of the sample tube using chemical and/or radiological
scanning instruments. Record results in a bound logbook. Using a sampling
knife, shave approximately 0.5 in. of soil from each end of the tube. Fill the
space with melted paraffin wax, then place a cap over each end of the tube.
The purpose of the wax is to prevent the shifting of soil in the tube during
shipment to the geotechnical laboratory.

6. Attach a sample label to the tube; then place custody seals over each end cap.

7. Place sample tubes in a vertical position for transport to the geotechnical
laboratory to preserve the soil compaction characteristics. If samples must be
shipped, it is best to mark the sample box as "FRAGILE." Also, denote on the
outside of the box which end is "UP." If no chemical analyses are being
performed, there is no need to chill the sample.

8. Transfer any soil left over from the sampling into a waste container. Prior to
leaving the site, all waste containers should be sealed, labeled, and handled
appropriately.

9. Have a professional surveyor survey the coordinates of the sampling point to
preserve the exact sampling location.

4.3.3.4.2   Deep Soil Sampling

Soil samples collected at depths greater than 5 ft are generally referred to as
"deep." These samples are most commonly collected by driving a split-tube or solid-
tube sampler (Figure 4.72) or hydraulically pushing a thin-walled tube (Figure 4.73)
into the ground with the assistance of an auger drill rig. These sampling methods
are similar to those described for shallow soil sampling (Sections 4.3.3.4.1.5 and
4.3.3.4.1.6) except samples are collected from deeper intervals.

Table 4.16 summarizes the effectiveness of the two recommended sampling
methods. A number "1" in the table indicates that a particular procedure is most
effective in collecting samples for a particular laboratory analysis, sample type, or
sampling depth. A number "2" indicates that the procedure is acceptable, but less
effective, while an empty cell indicates that the procedure is not recommended. For
example, Table 4.16 indicates that the split-tube and solid-tube samplers are both
effective in collecting soil samples for volatile organics and all other chemical
analyses, while the thin-walled tube sampler is most effectively used to collect
geotechnical samples. The following sections provide further details and SOPs for
these deep soil sampling methods.

4.3.3.4.2.1 Split-Tube or Solid-Tube Method — The split-tube or solid-tube
method used to collect deep soil samples is identical to that described for shallow

204 SAMPLING AND SURVEYING RADIOLOGICAL ENVIRONMENTS

Table 4.16 Rating Table for Deep Soil Sampling Methods

	Laboratory Analyses								Sample Type			Depth	
	Radionuclides	Volatiles	Semivolatiles	Metals	Pesticides	PCBs	TPH	Geotechnical	Grab	Composite (Vertical)	Composite (Areal)	Deep (> 5.0 ft)	Lithology Description
Split tube/ solid tube	1/1	1/1	1/1	1/1	1/1	1/1	1/1		1/1	1/2	2/2	1/1	1/2
Thin-walled tube								1	1			1	

Note: 1 = preferred method; 2 = acceptable method; empty cell = method not recommended.

soil sampling (Section 4.3.3.4.1.5), with the exception that samples are collected from a depth greater than 5 ft.

4.3.3.4.2.2 Thin-Walled Tube Method — The thin-walled tube method used to collect deep soil samples is identical to the procedure described for shallow soil sampling (Section 4.3.3.4.1.6), with the exception that samples are collected from a depth greater than 5 ft.

4.3.3.5 Sediment Sampling

This section provides the reader with guidance on selecting sediment sampling methods for remedial investigation studies. The criteria used in selecting the most appropriate method include the analyses to be performed on the sample, the type of sample being collected (grab or composite), and the sampling depth. SOPs have been provided for each of the methods to facilitate implementation.

One objective of a remedial investigation should be to determine if chemical and/or radiological contamination is present in the sediment of nearby surface water units such as streams, rivers, surface water drainages, ponds, lakes, retention basins, and/or tanks. It is particularly important to characterize the sediment in streams, rivers, and other surface water drainages since they provide avenues for rapid contaminant migration, and provide points where receptors are readily exposed to contamination. Ponds, lakes, retention basins, and tanks do not provide the same opportunity for rapid contaminant migration; however, they similarly provide exposure points for receptors.

The seven-step DQO process (Section 4.1.1.5) should be used to define the sampling approach, required number of samples, analyses to be performed, and analytical performance requirements.

The following sections present preferred sampling methods and procedures for collecting sediment samples from streams, rivers, surface water drainages, ponds, lakes, retention basins, and tanks. Sampling tools used to collect sediment for laboratory analysis should be made of Teflon and/or stainless steel.

4.3.3.5.1   Streams, Rivers, and Surface Water Drainages

Although a number of sophisticated sampling devices are available to collect sediment samples, not all of these tools are effective in collecting samples through the shallow, fast-moving water that typifies streams, rivers, and surface water drainages. The methods that have proved to be the most effective when sampling these environments include the scoop or dipper method, slide hammer method, and box sampler method.

Of these three methods, the scoop or dipper method is the easiest to implement since it simply involves pushing the sampler into the sediment, then either transferring the sediment directly into a sample jar or transferring the sediment into a stainless steel bowl and compositing with other sampling locations prior to filling the sample jar. This technique is only effective in collecting samples of the top 0.5 ft of sediment at locations where the water depth is less than 2 ft.

The slide hammer method involves beating a sampling tube into the sediment. This technique removes a core of sediment for analytical testing. Since the sampling tube is typically lined with sampling sleeves, either individual sleeves can be sent to the laboratory as grab samples or the soil can be removed from the sleeves and composited prior to filling a sample jar.

The box sampler method utilizes a spring-loaded sample box attached to a sampling pole to collect grab sediment samples. This method involves pushing the sampling box into the sediment, and releasing the spring-loaded sample jaws. After the sampler is retrieved, sediment from the sampling box is transferred into a sample jar. Similar to the scoop or dipper method, this method is only effective in collecting samples from the top 0.5 ft of sediment.

Depending on the depth of the water overlying the sediment, samplers may need a pair of waders, a raft, or a boat to access the sampling point. If waders are used, the sampler should face upstream while collecting the sample to assure that the sampler's boots do not contaminate the sample. Similarly, samples should be collected from the upstream side of the raft or boat. As a general rule, the sample located farthest downstream should always be the first collected. Sampling should then proceed upstream. By collecting samples in this manner, any sediment disturbed by the samplers will not contaminate downstream sampling points.

Table 4.17 summarizes the effectiveness of each of the three recommended sampling methods. A number "1" in the table indicates that a particular procedure is most effective in collecting samples for a particular laboratory analysis, sample type, or sampling depth. A number "2" indicates that the procedure is acceptable, but less effective, while an empty cell indicates that the procedure is not recommended. For example, Table 4.17 indicates that the slide hammer method is most effective in collecting sediment samples for volatile organics. While the scoop or dipper, and box sampler are acceptable sampling methods for volatile organic analysis, they are less effective than the slide hammer method.

Table 4.17 Rating Table for Sediment Sampling Methods for Streams, Rivers, and Surface Water Drainages

	Laboratory Analyses							Sample Type			Depth		
	Radionuclides	Volatiles	Semivolatiles	Metals	Pesticides	PCBs	TPH	Geotechnical	Grab	Composite (Vertical)	Composite (Areal)	Surface (0.0–0.5 ft)	Shallow (0.0–5.0 ft)	Lithology Description
Scoop or dipper	2	2	2	2	2	2	2		2		2	2		2
Slide-hammer	1	1	1	1	1	1	1		1	1	1	2	1	2
Box sampler	1	2	1	1	1	1	1		1		1	1		1

Note: 1 = preferred method; 2 = acceptable method; empty cell = method not recommended.

4.3.3.5.1.1 Scoop or Dipper Method — The scoop or dipper is the simplest sampling tool for collecting grab sediment samples from streams, rivers, and surface water drainages, and is available in many shapes and sizes. Figure 4.66 presents several varieties of scoop samplers. A dipper is basically a deep cup attached to a long handle. This method involves lowering a Teflon or stainless steel scoop or dipper by hand through the surface water and pushing the sampler deep into the underlying sediment. As the sampler is retrieved, the sediment is transferred into a sample jar. This technique is generally effective in collecting samples of the top 0.5 ft of sediment, at locations where the water is less than 2 ft in depth.

This method is most effective when collecting samples from water bodies with relatively slow flow velocities, since a significant amount of the finer-grained sediment tends to be lost when sampling higher-energy environments. The dipper typically works more effectively than the scoop at preventing the loss of fine-grained sediment when retrieving the sample; however, since the dipper has no cutting edge, it is only effective in sampling soft sediment. The scoop or dipper method is most commonly used for preliminary sampling activities. If contaminants are identified during the preliminary sampling, the slide hammer (Section 4.3.3.5.1.2) or box sampler (Section 4.3.3.5.1.3) methods should be considered to define the distribution of contaminants more accurately.

When collecting a grab sample, sediment is transferred from the sampler directly into a sample jar. If a composite sample is collected, the sediment to be composited is transferred into a stainless steel bowl and homogenized with a stainless steel spoon prior to filling a sample jar.

For most sampling programs, four people are sufficient for this sampling procedure. Two are needed for sample collection, labeling, and documentation; a third is needed for health and safety; and a fourth is needed for waste management and equipment decontamination.

The following equipment and procedure can be used to collect sediment samples for chemical and/or radiological analysis.

Equipment
1. Teflon or stainless steel scoop or dipper
2. Stainless steel bowl
3. Stainless steel spoon
4. Sample jars
5. Sample labels
6. Cooler packed with Blue Ice (Blue Ice is only required for nonradiological analyses)
7. Trip blank and coolant blank (only required for volatile organic and nonradiological analyses, respectively)
8. Sample logbook
9. Chain-of-custody forms
10. Chain-of-custody seals
11. Permanent ink marker
12. Health and safety and radiological scanning instruments
13. Health and safety clothing
14. Waste container
15. Sampling table
16. Plastic waste bags

Sampling Procedure
1. In preparation for sampling, confirm that all necessary preparatory work has been completed, including obtaining property access agreements; meeting health and safety, decontamination, and waste disposal requirements; and check calibration of all radiological and health and safety scanning instruments.
2. Approach the sampling point from downstream, being careful not to disturb the underlying sediment.
3. Push the scoop or dipper firmly downward into the sediment, then lift upward. Quickly raise the sampler out of the water in an effort to reduce the amount of sediment lost to the water current. If a grab sample is being collected, transfer the sediment from the scoop or dipper directly into a sample jar. If a composite sample is being collected, transfer the sediment from each composite interval or location into a stainless steel bowl and homogenize with a stainless steel spoon prior to filling a sample jar.
4. Scan the sediment in the jar using chemical and/or radiological scanning instruments. Record results in a bound logbook.
5. After the jar is capped, attach a sample label and custody seal to the jar and immediately place it into a sample cooler or radiologically shielded container. Samples for chemical analysis should be packed in Blue Ice.
6. See Chapter 5 for details on preparing sample jars and coolers for sample shipment.
7. Transfer any sediment left over from the sampling into a waste container. Prior to leaving the site, all waste containers should be sealed, labeled, and handled appropriately.
8. Have a professional surveyor survey the coordinates of the sampling point to preserve the exact sampling location.

4.3.3.5.1.2 Slide Hammer Method — The slide hammer is an effective tool for collecting core samples of stream and river sediment. This tool consists of a stainless steel core barrel, extension rod, and slide hammer (see Figure 4.69). Stock core barrels have an I.D. of 2 or 2.5 in., and are 1 to 3 ft in length. However, the core barrel can be special-ordered to meet project specific volume requirements. The top of the barrel is threaded so that it can be screwed into an extension rod to allow sampling through deeper water.

The barrel is constructed to accept sample liners (see Figure 4.70), which are commonly used to facilitate the removal of sediment from the barrel without disturbing the sample. The use of liners is not a necessity; however, they are recommended when grab samples are to be analyzed for volatile organics. Without sample liners, sediment must be extracted from the barrel and transferred into a sample jar. In this process, volatile organics can be lost into the ambient air. In contrast, sample liners can be quickly removed from the barrel and sealed with airtight Teflon caps. After labeling and custody-sealing the liners, they can be shipped directly to the laboratory for analysis. When collecting a composite sample, sediment from the intervals or locations to be composited is transferred into a stainless steel bowl and homogenized prior to filling a sample jar.

Extension rods can be ordered in various lengths to allow sampling through various depths of water. The rods are screwed into the core barrel at one end, and into the slide hammer at the other end. The slide hammer is used to beat the sampler into the sediment. The hammer is available in various weights to accommodate the needs of the sampler. If samples are to be collected from locations where water depths exceed several feet, a raft or boat will be required to assist the sampling procedure.

For most sampling programs, four people are sufficient for this sampling procedure. Two are needed for sample collection, labeling, and documentation; a third is needed for health and safety; and a fourth is needed for waste management and equipment decontamination. If a raft or boat is used to assist the sampling procedure, at least one additional person will be needed.

The following equipment and procedure can be used to collect sediment samples for chemical and/or radiological analysis.

Equipment
1. Slide hammer and extension rods
2. Stainless steel sample sleeves
3. Teflon end-caps for sample sleeves
4. Stainless steel bowl
5. Stainless steel spoon
6. Stainless steel knife
7. Sample jars
8. Sample labels
9. Cooler packed with Blue Ice (Blue Ice is only required for nonradiological analyses)
10. Trip blank and coolant blank (only required for volatile organic and nonradiological analyses, respectively)
11. Sample logbook

12. Chain-of-custody forms
13. Chain-of-custody seals
14. Permanent ink marker
15. Health and safety and radiological scanning instruments
16. Health and safety clothing
17. Waders, raft, or boat
18. Sampling table
19. Waste container
20. Plastic waste bags

Sampling Procedure
1. In preparation for sampling, confirm that all necessary preparatory work has been completed, including obtaining property access agreements; meeting health and safety, decontamination, and waste disposal requirements; and checking calibration of all radiological and health and safety scanning instruments.
2. Approach the sampling point from downstream, being careful not to disturb the underlying sediment.
3. Lower the sampler through the water, then beat the core barrel to the desired depth, and record the blow counts in a sample logbook.
4. Remove the core barrel from the hole by either rocking it from side to side several times before lifting or reverse beating the sampler from the hole.
5. To collect a grab sample, unscrew the core barrel from the sampler and slide the sample sleeves out onto the sampling table. Using a stainless steel knife, separate the sample sleeves. Scan the sediment exposed at the ends of the sample sleeves using chemical and/or radiological scanning instruments. Record results in a bound logbook. Place Teflon caps over the ends of the sleeves to be sent to the laboratory. If sampling sleeves are not being used, spoon sediment from the core barrel directly into a sample jar. Scan the sediment in the jar using chemical and/or radiological scanning instruments. Record results in a bound logbook.
   To collect a composite sample, sample sleeves are not needed; rather, sediment from each of the intervals to be composited should be transferred into a stainless steel bowl and homogenized prior to filling a sample jar.
6. After the jar is capped, attach a sample label and custody seal to the jar and immediately place it into a sample cooler or radiologically shielded container. Samples for chemical analysis should be packed in Blue Ice.
7. See Chapter 5 for details on preparing sample jars and coolers for sample shipment.
8. Transfer any soil left over from the sampling into a waste container. Prior to leaving the site, all waste containers should be sealed, labeled, and handled appropriately.
9. Have a professional surveyor survey the coordinates of the sampling point to preserve the exact sampling location.

4.3.3.5.1.3 Box Sampler Method — The stainless steel box sampler is a very effective tool for collecting grab samples of stream and river sediment. This sampler is composed of a sample box, spring-loaded sample jaws, and a pole containing a spring release mechanism (Figure 4.74). After the sample box is pushed firmly into the sediment, the jaws are released to seal off the bottom of the box. The closed

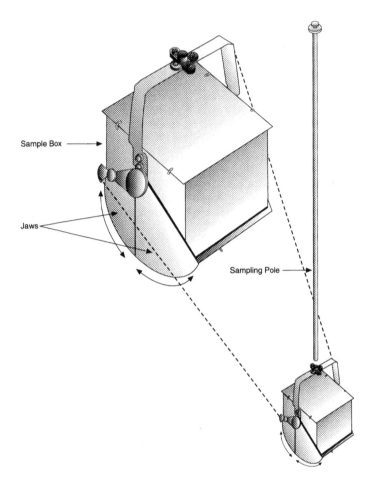

Sample Box

Jaws

Sampling Pole

Figure 4.74 Box sampler. (From Byrnes, M.E., *Field Sampling Methods for Remedial Investigations*, Lewis Publishers, Boca Raton, FL, 1994. With permission.)

box is then retrieved from the water. The advantage of this sampling method is that fine-grained sediment is not stripped from the sample as it is removed from the water.

For most sampling programs, four people are sufficient for this sampling procedure. Two are needed for sample collection, labeling, and documentation; a third is needed for health and safety; and a fourth is needed for waste management and equipment decontamination. If a raft or boat is used to assist the sampling procedure, at least one additional person will be needed.

The following equipment and procedure can be used to collect sediment samples for chemical and/or radiological analysis.

Equipment
1. Stainless steel box sampler
2. Stainless steel bowl
3. Stainless steel spoon

4. Sample jars
5. Sample labels
6. Cooler packed with Blue Ice (Blue Ice is only required for nonradiological analyses)
7. Trip blank and coolant blank (only required for volatile organic and nonradiological analyses, respectively)
8. Sample logbook
9. Chain-of-custody forms
10. Chain-of-custody seals
11. Permanent ink marker
12. Health and safety and radiological scanning instruments
13. Health and safety clothing
14. Waders, raft, or boat
15. Sampling table
16. Waste containers
17. Plastic waste bags

Sampling Procedure
1. In preparation for sampling, confirm that all necessary preparatory work has been completed, including obtaining property access agreements; meeting health and safety, decontamination, and waste disposal requirements; and checking the calibration of all radiological and health and safety scanning instruments.
2. Approach the sampling point from downstream, being careful not to disturb the underlying sediment.
3. Hold the sampling pole so the open sampler jaws are positioned several inches above the surface of the sediment; then firmly thrust the sampler downward. Depress the button at the top of the sampling pole to release the spring-loaded jaws.
4. If a grab sample is being collected, transfer the sediment from the box sampler directly into a sample jar. If a composite sample is being collected, transfer the sediment from the locations to be composited into a stainless steel bowl and homogenize with a stainless steel spoon prior to filling a sample jar. Scan the sediment in the jar using chemical and/or radiological scanning instruments. Record results in a bound logbook.
5. After the jar is capped, attach a sample label and custody seal to the jar and immediately place it into a sample cooler or radiologically shielded container. Samples for chemical analysis should be packed in Blue Ice.
6. See Chapter 5 for details on preparing sample jars and coolers for sample shipment.
7. Transfer any soil left over from the sampling into a waste container. Prior to leaving the site, all waste containers should be sealed, labeled, and handled appropriately.
8. Have a professional surveyor survey the coordinates of the sampling point to preserve the exact sampling location.

4.3.3.5.2 Ponds, Lakes, Retention Basins, and Tanks

The sampling methods that have proved to be the most effective when sampling sediment from ponds, lakes, retention basins, and tanks include the scoop or dipper, slide hammer, box sampler, and dredge sampler. To collect sediment samples from highly radioactive tanks, the Light-Duty Utility Arm described in Section 4.2.2.3.1.1 will need to be used. The scoop or dipper, slide hammer, and box sampler methods are all effective when collecting sediment samples from around the edges of ponds, lakes, and retention basins, where the water is relatively shallow. On the other hand, the box sampler and dredge sampler, used in combination with a wire line, are effective methods for collecting sediment samples through deep water.

Of these four methods, the scoop or dipper method is the easiest to implement since it simply involves lowering a Teflon or stainless steel scoop or dipper by hand through the surface water and pushing the sampler deep into the underlying sediment. As the sampler is retrieved, the sediment is transferred into a sample jar. This technique is generally effective in collecting samples of the top 0.5 ft of sediment at locations where the water depth is less than 2 ft.

The slide hammer method involves beating a sampling tube into the sediment. This technique removes a core of sediment for analytical testing. Since the sampling tube is typically lined with sampling sleeves, either individual sleeves can be sent to the laboratory as grab samples or the soil can be removed from the sleeves and composited prior to filling a sample jar. Of the four sampling methods, only the slide hammer is effective in collecting sediment samples deeper than the top 0.5 ft. When used in combination with extension rods, the slide hammer can be used to collect a 2- to 3-ft sediment core through 10 to 15 ft of water.

The box sampler method utilizes a spring-loaded sample box attached to either a sampling pole or wire line to collect grab sediment samples. This method involves pushing the box sampler into the sediment, and releasing the spring-loaded sampler jaws. After the sampler is retrieved, sediment from the sampling box is transferred into a sample jar. This method is only effective in collecting samples from the top 0.5 ft of sediment. When using a sampling pole, one can collect a sediment sample through 5 ft of water. When using a wire line, there is no limit to the depth of water through which a sediment sample can be collected.

The dredge sampler method is composed of two jaws connected by a lever, and is lowered through the water using a wire line. As the sampler is quickly lowered through the water, the jaws open and embed themselves in the underlying sediment. As the wire line is raised, the jaws close to capture a sample. Sediment from the dredge is then transferred into a sample jar using a stainless steel spoon. Similar to the scoop or dipper and box sampler, this method is only effective in collecting samples from the top 0.5 ft of sediment.

Depending on the depth of the water overlying the sediment, samplers will need either a pair of waders, a raft, or a boat to access the sampling point. Table 4.18 summarizes the effectiveness of each of the four recommended sampling methods. A number "1" in the table indicates that a particular procedure is most effective in collecting samples for a particular laboratory analysis, sample type, or sampling depth. A number "2" indicates that the procedure is acceptable but less effective, while an empty cell indicates that the procedure is not recommended. For example, Table 4.18 indicates that the slide hammer is the most effective method for collecting sediment samples for volatile organic analysis. While the scoop or dipper, box sampler, and dredge sampler are acceptable sampling methods for volatile organic analysis, they are less effective than the slide hammer method.

Table 4.18 Rating Table for Sediment Sampling Methods for Ponds, Lakes, and Retention Basins

	Laboratory Analyses							Sample Type			Depth		
	Radionuclides	Volatiles	Semivolatiles	Metals	Pesticides	PCBs	TPH	Geotechnical	Grab	Composite (Vertical)	Composite (Areal)	Surface (0.0–0.5 ft)	Shallow (0.0–5.0 ft)	Lithology Description
Scoop or dipper	2	2	2	2	2	2	2		2		2	2		2
Slide-hammer	1	1	1	1	1	1	1		1	1	1	2	1	2
Box sampler	1	2	1	1	1	1	1		1		1	1		1
Dredge sampler	1	2	1	1	1	1	1		1		1	1		1

Note: 1 = preferred method; 2 = acceptable method; empty cell = method not recommended.

4.3.3.5.2.1 Scoop or Dipper Method — The scoop or dipper method used to collect sediment samples from ponds, lakes, and retention basins is identical to the procedure described for streams, rivers, and surface water drainage sampling (see Section 4.3.3.5.1.1).

4.3.3.5.2.2 Slide Hammer Method — The slide hammer method used to collect sediment samples from ponds, lakes, and retention basins is identical to the procedure described for streams, rivers, and surface water drainage sampling (see Section 4.3.3.5.1.2).

4.3.3.5.2.3 Box Sampler Method — The box sampler method used to collect sediment samples from ponds, lakes, and retention basins is identical to the procedure described for streams, rivers, and surface water drainage sampling (see Section 4.3.3.5.1.3), with the following modifications:

- A wire line may be used to lower the sampler through the water, as opposed to the sampling pole.
- Add "wire line" and "life jackets" to the equipment list.
- Modify Step 3 to read "Hold the sampling pole so the open sampler jaws are positioned several inches above the surface of the sediment, then firmly thrust the sampler downward. Depress the button at the top of the sampling pole to release the spring-loaded jaws. When using a cable to lower the sampler, as opposed to a sampling pole, allow the sampler to free fall through the last 5 to 10 ft of water to assure that the sampler jaws become deeply embedded into the sediment. Then slide a trip weight down the cable to trip the spring-loaded jaws."

4.3.3.5.2.4 Dredge Sampler Method — The dredge sampler is a very common and effective tool in collecting grab samples of sediment in ponds, lakes, and retention basins. This sampler is composed of two jaws connected by a lever (Figure 4.75). To use this sampling method, the dredge is attached to a cable and allowed to free-fall through the water to assure that the sampler jaws deeply embed themselves into the underlying sediment. As the cable is retrieved, the jaws to the sampler are forced closed. As with the box sampler, this technique is effective in preventing the loss of fine-grained sediment as the sample is retrieved.

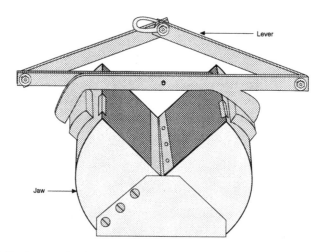

Figure 4.75 Dredge sampler. (From Byrnes, M.E., *Field Sampling Methods for Remedial Investigations*, Lewis Publishers, Boca Raton, FL, 1994. With permission.)

For most sampling programs, five people are sufficient for this sampling procedure. Two are needed for sample collection, labeling and documentation; a third is needed for health and safety; a fourth is needed for waste management and equipment decontamination; and a fifth is needed to maneuver the raft or boat.

The following equipment and procedure can be used to collect sediment samples for chemical and/or radiological analysis.

Equipment
1. Stainless steel dredge sampler
2. Wire line
3. Stainless steel spoon
4. Sample jars
5. Sample label
6. Cooler packed with Blue Ice (Blue Ice is only required for nonradiological analyses)
7. Trip blank and coolant blank (only required for volatile organic and nonradiological analyses, respectively)
8. Sample logbook
9. Chain-of-custody forms
10. Chain-of-custody seals

11. Permanent ink marker
12. Health and safety and radiological scanning instruments
13. Health and safety clothing (including life jackets)
14. Boat large enough for five people and sampling equipment
15. Sampling table
16. Waste container
17. Plastic waste bags

Sampling Procedure
 1. In preparation for sampling, confirm that all necessary preparatory work has been completed, including obtaining property access agreements; meeting health and safety, decontamination, and waste disposal requirements; and checking the calibration of all radiological and health and safety scanning instruments.
 2. Maneuver the raft or boat over the sampling location.
 3. Lower the dredge sampler to the bottom of the pond or lake using a wire line. The faster the sampler is dropped, the deeper the sampler will be embedded into the sediment. When the sampler hits bottom, allow the line to go slack for a few seconds, then retrieve.
 4. If a grab sample is being collected, transfer the sediment from the dredge directly into a sample jar. If a composite sample is being collected, transfer the sediment to be composited into a stainless steel bowl and homogenize with a stainless steel spoon prior to filling a sample jar. Scan the sediment in the jar using chemical and/or radiological scanning instruments. Record results in a bound logbook.
 5. After the jar is capped, attach a sample label and custody seal to the jar and immediately place it into a sample cooler or radiologically shielded container. Samples for chemical analysis should be packed in Blue Ice.
 6. See Chapter 5 for details on preparing sample jars and coolers for sample shipment.
 7. Transfer any sediment left over from the sampling into a waste container. Prior to leaving the site, all waste containers should be sealed, labeled, and handled appropriately.
 8. Have a professional surveyor survey the coordinates of the sampling point to preserve the exact sampling location.

4.3.3.6 *Surface Water and Liquid Waste Sampling*

This section provides the reader with guidance on selecting surface water and liquid waste sampling methods for site characterization. The criteria used in selecting the most appropriate method include the analyses to be performed on the sample, the type of sample being collected (grab, composite, or integrated), and the sampling depth. SOPs have been provided for each of the methods to facilitate implementation.

One objective of an initial site characterization study should be to determine if contamination is present in nearby surface water and/or liquid waste units such as streams, rivers, surface and storm sewer drainages, ponds, lakes, retention basins, and tanks. It is particularly important to characterize the surface water in streams, rivers, and other surface water drainages since they provide avenues for rapid contaminant migration, and provide points where receptors are readily exposed to contamination. Ponds, lakes, retention basins, and tanks do not provide the same opportunity for rapid contaminant migration; however, they similarly provide exposure points for receptors.

The seven-step DQO process (Section 4.1.1.5) should be used to define the sampling approach, required number of samples, analyses to be performed, and analytical performance requirements.

The depth at which surface water and/or liquid waste samples are collected should be based on the suspected concentration and specific gravity of the contaminants of concern. If historical information leads the investigator to believe that contamination could be layered, due to varying specific gravities of the contaminants, collecting either a vertical composite sample or several grab samples from different depth intervals should be considered (Figure 4.76).

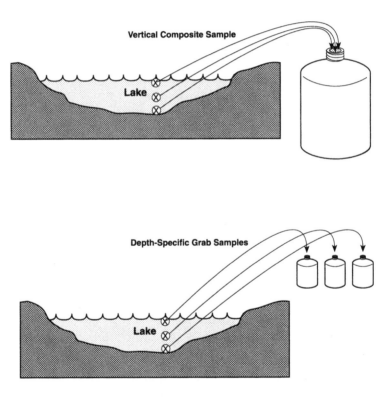

Figure 4.76 Vertical grab or composite surface water sampling. (From Byrnes, M.E., *Field Sampling Methods for Remedial Investigations*, Lewis Publishers, Boca Raton, FL, 1994. With permission.)

When preparing to collect a surface water sample, a presample should be collected in a glass jar, and should be analyzed for pH, temperature, and conductivity. Additional measurements that should be considered include turbidity, alkalinity, dissolved oxygen, and oxidation-reduction potential.

The following sections present preferred sampling methods and procedures for collecting surface water and/or liquid waste samples from streams, rivers, and surface water drainages, and ponds, lakes, retention basins, and tanks. Sampling tools used to collect samples for laboratory analysis should be made of Teflon, stainless steel, or glass.

4.3.3.6.1 Streams, Rivers, and Surface Water Drainages

Although a number of sophisticated sampling devices are available to collect surface water samples, not all of these tools are effective in collecting samples in the shallow, fast-moving water which typifies streams, rivers, and surface water drainages. Since the movement of water in these environments tends to homogenize the water column naturally, in most cases samples are collected from the water surface. However, if vertical fractionation is suspected, deeper sampling can be performed.

The methods that have proved to be the most effective when sampling these environments include the bottle submersion method, dipper method, and extendable tube sampler method.

The bottle submersion and extendable tube sampler are all effective methods for collecting grab samples. Of these methods, the bottle submersion method is the easiest to implement, since it simply involves submerging a sample bottle beneath the water surface. With this method, a telescoping extension rod is often used to hold the sample bottle as it is lowered into the water. The dipper method is also very implementable since it simply involves lowering a stainless steel dipper below the water surface, then transferring the water into sample bottles.

The extendable tube sampler utilizes a sampling tube and a vertical extension rod. This method utilizes a check valve to open and close the sampling tube when the desired sampling depth is reached. After the sampling tube is retrieved, water is transferred into a sample bottle. This method is effective in collecting samples as deep as 5 ft below the water surface.

Any of the three methods can be used to collect areal, composite, or integrated samples, where an areal composite sample is collected by compositing a number of grab samples from different locations and an integrated sample involves collecting a portion of a sample from the same location several times over a period of time. Depth composite samples can be collected using the extendable tube sampler. A depth composite sample is collected by compositing several grab samples, each representing a different depth in the water column.

Table 4.19 summarizes the effectiveness of each of the three recommended sampling methods. A number "1" in the table indicates that a particular procedure is most effective in collecting samples for a particular laboratory analysis, sample type, or sampling depth. A number "2" indicates that the procedure is acceptable but less effective, while an empty cell indicates that the procedure is not recommended. For example, Table 4.19 indicates that bottle submersion is the preferred procedure for collecting samples for volatile organic analysis. While the dipper and

Table 4.19 Rating Table for Surface Water Sampling Methods for Streams, Rivers, and Drainages

	Laboratory Analyses							Sample Type			Depth		
	Radionuclides	Volatiles	Semivolatiles	Metals	Pesticides	PCBs	TPH	Grab	Composite (Vertical)	Composite (Areal)	Integrated	Surface (0.0–0.5 ft)	Shallow (0.0–5.0 ft)	Deep (>5.0 ft)
Bottle submersion	1	1	1	1	1	1	1	1		1	1	1		
Dipper	1	2	1	1	1	1	1	1		1	1	1		
Extendable tube sampler	1	2	1	1	1	1	1	1	1	1	2	1	1	

Note: 1 = preferred method; 2 = acceptable method; empty cell = method not recommended.

extendable tube sampler are acceptable methods for volatile organic analysis, they are less desirable since the water from the sampler must be poured into sample bottles, which tends to aerate the sample.

Whichever sampling method is selected, sample bottles for volatile organic analysis should always be the first to be filled. Bottles for radiological and other analyses should be filled consistent with their relative importance to the sampling program. As a general rule, the sample located farthest downstream should always be the first to be collected. Sampling should then proceed upstream. By collecting samples in this manner, any sediment disturbed by the samplers will not contaminate downstream sampling points. Depending on the depth of the water, samplers may need a pair of waders, a raft, or a boat to access the sampling point. If waders are used, the sampler should face upstream while collecting the sample to assure that the waders do not contaminate the sample. Similarly, samples should be collected from the upstream side of a raft or boat.

4.3.3.6.1.1 Bottle Submersion Method — The bottle submersion method is the simplest and one of the most commonly used surface water sampling methods for collecting samples from streams, rivers, and surface water drainages. This method utilizes a water sample bottle and an optional telescoping extension rod with an adjustable beaker clamp. This method is only effective in collecting grab samples from the top few inches of the water column.

The following procedure can be used to collect grab surface water samples, and is written to include the use of a telescoping extension rod. If an extension rod is not available, the bottles can be lowered into the water by hand. Since this procedure is both simple and effective, it is recommended that initial surface water sampling be performed using either this method or the dipper method. If contaminants are identified from this preliminary sampling, the extendable tube sampler method should be considered to define the distribution of contaminants more accurately. For a summary of the effectiveness and limitations of this sampling method, see Table 4.19.

For most sampling programs, four people are sufficient for this sampling procedure. Two are needed for field testing, sample collection, labeling, and documentation; a third is needed for health and safety; and a fourth is needed for miscellaneous tasks such as waste management and equipment decontamination.

The following equipment and procedure can be used to collect surface water samples for chemical and/or radiological analysis.

Equipment
1. Telescoping extension rod with adjustable beaker clamp
2. Sample bottles
3. Sample preservatives
4. pH, temperature, and conductivity meters
5. Sample labels
6. Cooler packed with Blue Ice (Blue Ice is only required for nonradiological analyses)
7. Trip blank and coolant blank (only required for volatile organic and nonradiological analyses, respectively)
8. Sample logbook
9. Chain-of-custody forms
10. Chain-of-custody seals
11. Permanent ink marker
12. Health and safety and radiological scanning instruments
13. Health and safety clothing (including life jackets)
14. Waste container
15. Sampling table
16. Plastic waste bags

Sampling Procedure
1. In preparation for sampling, confirm that all necessary preparatory work has been completed, including obtaining property access agreements; meeting health and safety, decontamination, and waste disposal requirements; and checking the calibration of all radiological and health and safety scanning instruments.
2. Prior to sample collection, fill a clean glass jar with sample water, and measure the pH, temperature, and conductivity of the water. Record this information in a sample logbook.
3. Secure a clean sample bottle to the end of the telescoping extension rod using a beaker clamp. Remove the bottle cap just prior to sampling.
4. While standing on the bank of the stream, river, or surface water drainage, extend the rod out over the water and lower the sample bottle just below the water surface. If the analyses to be performed require the sample to be preserved, this should be performed prior to filling the sample bottle. Scan the sample using chemical and/or radiological scanning instruments, and record results in a bound logbook.
5. After the bottle is capped, attach a sample label and custody seal to the bottle and immediately place it into a sample cooler or radiologically shielded container. Samples for chemical analysis should be packed in Blue Ice.

6. See Chapter 5 for details on preparing sample bottles and coolers for sample shipment.
7. Transfer any water left over from the sampling into a waste container. Prior to leaving the site, all waste containers should be sealed, labeled, and handled appropriately.
8. Have a professional surveyor survey the coordinates of the sampling point to preserve the exact sampling location.

4.3.3.6.1.2 Dipper Method — The dipper method is a simple but effective method for collecting grab samples of surface water from streams, rivers, and surface water drainages for both chemical and radiological analysis. With this method, a Teflon or stainless steel dipper is used to collect a water sample, which is then transferred into a sample bottle. This method is only effective in collecting grab samples from the top few inches of the water column.

Since this procedure is both simple and effective, it is recommended that initial surface water sampling be performed using either this method or the bottle submersion method. If contaminants are identified from this preliminary sampling, the extendable tube sampler method should be considered to define the distribution of contaminants more accurately. For a summary of the effectiveness and limitations of this sampling method, see Table 4.19.

For most sampling programs, four people are sufficient for this sampling procedure. Two are needed for field testing, sample collection, labeling, and documentation; a third is needed for health and safety; and a fourth is needed for waste management and equipment decontamination.

The following equipment and procedure can be used to collect surface water samples for chemical and/or radiological analysis.

Equipment
1. Teflon or stainless steel dipper
2. Sample bottles
3. Sample preservatives
4. pH, temperature, and conductivity meters
5. Sample labels
6. Cooler packed with Blue Ice (Blue Ice is only required for nonradiological analyses)
7. Trip blank and coolant blank (only required for volatile organic and nonradiological analyses, respectively)
8. Sample logbook
9. Chain-of-custody forms
10. Chain-of-custody seals
11. Permanent ink marker
12. Health and safety and radiological scanning instruments
13. Health and safety clothing (including life jackets)
14. Waste container
15. Sampling table
16. Plastic waste bags

Sampling Procedure

1. In preparation for sampling, confirm that all necessary preparatory work has been completed, including obtaining property access agreements; meeting health and safety, decontamination, and waste disposal requirements; and checking the calibration of all radiological and health and safety scanning instruments.

2. Prior to sample collection, fill a clean glass jar with sample water, and measure the pH, temperature, and conductivity of the water. Record this information in a sample logbook.

3. Extend the dipper out over the water and lower it just below the water surface, being careful not to disturb the underlying sediment.

4. When the dipper is full, carefully transfer the water into a sample bottle. If the analyses to be performed require the sample to be preserved, this should be performed prior to filling the sample bottle. Scan sample using chemical and/or radiological scanning instruments, and record results in a bound logbook.

5. After the bottle is capped, attach a sample label and custody seal to the bottle and immediately place it into a sample cooler or radiologically shielded container. Samples for chemical analysis should be packed in Blue Ice.

6. See Chapter 5 for details on preparing sample bottles and coolers for sample shipment.

7. Transfer any water left over from the sampling into a waste container. Prior to leaving the site, all waste containers should be sealed, labeled, and handled appropriately.

8. Have a professional surveyor survey the coordinates of the sampling point to preserve the exact sampling location.

4.3.3.6.1.3 Extendable Tube Sampler Method — The extendable tube sampler is an effective tool for collecting grab or composite samples as deep as 5 ft below the water surface. The sampler is composed of a sampling tube, which contains a ball check-valve with plunger-type inner rod, and a vertical sampling pole with an inner rod to open and close the check valve (Figure 4.77).

This method is particularly effective for characterizing streams and rivers that are suspected of having stratified zones of contamination, since they can collect grab samples from various depth intervals within a water column. While this method is most effective in collecting grab samples, it can also be used to collect composite samples by combining a number of grab samples together in one sample bottle. A composite sample should not be collected when the analyses to be performed on the sample include volatile organics, since volatilization is facilitated by the compositing process. For a summary of the effectiveness and limitations of this sampling method, see Table 4.19.

For most sampling programs, four people are sufficient for this sampling procedure. Two are needed for field testing, sample collection, labeling, and documentation; a third is needed for health and safety; and a fourth is needed for waste management and equipment decontamination. If a raft or boat is used to assist the sampling procedure, at least one additional person will be needed.

The following equipment and procedure can be used to collect surface water samples for chemical and/or radiological analysis.

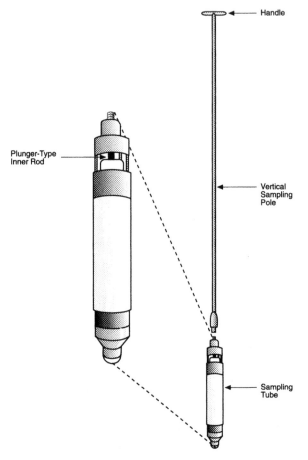

Figure 4.77 Extendable tube sampler. (From Byrnes, M.E., *Field Sampling Methods for Remedial Investigations*, Lewis Publishers, Boca Raton, FL, 1994. With permission.)

Equipment
1. Extendable tube sampler
2. Sample bottles
3. Sample preservatives
4. pH, temperature, and conductivity meters
5. Sample labels
6. Cooler packed with Blue Ice (Blue Ice is only required for nonradiological analyses)
7. Trip blank and coolant blank (only required for volatile organic and nonradiological analyses, respectively)
8. Sample logbook
9. Chain-of-custody forms
10. Chain-of-custody seals
11. Permanent ink marker
12. Health and safety and radiological scanning instruments
13. Health and safety clothing (including life jackets)

14. Waders, raft, or boat
15. Waste container
16. Sampling table
17. Plastic waste bags

Sampling Procedure

1. In preparation for sampling, confirm that all necessary preparatory work has been completed, including obtaining property access agreements; meeting health and safety, decontamination, and waste disposal requirements; and checking the calibration of all radiological and health and safety scanning instruments.
2. Prior to sample collection, fill a clean glass jar with sample water and measure the pH, temperature, and conductivity of the water. Record this information in a sample logbook.
3. To collect a grab sample, lower the sampler to the desired sampling interval, and open the ball check-valve to allow the sampling tube to fill with water. When the sampler is full, close the check-valve and retrieve the sample. The water is then transferred into a sample bottle through the bottom dump valve. A composite sample can be collected by combining a number of grab samples over either a vertical or horizontal sampling area. Scan sample using chemical and/or radiological scanning instruments, and record results in a bound logbook.
4. If the analyses to be performed require the sample to be preserved, this should be performed prior to filling the sample bottle.
5. After the bottle is capped, attach a sample label and custody seal to the bottle and immediately place it into a sample cooler or radiologically shielded container. Samples for chemical analysis should be packed in Blue Ice.
6. See Chapter 5 for details on preparing sample bottles and coolers for sample shipment.
7. Transfer any water left over from the sampling into a waste container. Prior to leaving the site, all waste containers should be sealed, labeled, and handled appropriately.
8. Have a professional surveyor survey the coordinates of the sampling point to preserve the exact sampling location.

4.3.3.6.2 Ponds, Lakes, Retention Basins, and Tanks

The depth at which surface water and/or liquid waste samples are collected should be based on the suspected concentration and specific gravity of the contaminants of concern. If historical information leads the investigator to believe that contamination could be layered, due to varying specific gravities of the contaminants, collecting either a vertical composite sample or several grab samples from different depth intervals should be considered (see Figure 4.76).

Although a number of sophisticated sampling devices are available to collect surface water and/or liquid waste samples, not all of these tools are effective in collecting samples from ponds, lakes, retention basins, and tanks. The methods that have proved to be the most effective when sampling these environments include the bottle submersion method, dipper method, extendable tube sampler method, bailer method, Kemmerer bottle method, Kabis sampler, and bomb sampler method. The first three of these methods are identical to those used to collect samples from streams, rivers, and surface water drainages (see Section 4.3.3.6.1), with a few minor modifications.

All seven of these methods are effective in collecting grab samples. Of these methods, the bottle submersion method is the easiest to implement since it simply involves submerging a sample bottle just beneath the liquid surface. A telescoping extension rod is often used with this method to hold the sample bottle as it is lowered into the liquid. The dipper method is also very easy to implement since it simply involves lowering a dipper below the liquid surface, then transferring the liquid into sample bottles. These two methods are only effective in collecting samples from the water surface.

The extendable tube sampler utilizes a sampling tube and a vertical extension rod. This method utilizes a check valve to open and close the sampling tube when the desired sampling depth is reached. After the sampling tube is retrieved, liquid is transferred into a sample bottle. This method is effective in collecting samples as deep as 5 ft below the liquid surface.

The bailer is composed of a sampling tube, pouring spout, and a bottom check valve. The bailer is lowered by hand just below the liquid surface. When the bailer is full, it is retrieved and the liquid is transferred into a sample bottle. This method collects a sample representative of the top several feet of the liquid column.

The Kemmerer bottle is composed of a vertical sampling tube, center rod, trip head, and bottom plug. The bottle is lowered through the liquid by means of a sampling line to the desired sampling depth. The sampling tube is then opened with the assistance of a trip weight. When the sampler is full, the liquid is retrieved and transferred into a sample bottle. The depth at which a sample can be collected with this method is only restricted by the length of the sampling line.

The Kabis sampler is composed of a bullet-shaped sampling tube, threaded lid, fill tube, exhaust tube, and support ring. The sampler is lowered to the desired sampling depth using a support line. After pressure equilibration, water enters the sampler through the fill tube while air exits through the exhaust tube. When all of the air has been replaced by water, the sampler is retrieved. The sample bottle is then removed from the sampler. The depth at which a sample can be collected with this method is only restricted by the length of the sampling line.

The bomb sampler is composed of a sampling tube, center rod, and support ring. The sampler is lowered to the desired sampling depth using a support line. The center rod is then lifted using a second line, which allows the sampler to fill with liquid. The liquid is then retrieved and transferred into a sample bottle. The depth at which a sample can be collected with this method is only restricted by the length of the sampling line.

Vertical composite samples can be collected using either the extendable tube sampler, Kemmerer bottle, or bomb sampler. A vertical composite sample is collected by compositing several grab samples; each represents a different depth in the liquid column (see Figure 4.76). Any of the seven methods can be used to collect areal composite samples where grab samples from different locations are composited together (Figure 4.78). Any of the seven methods can also be used to collect an integrated sample where a portion of a sample is collected from the same location several times over a selected time period.

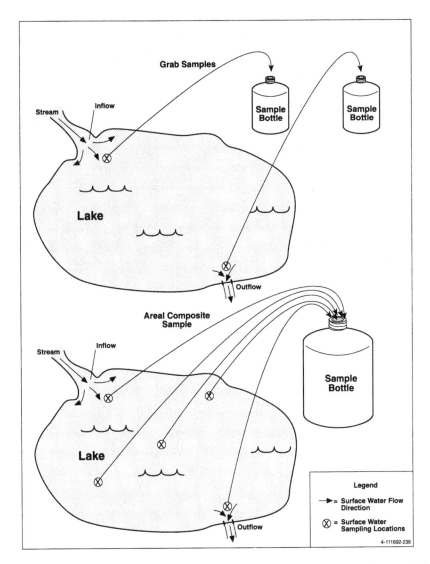

Figure 4.78 Collecting areal composite or grab surface water samples. (From Byrnes, M.E., *Field Sampling Methods for Remedial Investigations*, Lewis Publishers, Boca Raton, FL, 1994. With permission.

Table 4.20 summarizes the effectiveness of each of the seven sampling methods. A number "1" in the table indicates that a particular procedure is most effective in collecting samples for a particular laboratory analysis, sample type, or sampling depth. A number "2" indicates that the procedure is acceptable but less effective, while an empty cell indicates that the procedure is not recommended. For example, Table 4.20 indicates that bottle submersion and Kabis sampler are the preferred methods for collecting samples for volatile organic analysis. While the dipper, bailer, extendable tube sampler, Kemmerer bottle, and bomb sampler are acceptable methods

Table 4.20 Rating Table for Surface Water Sampling Methods for Ponds, Lakes, Retention Basins, and Tanks

	Laboratory Analyses							Sample Type				Depth		
	Radionuclides	Volatiles	Semivolatiles	Metals	Pesticides	PCBs	TPH	Grab	Composite (Vertical)	Composite (Areal)	Integrated	Surface (0.0–0.5 ft)	Shallow (0.0–5.0 ft)	Deep (>5.0 ft)
Bottle submersion	1	1	1	1	1	1	1	1		1	1	1		
Dipper	1	2	1	1	1	1	1	1		1	1	1		
Extendable tube sampler	1	2	1	1	1	1	1	1	1	2	1	2	1	
Bailer	1	2	1	1	1	1	1	1		1	1	1	2[a]	
Kemmerer bottle	1	2	1	1	1	1	1	1	1	2	1		2	1
Kabis sampler	1	1	1	1	1	1	1	1	1	2	1		2	1
Bomb sampler	1	2	1	1	1	1	1	1	1	2	1		2	1

[a] Able to collect a sample equal to the length of the bailer.

Note: 1 = preferred method; 2 = acceptable method; empty cell = method not recommended.

for volatile organic analysis, they are less desirable since the liquid from the sampler must be poured into sample bottles, which tends to aerate the sample.

Whichever sampling method is selected, sample bottles for volatile organic analysis should always be the first to be filled. Bottles for the remaining parameters should be filled consistent with their relative importance to the sampling program.

4.3.3.6.2.1 Bottle Submersion Method — The bottle submersion method used to collect grab liquid samples from ponds, lakes, retention basins, and tanks is identical to the procedure described for stream, river, and water drainage sampling (Section 4.3.3.6.1.1), with the following modifications:

- If contaminants are identified from the preliminary sampling, the Kemmerer bottle, Kabis sampler, and bomb sampler should be considered to delineate clearly the vertical distribution of contaminants.
- If a raft or boat is used to assist the sample collection, a fifth person will be needed on the sampling team. This person's responsibilities are to maneuver and steady the boat.
- Add "waders, raft, or boat" to the equipment list.
- Modify Step 4 to read "Extend the extension rod out over the water or liquid waste and lower the sample bottle just below the liquid surface. If the analyses to be performed require the sample to be preserved, this should be performed prior to filling the sample bottle."

4.3.3.6.2.2 Dipper Method — The dipper method used to collect grab surface water samples from ponds, lakes, retention basins, and tanks is identical to the procedure described for stream, river, and water drainage sampling (Section 4.3.3.6.1.2), with the following modifications:

- If contaminants are identified from the preliminary sampling, the Kemmerer bottle, Kabis sampler, and bomb sampler should be considered to delineate the vertical distribution of contaminants clearly.
- If a raft or boat is used to assist the sample collection, a fifth person will be needed on the sampling team. This person's responsibilities are to maneuver and steady the boat.
- Add "waders, raft, or boat" to the equipment list.
- Modify Step 3 to read, "Lower the dipper below the liquid surface."

4.3.3.6.2.3 Extendable Tube Sampler Method — The extendable tube sampler method used to collect grab or composite liquid samples from ponds, lakes, retention basins, and tanks is identical to the procedure described for stream, river, and water drainage sampling (Section 4.3.3.6.1.3), with the following modification:

- If contaminants are identified from the preliminary sampling, the Kemmerer bottle, Kabis sampler, and bomb sampler should be considered to delineate clearly the vertical distribution of contaminants.

4.3.3.6.2.4 Bailer Method — A bailer is most commonly used to collect ground-water samples; however, it can also be used to collect liquid samples from ponds, lakes, and retention basins and tanks. A standard bailer is composed of a bailer body, which is available in various lengths and diameters, a pouring spout, and a bottom check valve which contains a check ball (Figure 4.79).

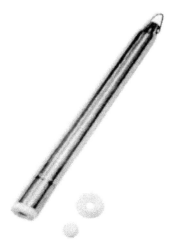

Figure 4.79 AMS TaperTop Bailer.

As the bailer is lowered into the liquid, water and/or liquid waste flows into the bailer through the bottom check valve. When the sampler is retrieved, a check ball prevents liquid from escaping through the bottom of the sampler. The sample is then poured from the bailer through a spout into sample bottles. Bailers used to collect liquid samples for laboratory analysis should be made of Teflon or stainless steel. Depending on the depth of the liquid at the sampling point, samplers may need a pair of waders, raft, or boat to access the sampling point.

Some common modifications to the bailer include the use of extension couples to increase the length of the bailer, and a controlled flow bottom assembly. The bottom assembly allows the bailer to be emptied through the bottom of the sampler, which reduces the opportunity for volatilization to occur. For more information about bailers, visit the AMS Web page at www.ams-samplers.com.

For most sampling programs, four people are sufficient for this sampling procedure. Two are needed for field testing, sample collection, labeling, and documentation; a third is needed for health and safety; and a fourth is needed for miscellaneous tasks such as managing wastewater drums and equipment decontamination. If a raft or boat is used to assist the sampling procedure, at least one additional person will be needed.

The following equipment and procedure can be used to collect liquid samples for chemical and/or radiological analysis.

Equipment
1. Teflon or stainless steel bailer
2. Sample bottles
3. Sample preservatives
4. pH, temperature, and conductivity meters
5. Sample labels
6. Cooler packed with Blue Ice (Blue Ice is only required for nonradiological analyses)
7. Trip blank and coolant blank (only required for volatile organic and nonradiological analyses, respectively)
8. Sample logbook
9. Chain-of-custody forms
10. Chain-of-custody seals
11. Permanent ink marker
12. Health and safety and radiological scanning instruments
13. Health and safety clothing (including life jackets)
14. Waders, raft, or boat
15. Waste containers
16. Sampling table
17. Plastic waste bags

Sampling Procedure
1. In preparation for sampling, confirm that all necessary preparatory work has been completed, including obtaining property access agreements; meeting health and safety requirements, decontamination, and waste disposal requirements; and checking the calibration of all radiological and health and safety scanning instruments.

2. Prior to sample collection, fill a clean glass jar with sample liquid and measure the pH, temperature, and conductivity of the water. Record this information in a sample logbook.

3. When properly positioned over the sampling point, hold the bailer just above the pouring spout, and slowly lower it just deep enough to fill the bailer. Retrieve the sampler and carefully transfer the liquid into a sample bottle. Scan sample using chemical and/or radiological scanning instruments, and record results in a bound logbook.

4. If the analyses to be performed require the sample to be preserved, do so just prior to filling the sample bottle.

5. After the bottle is capped, attach a sample label and custody seal to the bottle and immediately place it into a sample cooler or radiologically shielded container. Samples for chemical analysis should be packed in Blue Ice.

6. See Chapter 5 for details on preparing sample bottles and coolers for sample shipment.

7. Transfer any liquid left over from the sampling into a waste container. Prior to leaving the site, all waste containers should be sealed, labeled, and handled appropriately.

8. Have a professional surveyor survey the coordinates of the sampling point to preserve the exact sampling location.

4.3.3.6.2.5 Kemmerer Bottle Method — The Kemmerer bottle sampler method is effective for collecting at-depth grab samples of liquid from ponds, lakes, retention basins, and tanks. The sampler is composed of a vertical sampling tube, center rod, head plug, and bottom plug (Figure 4.80). A line attached to the top of the sampler is used to lower the sampler to the desired sampling depth. The head plug and bottom plug are then tripped open by sliding a trip weight down the sampling line. When the sampling tube is full of liquid, the sampler is retrieved and the liquid is transferred into a sample bottle.

The sampling line should be monofilament, such as common fishing line, and should be discarded between sampling points. The line should be cut to a length long enough to reach the desired sampling depth, and it must be strong enough to lift the weight of the Kemmerer bottle when it is full of liquid. Prior to using the sampling line, it should first be decontaminated in the same manner as other sampling equipment (see Chapter 9).

The effective sampling depth of this sampler is only limited by the length of the sampling line. Since this method is often used to collect deep liquid samples, a raft or boat may be required to assist the sampling procedure.

When samples are to be collected from depths of 5 ft or less, the extendable tube sampler (Section 4.3.3.6.2.3) is recommended over the Kemmerer bottle, since it is easier to use.

This method is particularly effective for characterizing ponds, lakes, retention basins, and tanks, which may contain vertically stratified contaminant layers. To characterize stratified conditions, the Kemmerer bottle, Kabis sampler, or bomb sampler (Sections 4.3.3.6.2.5, 4.3.3.6.2.6, and 4.3.3.6.2.7) can be used to collect liquid samples from discrete intervals throughout the liquid column.

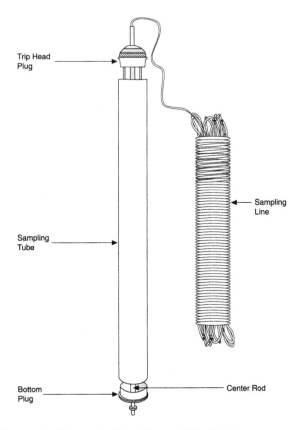

Figure 4.80 Kemmerer bottle sampler. (From Byrnes, M.E., *Field Sampling Methods for Remedial Investigations*, Lewis Publishers, Boca Raton, FL, 1994. With permission.)

The Kemmerer bottle is effective in collecting depth and areal composite liquid samples as well as integrated samples. A depth composite sample is acquired by compositing several grab samples, each representing a different depth in the liquid column. An areal composite sample is collected by compositing liquid samples from different locations, whereas an integrated sample is acquired by collecting a portion of a sample from the same location several times over an extended period of time.

For most sampling programs, five people are sufficient for this sampling procedure. Two are needed for field testing, sample collection, labeling, and documentation; a third is needed for health and safety; a fourth is needed for waste management and equipment decontamination; and a fifth may be needed to operate the boat or maneuver the raft (if required).

The following equipment and procedure can be used to collect surface liquid samples for chemical and/or radiological analysis.

Equipment
1. Teflon or stainless steel Kemmerer bottle
2. Sampling line

 3. Sample bottles
 4. Sample preservatives
 5. pH, temperature, and conductivity meters
 6. Sample labels
 7. Cooler packed with Blue Ice (Blue Ice is only required for nonradiological analyses)
 8. Trip blank and coolant blank (only required for volatile organic and nonradiological analyses, respectively)
 9. Sample logbook
10. Chain-of-custody forms
11. Chain-of-custody seals
12. Permanent ink marker
13. Health and safety and radiological scanning instruments
14. Health and safety clothing (including life jackets)
15. Waders, raft, or boat
16. Waste container
17. Sampling table
18. Plastic waste bags

Sampling Procedure
 1. In preparation for sampling, confirm that all necessary preparatory work has been completed, including obtaining property access agreements; meeting health and safety, decontamination, and waste disposal requirements; and checking the calibration of all radiological and health and safety scanning instruments.
 2. Prior to sample collection, fill a clean glass jar with sample liquid from the desired sampling depth, and measure the pH, temperature, and conductivity of the liquid. Record this information in a sample logbook.
 3. When properly positioned over the sampling point, slowly lower the Kemmerer bottle to the desired sampling depth.
 4. Slide the trip weight down the sampling line to trip open the sample tube. When the sampler is full of liquid, retrieve it and transfer the liquid into a sample bottle. Scan sample using chemical and/or radiological scanning instruments, and record results in a bound logbook.
 5. If the analyses to be performed require the sample to be preserved, do this prior to filling the sample bottle.
 6. After the bottle is capped, attach a sample label and custody seal to the bottle and immediately place it into a sample cooler or radiologically shielded container. Samples for chemical analysis should be packed in Blue Ice.
 7. See Chapter 5 for details on preparing sample bottles and coolers for sample shipment.
 8. Transfer any liquid left over from the sampling into a waste container. Prior to leaving the site, all waste containers should be sealed, labeled, and handled appropriately.
 9. Have a professional surveyor survey the coordinates of the sampling point to preserve the exact sampling location.

4.3.3.6.2.6 Kabis Sampler Method — The Kabis sampler method used to collect grab liquid samples from ponds, lakes, retention basins, and tanks is identical to the procedure described for groundwater sampling (Section 4.3.3.7.2.3.2) with the following exceptions:

- Add "waders, raft, or boat" and "life preservers" to the equipment list
- Delete Steps 2 and 3 since there is no need to collect water level measurements or perform well development.

4.3.3.6.2.7 Bomb Sampler Method — The bomb sampler method is effective for collecting at-depth grab samples of water or liquid waste from ponds, lakes, retention basins, and tanks. The sampler is composed of a sampling tube, center rod, and support ring (Figure 4.81). A line attached to the support ring is used to lower the sampler to the desired sampling depth. A second line is attached to the top of the spring-loaded center rod, and is used to open and close the sampling tube.

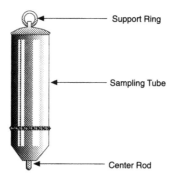

Figure 4.81 Bomb sampler. (From Byrnes, M.E., *Field Sampling Methods for Remedial Investigations*, Lewis Publishers, Boca Raton, FL, 1994. With permission.)

The support and sampling line should be monofilament, such as common fishing line, and should be discarded between sampling points. The line should be cut to a length long enough to reach the desired sampling depth, and must be strong enough to lift the weight of the bomb sampler when it is full of liquid. Prior to using the sampling line, it should first be decontaminated in the same manner as other sampling equipment (Chapter 9).

After lowering the sampler to the desired sampling depth, the sampling line is lifted to allow the sampling tube to fill with liquid. When the sampling line is released, the center rod drops to reseal the sampling tube. The sampler is then retrieved and liquid is transferred into a sample bottle by placing the bottle beneath the center rod, and lifting up on the sampling line. Bomb samplers used to collect liquid samples for chemical or radiological analysis should be made of Teflon or stainless steel.

The effective sampling depth of this sampler is only limited by the length of the sampling line. Since this method is used primarily to collect deep liquid samples, a raft or boat may be required to assist the sampling procedure. When samples are to be collected from depths of 5 ft or less, the extendable tube sampler is recommended over the bomb sampler, since it is easier to use.

The bomb sampler is effective in collecting deep grab and areal composite liquid samples, as well as integrated samples. A depth composite sample is acquired by compositing several grab samples, each representing a different depth in the liquid column. An areal composite sample is collected by compositing liquid samples from different locations, whereas an integrated sample is acquired by collecting a portion of a sample from the same location, several times over a selected time period.

For most sampling programs, five people are sufficient for this sampling procedure. Two are needed for field testing, sample collection, labeling, and documentation; a third is needed for health and safety; a fourth is needed for waste management and equipment decontamination; and a fifth may be needed to operate the boat or maneuver the raft.

The following equipment and procedure can be used to collect liquid samples for chemical and/or radiological analysis.

Equipment
1. Teflon or stainless steel bomb sampler
2. Sampling line
3. Sample bottles
4. Sample preservatives
5. pH, temperature, and conductivity meters
6. Sample labels
7. Cooler packed with Blue Ice (Blue Ice is only required for nonradiological analyses)
8. Trip blank and coolant blank (only required for volatile organic and nonradiological analyses, respectively)
9. Sample logbook
10. Chain-of-custody forms
11. Chain-of-custody seals
12. Permanent ink marker
13. Health and safety and radiological scanning instruments
14. Health and safety clothing (including life jackets)
15. Waders, raft, or boat
16. Waste container
17. Sampling table
18. Plastic waste bags

Sampling Procedure
1. In preparation for sampling, confirm that all necessary preparatory work has been completed, including obtaining property access agreements; meeting health and safety, decontamination, and waste disposal requirements; and checking the calibration of all radiological and health and safety scanning instruments.
2. Prior to sample collection, fill a clean glass jar with liquid from the desired sampling depth, and measure the pH, temperature, and conductivity. Record this information in a sample logbook.
3. When properly positioned over the sampling point, slowly lower the bomb sampler to the desired sampling depth.
4. Lift up on the sampling line and allow the sampling tube to fill with liquid. When the sampler is full, release the sampling line to reseal.

5. Retrieve the sampler, and transfer the liquid into a sample bottle. Scan sample using chemical and/or radiological scanning instruments, and record results in a bound logbook.
6. If the analyses to be performed require the sample to be preserved, do this prior to filling the sample bottle.
7. After the bottle is capped, attach a sample label and custody seal to the bottle and immediately place it into a sample cooler or radiologically shielded container. Samples for chemical analysis should be packed in Blue Ice.
8. See Chapter 5 for details on preparing sample bottles and coolers for sample shipment.
9. Transfer any liquid left over from the sampling into a waste container. Prior to leaving the site, all waste containers should be sealed, labeled, and handled appropriately.
10. Have a professional surveyor survey the coordinates of the sampling point to preserve the exact sampling location.

4.3.3.7 Groundwater Sampling

This section provides the reader with guidance on selecting the most appropriate groundwater sampling method for the site under investigation. The criteria used to select the most appropriate method include the analyses to be performed on the sample, the type of sample to be collected (grab, composite, or integrated), and the sampling depth. SOPs have been provided for each of the recommended sampling methods to facilitate implementation. The following groundwater characterization strategies are provided as a supplement to general guidance provided in Section 4.1.1.5.

Whenever contamination is identified in soil, there is always the possibility of contaminants migrating to groundwater. This migration is possible through the transport mechanism of water percolating through the soil, while the rate of migration is controlled by soil physical properties, such as pore size, and geochemical properties, such as distribution coefficient (Kd). Once contaminants reach the groundwater, they commonly disperse into the saturated formation. Depending on their physical and/or chemical properties, contaminants can concentrate near the top or bottom of the aquifer, or may evenly distribute themselves throughout the aquifer. Light, nonaqueous-phase liquids (LNAPLs), such as benzene, toluene, and xylene, have specific gravities less than water and therefore concentrate near the top of the aquifer. Similarly, dissolved radionuclides and metals tend to show the highest concentration near the top of the aquifer as well, since this is closest to the source of contamination. Dense, nonaqueous-phase liquids (DNAPLs), such as PCE, TCE, and vinyl chloride, have specific gravities greater than water and therefore concentrate near the base of the aquifer.

When assessing the groundwater conditions at a site, serious consideration should be given to using the direct push method in combination with a mobile laboratory to perform a preliminary groundwater assessment prior to installing permanent monitoring wells (see Section 4.3.3.7.1). The advantages of the direct push method are that groundwater samples can be collected quickly and inexpensively when compared with collecting the same data through monitoring well installation and sampling. Since a mobile laboratory can typically provide analytical results within hours after sampling, the direct push method facilitates the "observational approach," where the results from samples analyzed in the field are used to guide the characterization effort. Once preliminary groundwater characterization is complete, groundwater wells can be precisely positioned for long-term monitoring, aquifer testing, and/or remediation.

If for some reason the direct push method is not selected to perform preliminary groundwater characterization, one groundwater monitoring well should be installed downgradient from each suspected source of contamination, and one well upgradient (Figure 4.82). These wells should be built to screen the first encountered water-bearing unit. When the analytical results are obtained from samples collected from these wells, any contaminants identified in the downgradient samples that are not also found in the upgradient sample are contaminants derived from the site.

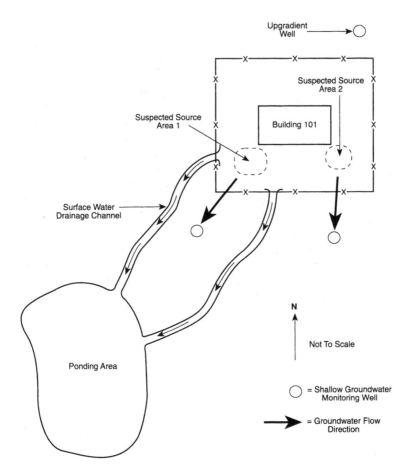

Figure 4.82 Common initial groundwater well configuration when the general groundwater flow direction is known. (From Byrnes, M.E., *Field Sampling Methods for Remedial Investigations*, Lewis Publishers, Boca Raton, FL, 1994. With permission.)

Although shallow groundwater flow contours often mimic topographic contours, this is not always the case. When groundwater flow direction at a site is unknown, a minimum of three monitoring wells or piezometers should be positioned to form a triangle around the site, one of which should be positioned topographically downslope from the largest suspected source of contamination (Figure 4.83). This arrangement allows for an accurate determination of groundwater flow direction and gradient. After groundwater flow direction has been determined, one well should be installed downgradient from each suspected contaminant source. The wells should then be developed, purged, and sampled. These samples should be analyzed for all of the contaminants of concern.

If groundwater contamination is identified in the upper aquifer, it is important to define the source, and lateral and vertical extent of contaminant migration. This is most effectively accomplished by using the direct push method to characterize the upper and lower aquifer thoroughly, followed by the installation of long-term monitoring wells. The next section provides examples of how this type of investigation is performed.

4.3.3.7.1   Direct Push Method

The direct push method utilizes a hydraulic press and slide hammer mounted on the rear end of a truck, to advance a sampling probe to a depth where a groundwater

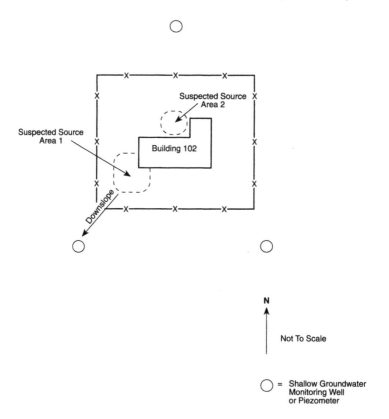

Figure 4.83 Common initial groundwater well configuration when the general groundwater flow direction is unknown. (From Byrnes, M.E., *Field Sampling Methods for Remedial Investigations*, Lewis Publishers, Boca Raton, FL, 1994. With permission.)

sample can be collected (Figures 4.84 and 4.85). Three sizes of trucks are available to perform this procedure. The size of the truck required is dependent on the depth of groundwater at the site. A lightweight van or truck is generally able to provide a reaction weight of 1000 to 2000 lb, which can advance a sampling probe 10 to 20 ft below the ground surface. The next larger truck provides a reaction weight of 3000 to 5000 lb, and is designed to sample as deep as 50 ft. The heaviest vehicle is the cone penetrometer which has a reaction weight of 10 to 30 tons, and has been successful at penetrating through as much as 200 ft of unconsolidated sediment.

Figure 4.84 AMS PowerProbe™ Direct Push Sampler.

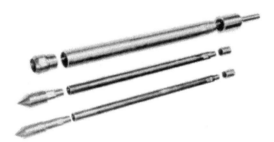

Figure 4.85 Example of Direct Push Sample Probes.

The primary advantages of using the direct push method to assist a groundwater characterization study include:

- There is very little investigation-derived waste generated with the procedure.
- Samples can be collected quickly.
- The procedure is much less expensive than collecting the same data by installing and sampling monitoring wells.
- The equipment gathers less public attention than a drill rig.
- The procedure produces little disturbance to the surrounding environment.

Until recently, groundwater characterization has been performed by drilling and installing numerous groundwater monitoring wells in and around areas suspected of being contaminated. Since drilling procedures generate larger volumes of waste soil, which must be drummed, stored, and ultimately disposed of, the less drilling that is required, the better. Although the need to install groundwater wells has not gone away, the direct push method can assist a groundwater investigation by selecting the optimum location for fewer wells.

The primary limitations of the direct push method include the sampling depth, volume of sample that can be retrieved, and difficulties in penetrating through soils that contain gravel. Groundwater samples can be collected one of five different ways when using the direct push method:

1. Lowering a <0.5-in.-diameter bailer down the inside of the probe multiple times to collect the volume of sample needed. This technique works well when only small volumes of water are needed.
2. Lowering a weighted sample vial under vacuum down the inside of the probe. A needle inside the probe punctures the septum and allows water to flow into the vial. This method works well for collecting small volumes of water.
3. A third method utilizes chambers in the probe which can be filled at depth, then brought to the surface. In most cases, the capacity of the chambers does not exceed 500 mL. To obtain larger volumes of water with this technique, the sampler must be advanced and retrieved repeatedly.
4. Lowering a sample tube down the inside of the probe and using a suction-lift pump to extract as much water as needed. This is the least preferred of all the available methods since it is not effective in collecting samples deeper than 25 ft, and is reported by the EPA to cause the volatilization of the sample and, possibly, to affect the pH.
5. A fifth method can be used to collect samples from formations with very low permeability. This technique involves running a screened tube down the inside of the sampling hole, removing the steel rods, and packing sand around the screened section. This mini-monitoring well can be left in place as long as is required for water to fill the hole. Tests have shown that analytical results from water samples collected from the temporary wells compare favorably with data from conventional wells, while the temporary wells are a fraction of the cost. In addition to installing mini-monitoring wells, the direct push method can also be used to install temporary piezometers for water level monitoring.

If groundwater contaminants at a site include both LNAPLs and DNAPLs, it is recommended that the groundwater samples be collected from both the top and bottom of the aquifer, since this is where these contaminants will concentrate. To characterize the distribution of contaminants more completely, samples can be collected at regular intervals throughout the depth of an aquifer. This type of depth interval sampling is commonly performed at locations close to the suspected contaminant source(s), and the results are used to determine the most appropriate sampling interval for more distant sampling points.

If contamination is identified in the upper aquifer, the direct push method can be used to collect samples from the next deeper aquifer, assuming it is within the depth penetration range of the sampler. The sampling intervals within the lower aquifer should be selected based on the types of contaminants identified in the upper aquifer.

To use the direct push method most cost-effectively, groundwater investigations should be performed in combination with a mobile laboratory and should be performed using the "observational approach." The observational approach utilizes the analytical data from each sampling point to decide where to position additional sampling points. This method is only possible when using a mobile laboratory that can provide analytical results shortly after sampling. This approach avoids the problem of collecting unnecessary or insufficient data.

An example of a successful groundwater investigation using the observational approach is illustrated in Figure 4.86. In this example, historical information led investigators to believe that buried tanks located south and west of Building 101 were potential sources of groundwater contamination at the site. The first step in this investigation involved collecting an initial row of groundwater samples in a "V" pattern, just inside the site property boundary, with the "V" pointing in the downgradient direction. Collecting initial groundwater samples from these locations will assure that any groundwater contamination leaving the site will be detected by one or more of these sampling points. Based on the results from this first phase of sampling, the observational approach is used to track the extent of the contaminant plume. This approach involves sampling outward from a contaminated sampling point in a grid pattern until the edge of the contaminant plume is defined.

Once the plume has been defined, monitoring wells should be installed for long-term monitoring purposes (see Figure 4.86). These wells are commonly positioned near the source and downgradient from the leading edge of the contaminant plume(s) to track the long-term migration of the contamination. Shallow and deep well pairs are recommended to track both the vertical and horizontal migration of contaminants. The number of wells required is based on the size of the contaminant plume(s) identified. If groundwater remediation is later determined to be necessary, additional wells may be needed near the source of contamination.

4.3.3.7.2 Monitoring Wells

The primary objective of installing monitoring wells is to provide an access point where groundwater samples can be repeatedly collected, and groundwater elevations measured. When installing monitoring wells, it is important to minimize the disturbance to the surrounding formation, and to construct a well from materials that will not interfere with the chemistry of the groundwater.

The primary components of a groundwater monitoring well are the well screen, sump, riser pipe, well cap, protective steel casing, and lock (Figure 4.87). The well screen is by far the most critical component of a well. A well screen must have slots that are large enough to allow groundwater and contaminants to flow freely into a well, yet small enough to prevent formation soils from entering the well. The most common lengths of screen used to construct monitoring wells are 2, 5, 10, 15, and 20 ft. However, the EPA generally discourages the use of well screens larger than 10 ft since larger sampling intervals tend to dilute the sample. Another potential problem with using long screen lengths is that two aquifers can unintentionally be screened in the same well. Such an error provides a conduit for cross-contamination between aquifers.

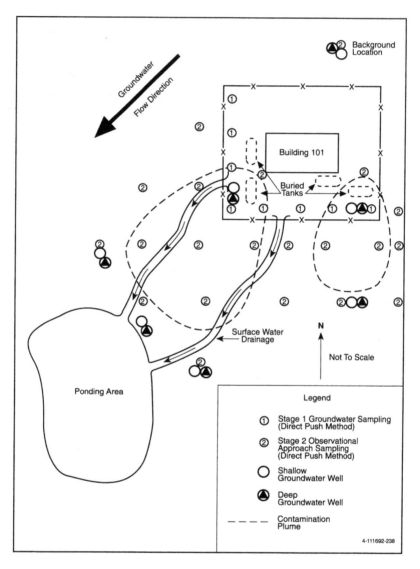

Figure 4.86 Example of observational approach using direct push method. (From Byrnes, M.E., *Field Sampling Methods for Remedial Investigations*, Lewis Publishers, Boca Raton, FL, 1994. With permission.)

When setting wells in an unconfined aquifer where dissolved radionuclides, metals, and LNAPLs may be present, the screen should be set so that it spans the vadose zone and the upper portion of the aquifer since this is where these contaminants will be found in their highest concentrations. For example, a 10-ft well screen should be set so that approximately 8 ft of the screen are below the mean static water level, and 2 ft are above. By constructing the well in this manner, any floating hydrocarbons that are present in the formation will be visible in the well, even with

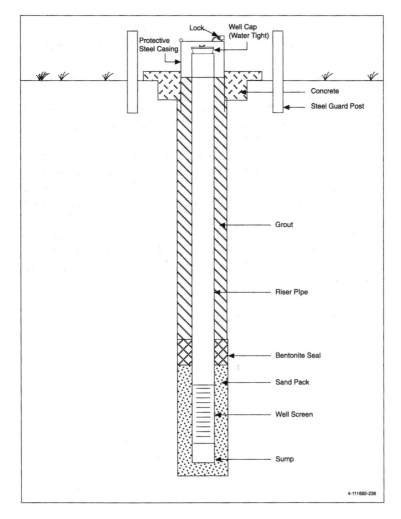

Figure 4.87 Primary components of a groundwater monitoring well. (From Byrnes, M.E., *Field Sampling Methods for Remedial Investigations*, Lewis Publishers, Boca Raton, FL, 1994. With permission.)

seasonal fluctuations in groundwater levels. Similarly, when using a well screen that is 15 or 20 ft in length, 3 and 4 ft of the screen should be set above the mean static water level, respectively. Screen lengths of 2 and 5 ft are used to sample specific intervals within the aquifer and are not designed to screen the static water level.

When the contaminants of concern are DNAPLs (e.g., TCE, PCE, vinyl chloride), the bottom of the well screen should be set at the bottom of the aquifer (Figure 4.88). In a relatively thick aquifer where discrete layering of contamination is suspected to occur, several casings can be nested together in one large-diameter well boring to allow the screening of several depth intervals (Figure 4.89).

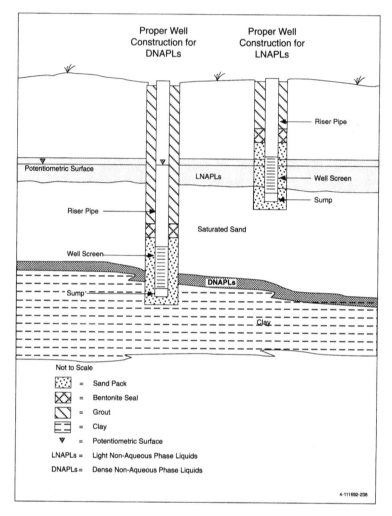

Figure 4.88 Common well construction for an unconfined aquifer when contaminants include DNAPLs and LNAPLs. (From Byrnes, M.E., *Field Sampling Methods for Remedial Investigations*, Lewis Publishers, Boca Raton, FL, 1994. With permission.)

When setting a well in a confined aquifer where the contaminants of concern are expected to be dissolved in the water, or floating hydrocarbons, the top of the screened interval is positioned at the top of the aquifer (Figure 4.90). Since confined aquifers experience overburden pressures, the static water level often rises well above the top of the aquifer. Consequently, these wells are very seldom built to screen the static water level. If the contaminants of concern have specific gravities greater than water, the bottom of the well screen is set at the bottom of the confined aquifer.

Well screens for use in environmental sampling are available in stainless steel, Teflon, and polyvinylchloride (PVC). Of these three materials, stainless steel is preferred since it is relatively inert, durable, and of moderate cost. In contrast, Teflon is a very inert substance; however, it is very expensive, and since it is not as strong

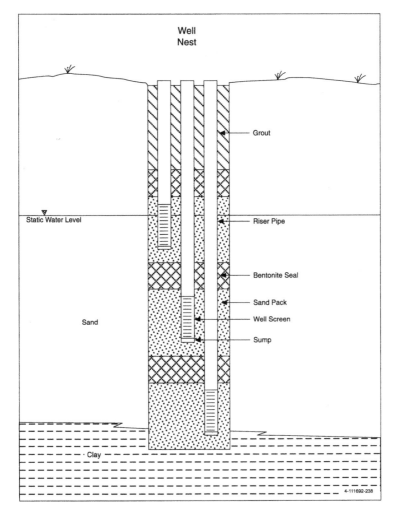

Figure 4.89 Groundwater monitoring well nest. (From Byrnes, M.E., *Field Sampling Methods for Remedial Investigations*, Lewis Publishers, Boca Raton, FL, 1994. With permission.)

as stainless steel it can be easily damaged during well installation. PVC is relatively inexpensive; however, studies have shown that it can release and absorb trace amounts of various organic constituents after prolonged exposure. The most commonly used screen slot sizes range from 0.01 to 0.03 in. for silty to coarse sandy formations, respectively. However, screens can be special-ordered with slots as small as 0.006 in., and as large as needed. To determine the appropriate screen slot size for use in a particular formation, a sieve analysis must be performed either in the field or in a laboratory to determine the grain size distribution of the formation. With this information, slot size calculations can be made using procedures outlined in Driscoll (1986).

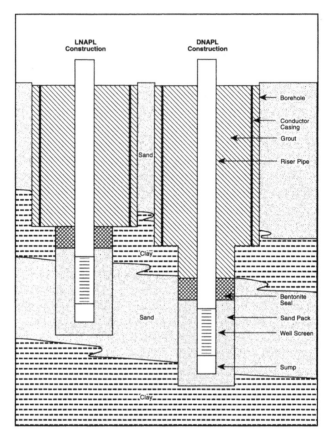

Figure 4.90 Common well construction for a confined aquifer when contaminants include DNAPLs and LNAPLs. (From Byrnes, M.E., *Field Sampling Methods for Remedial Investigations*, Lewis Publishers, Boca Raton, FL, 1994. With permission.)

A sump is threaded to the bottom of a well screen for the purpose of catching any fine-grained soil that enters the well through the screen. If a sump is not used, soil accumulation will occur within the well screen and will eventually plug up the well. Common well sumps range in lengths from 0.5 to 3 ft, and should be made of the same material as the well casing. Sumps must be cleaned out on a regular basis to assure that the screen remains clear.

A threaded riser pipe is attached to the well screen to complete the well to the ground surface. The riser pipe is cut near the ground surface to allow for either an aboveground or belowground completion. For an aboveground completion, the riser pipe is typically cut 1.5 to 2.0 ft above the ground surface. A locking protective steel casing is then grouted in place over the riser pipe. A concrete pad and guard posts are often positioned around a well for added protection (see Figure 4.87).

For a belowground completion, the riser pipe is cut several inches below the ground surface. A protective steel casing is then grouted in place over the riser pipe and completed flush with the ground surface (Figure 4.91). A watertight well cap

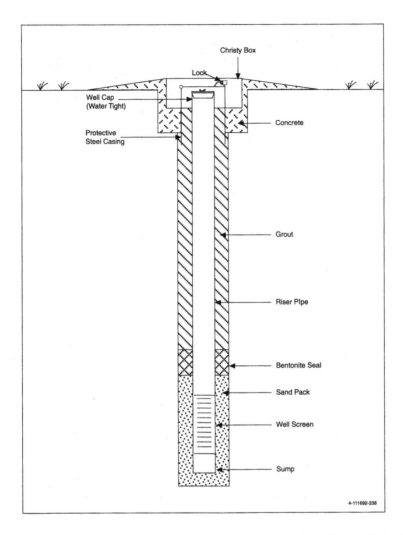

Figure 4.91 Example of a below-ground monitoring well completion. (From Byrnes, M.E., *Field Sampling Methods for Remedial Investigations*, Lewis Publishers, Boca Raton, FL, 1994. With permission.)

should always be used in a belowground completion to prevent surface water from entering the well.

A sand pack is built around the well screen to a level 2 to 3 ft above the top of the screen. The purpose of the sand pack is to reduce the amount of fine-grained formation soils entering the well screen. The sand pack is built above the top of the screen to allow for the potential settling of the sand pack over time. The grain size and size distribution of the sand pack used to build the well should be calculated using procedures outlined in Driscoll (1986), and should be compatible with the screen slot size. A 2- to 3-ft bentonite seal should be built over the sand pack using

246 SAMPLING AND SURVEYING RADIOLOGICAL ENVIRONMENTS

bentonite pellets. The remainder of the borehole should be filled with grout composed of Portland type I/II cement mixed with approximately 4 to 5% bentonite powder.

4.3.3.7.2.1 Well Development — Prior to collecting a groundwater sample from a well, it must first be properly developed and then purged. When a well is installed using an auger rig, it is not uncommon for clayey soils to smear along the walls of the borehole. Similarly, when the mud rotary technique is used, a mudcake often develops along the walls of the borehole. The development procedure restores the natural hydraulic conductivity and geochemical equilibrium of the aquifer near the well so that representative samples of formation water can be collected. In the development procedure, a well is surged and pumped long enough to clean out the well, and until the water stabilizes in pH, temperature, conductivity, turbidity, and clarity. Other measurements that should be considered when developing a well are dissolved oxygen and oxidation reduction potential.

Surging is the first step in the development procedure which involves lowering a surge block down the well using a wire line from a drill rig or development truck. The surge block is repeatedly raised and lowered over 2- to 3-ft intervals starting at the top of the screened interval and moving downward. This procedure pulls soil particles that are finer than the well screen into the well so that they can later be removed. This procedure helps to compact the sand pack around the well screen, and should be performed several times during the development operation to assure that most of the fine-grained soil particles have been removed from the immediate vicinity of the well screen. Following each surging step, a submersible pump and/or bailer is used to remove turbid water and sediment from the well.

A submersible pump is used to develop wells that can yield water at rates greater than several gallons per minute. In low-yielding wells, a large bottom-filling bailer is commonly used. When a submersible pump is used, the intake is first set near the bottom of the well screen, then near the center, and finally near the top. The pump is left at each interval until the water has cleared, and the pH, temperature, conductivity, and turbidity of the water have stabilized. For the most effective development, several different pumping rates are used at each development interval. The pumping rate for a submersible pump can be controlled by placing a restriction valve on the end of the discharge line, or by using a variable-speed pump.

For poorly producing aquifers, a bottom-filling bailer is used for well development. Since the bailer method does not work as well as the submersible pump at pulling fine-grained soil from the surrounding formation, the surging procedure is that much more important when developing these wells. After surging, the bailer is lowered to the bottom of the well using a wire line from a drill rig or development truck. When the bailer is full, it is retrieved to the ground surface, and the water is transferred into drums or a water holding tank. The bottom-filling bailer not only removes groundwater from the well, but it also works effectively in removing any silt that has accumulated in the sump. In poorly producing aquifers, it is tempting to want to add water to the well to assist the procedure; however, this should be avoided whenever possible because it can alter the groundwater chemistry. If a well will not develop without adding water, a field blank should be taken of the water added to the well, and it should be analyzed for the same parameters that the groundwater sample is to be analyzed for. Development methods using air should be avoided since they have the potential to alter the groundwater chemistry, and can damage the integrity of the well.

At regular intervals during the development procedure, a sample of the development water should be collected in a glass jar for pH, temperature, conductivity, and turbidity measurements. The results from these measurements are recorded on a Well Development Form (Chapter 5). A well is considered adequately developed when a minimum of three borehole volumes of water have been removed from the well, and the pH (±0.1), temperature (±0.5°C), conductivity (±10%), and clarity have stabilized, and turbidity measurements fall below 5 NTUs. A borehole volume is calculated using the formula:

$$V = \pi r^2 l$$

where

V = volume
π = 3.14
r = radius of the borehole
l = thickness of the water column

In clayey formations, it is not always possible to meet the <5-NTU turbidity requirement. In this situation, development should continue until the turbidity of the water has stabilized. If the above physical parameters stabilize quickly, a minimum of three borehole volumes of water must be removed to consider the well adequately developed. In fine-grained formations it is not uncommon for it to take five to ten borehole volumes for all the physical parameters to stabilize.

Immediately after the completion of well development, the depth to groundwater should be measured using a water level probe, and recorded. This information can later be used to estimate the transmissivity of a formation.

If a well is built using the appropriate screen slot size and sand pack, very little siltation should occur over time. In this instance, future sampling only requires a well to be purged prior to sampling. However, if fine-grained sediment begins building up inside a well sump and screen over time, the well will need to be redeveloped.

4.3.3.7.2.2 Well Purging — After development, a well should be allowed to set for several days, prior to purging and sampling. The well-purging procedure is identical to that for well development with the exception that surging is not performed. The objective of the purging procedure is to remove stagnant water from the well so that a representative water sample can be collected. The purging procedure should remove water throughout the screened interval to ensure that fresh formation water has replaced all stagnant water in the well. However, it is recommended that low pumping rates be used when purging to reduce the size of the cone of depression.

At regular intervals during the purging procedure, a sample of the purge water is collected in a glass jar for pH, temperature, conductivity, and turbidity measurements.

The results from these measurements are recorded on a Well Purging Form (Chapter 5). Other measurements that should be considered when purging a well include dissolved oxygen, and oxidation reduction potential. In the purging procedure, water should be removed from a well until the pH (±0.1), temperature (±0.5°C), conductivity (±10%), turbidity (<5 NTUs), and clarity have stabilized.

It is important not to overpurge a well, since this may cause dilution or concentration of contaminants where stratification or leachates occur. High purge rates should also be avoided, since purging a well dry causes formation water to cascade into the well, allowing volatilization to occur. Purge rates should also never exceed development rates since this may draw new sediment into the well.

Immediately after the completion of well purging, the depth to groundwater should be measured using a water level probe, and recorded on the Well Purging Form. This information can later be used to estimate the transmissivity of the formation. As a general rule, one should attempt to sample a well before the static water level has had time to equilibrate. Wells that are slow to recover should be allowed to set only long enough for a sufficient sample volume of water to enter the well.

For very low yielding wells (< 1 gpm), the above purging procedure is not practical since it would take days to complete. Rather, these wells should be bailed, or pumped dry twice, then sampled before groundwater equilibration.

4.3.3.7.2.3 Well Sampling — When collecting groundwater samples, bottles for volatile organic analysis should always be the first to be filled. This is critical since volatile organics are continuously being lost to the atmosphere during the sampling procedure. The remaining sample bottles should be filled in an order consistent with their relative importance to the sampling program. If the analyses to be performed require preservation, preservatives should be added to sample bottles prior to sample collection.

Shortly after a water sample has been collected from a well, it is common practice to take a final water level reading, and collect a final sample for pH, temperature, conductivity, and turbidity readings.

Collecting filtered and unfiltered groundwater samples for metals and/or radiological analysis should be considered, particularly for wells that remain cloudy or turbid throughout the development and purging procedure. Collecting samples in this way allows one to identify the dissolved concentration of metals and radionuclides. A pressurized filtration system utilizing a 0.45 μm millipore membrane filter should be used when filtering samples.

The most effective sampling tools for collecting groundwater samples include the bailer, Kabis sampler, bomb sampler, bladder pump, and submersible pump. The bailer is the least complicated of all the sampling tools and is commonly used to sample shallow wells. This tool is lowered down a well using a monofilament line. When the bailer is retrieved, the water is transferred into sample bottles. Similar to the bailer, the Kabis sampler and bomb sampler are lowered down the well using a monofilament line. These samplers are used to collect water or product samples from specific intervals in the water column. The bladder pump contains a Teflon bladder which fills and ejects water through a Teflon discharge line with the assistance of a compressed gas source such as a nitrogen bottle or compressor. Submersible pumps are powered by an electric motor. This motor rotates impellers, which force water up the discharge line.

Other groundwater sampling methods that are available but are not recommended include the suction-lift, and air-lift methods. These sampling procedures have the tendency to cause the oxidation and degassing of the samples being collected. Other problems associated with these techniques are that the suction-lift method has a tendency to affect the sample ph, and the air-lift method can damage the integrity of the sand pack around the well screen if high pressures are used.

Table 4.21 summarizes the effectiveness of each of the five recommended procedures. A number "1" in the table indicates that a particular procedure is most effective in collecting samples for a particular laboratory analysis, sample type, or sample depth. A number "2" indicates that the procedure is acceptable but less effective, while an empty cell indicates that the procedure is not recommended. For example, Table 4.21 indicates that the bailer, Kabis sampler, bomb sampler, and bladder pump are all effective methods for collecting samples for volatile organic analysis. Although the submersible pump method is an acceptable method for volatile organic analysis, it is less effective than the other methods.

Table 4.21 Rating Table for Groundwater Sampling Methods

	Laboratory Analyses							Sample Type		Depth	
	Radionuclides	Volatiles	Semivolatiles	Metals	Pesticides	PCBs	TPH	Grab	Composite (Vertical)	Integrated	Shallow (0.0–30 ft)	Deep (>30 ft)
Bailer	1	2	1	1	1	1	1	1		2	1	2
Kabis sampler	1	1	1	1	1	1	1	1	1	2	1	2
Bomb sampler	1	2	1	1	1	1	1	1	1	2	1	2
Bladder pump	1	1	1	1	1	1	1	1	2	1	1	1
Submersible pump	1	2	1	1	1	1	1	1	2	1	1	1

Note: 1 = preferred method; 2 = acceptable method; empty cell = method not recommended.

Whichever sampling method is selected, sample bottles for volatile organic analysis should always be the first to be filled. Bottles for radiological and other analyses should be filled consistent with their relative importance to the sampling program.

4.3.3.7.2.3.1 Bailer method — The bailer method is the simplest of all the groundwater sampling methods. A standard bailer is composed of a bailer body, which is available in various lengths and diameters, a pouring spout, and a bottom check valve, which contains a check ball (see Figure 4.79). As the bailer is lowered into groundwater, water flows into the sampler through the bottom check valve.

When the sampler is retrieved, the check ball seals the bottom of the bailer, which prevents water from escaping. Water is poured from the bailer through the pouring spout into sample bottles. Bailers used to collect water samples for chemical or radiological analysis should be made of Teflon or stainless steel.

Some common modifications to the bailer include the use of extension couples to increase the length of the bailer, and the use of a controlled flow bottom assembly. The bottom assembly allows the bailer to be emptied through the bottom of the sampler, which reduces the opportunity for volatilization to occur.

Some of the major advantages of the bailer are that it is easy to operate, portable, available in many sizes, and relatively inexpensive. The disadvantages are that the sampling procedure is labor-intensive, and aeration of the sample can be a problem when transferring water into sample bottles.

The bailer should be lowered down a well using a monofilament line, such as common fishing line. The selected line should be cut to a length long enough to reach the groundwater, and it must be strong enough to lift the weight of the bailer when it is full of water. The line should be decontaminated in the same manner as the sampling bailer (see Chapter 9). The two most effective means of lowering a bailer and sampling line down a well are using the hand-over-hand or tripod and reel methods.

To implement the hand-over-hand method, one end of the sampling line is tied to the top of the bailer, and the other end to the sampler's wrist. With the sampler's arms fully extended horizontally, the slack in the line is removed by winding it between the sampler's thumbs. To lower the bailer down the well, the sampler simply allows the line to unwind. To retrieve the bailer, the line is rewound. This method works very effectively for sampling wells that are less than 30 ft in depth, and when a small 2-in.-diameter bailer is used.

To implement the tripod and reel method, one end of the sampling line is tied to the top of the bailer, and the other end is tied to a reel. To lower the bailer down the well, the line is allowed to unwind from the reel. The handle on the reel is then used to rewind the line when the bailer is retrieved. This method is most effective in collecting samples from a depth less than 30 ft, and can handle a bailer as large as 4 in. in diameter.

For most sampling programs, four people are sufficient for the purging and sampling procedure. Two are needed for field testing, sample collection, labeling, and documentation; a third is needed for health and safety; and a fourth is needed for miscellaneous tasks such as managing wastewater containers, and equipment decontamination.

The following equipment and procedure can be used to collect groundwater samples for chemical and/or radiological analysis.

Equipment
1. Teflon or stainless steel bailer
2. Monofilament line
3. Tripod and reel (not needed if the hand-over-hand method is selected)
4. Water level probe
5. Sample bottles

6. Sample preservatives
7. pH, temperature, conductivity, and turbidity meters
8. Wide-mouth glass jar
9. Sample labels
10. Cooler packed with Blue Ice (Blue Ice is only required for nonradiological analyses)
11. Trip blank and coolant blank (only required for volatile organic and nonradiological analyses, respectively)
12. Sample logbook
13. Chain-of-custody forms
14. Chain-of-custody seals
15. Permanent ink marker
16. Health and safety and radiological scanning instruments
17. Health and safety clothing
18. Sampling table
19. Waste container
20. Plastic waste bags

Sampling Procedure
1. In preparation for sampling, confirm that all the necessary preparatory work has been completed, including obtaining property access agreements; meeting health and safety, decontamination, and waste disposal requirements; and checking the calibration of all radiological and health and safety scanning instruments.
2. Prior to sampling, a groundwater well must be properly developed and purged (see Sections 4.3.3.7.2.1 and 4.3.3.7.2.2). If a well has been previously developed, there is no need to repeat this procedure unless accumulated sediment is blocking the well screen, or the well was inadequately developed the first time. A well must be purged each time it is sampled.
3. When well purging is complete, collect a final water level measurement, and record this information on the Well Purging and Sampling Form (see Chapter 5).
4. Cut a length of monofilament line long enough to reach the water table. Tie one end of the line to the top of the bailer, and the other end to either the sampling reel or sampler's wrist, depending on whether the tripod and reel or hand-over-hand method is used.
5. Lower the bailer quickly down the well to a depth just above the water table. At this point, slowly lower the bailer through the water just deep enough to fill it with water. When the bailer is full, slowly raise it out of the water, then retrieve it quickly to the ground surface. Collecting a groundwater sample in this manner creates little disturbance of the water column, which in turn reduces the loss of volatile organics and minimizes the turbidity of the water sample.
6. Transfer the water from the bailer carefully into the appropriate sample bottle(s). If the analyses to be performed require the sample to be preserved, this should be performed prior to filling the sample bottle.
7. After the bottle is capped, attach a sample label and custody seal to the bottle and immediately place it into a sample cooler or radiologically shielded container. Samples for chemical analysis should be packed in Blue Ice.
8. See Chapter 5 for details on preparing sample bottles and coolers for sample shipment.

9. Fill a glass jar with sample water and collect and record a final pH, temperature, conductivity, and turbidity measurement, in addition to collecting a final water level measurement. Record this information on the Well Purging and Sampling Form.

10. Replace the well cap, lock the well, and containerize any waste in a waste container. Prior to leaving the site, all waste containers should be sealed, labeled, and handled appropriately.

4.3.3.7.2.3.2 Kabis sampler method — The Kabis sampler is manufactured by Sibak Industries and is an effective tool for collecting groundwater samples from discrete depth intervals. This sampler is particularly effective in helping delineate the vertical boundaries of a groundwater contaminant plume. The Kabis sampler is a passive type sampler with no electrical or mechanical components (Figure 4.92). The sampler includes a bullet-shaped stainless steel container. The inside lid of the container is equipped to attach as many as three 40-mL sample bottles, or one 0.25-, 0.5-, or 1.0-L sample bottle (Figure 4.93).

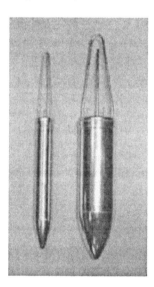

Figure 4.92 Kabis sampler. (Courtesy of Sibak Industries.)

Two thin tubes extend upward from the lid of the sampler to different heights. The difference in height creates a hydraulic gradient through the sampler (the Model-II KABIS Sampler, which accepts three 40-mL sample bottles, has three fill tubes, and one exhaust tube). The orifice diameter of the tubes prohibits sample entry while the sampler is descending through the water column. This occurs because friction due to the surface tension of water prohibits water entry/air loss through the narrow tubes during descent.

Once the sampler reaches the desired sampling depth, water pushes through the shorter tube and spills into the sample bottle while air exits the longer exhaust tube. The internal volume of the stainless steel container is designed to accept three sample

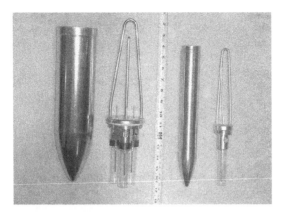

Figure 4.93 Internal components to Kabis sampler. (Courtesy of Sibak Industries.)

container volumes; thus the bottles are triple rinsed according to standard laboratory procedures. When all the air has been replaced by water, the sampler is retrieved. No water from upper levels in the well can displace the water already in the full sampler during retrieval. Once out of the well, the lid of the sampler is removed and the sample bottle is removed. The bottle is capped using a patented conic lid that removes all head space. The water in the stainless steel container can then be used to collect standard field parameters (pH, temperature, conductivity, turbidity). Additional details on the Kabis sampler can be obtained from the Sibak Web page at www.sibak.com.

The following procedure identifies how the Kabis sampler should be used to collect groundwater samples. For most sampling programs, four people are sufficient for implementing this sampling procedure. Two are needed for sample collection, labeling, recording sampling location, and documentation; a third is needed for health and safety; and a fourth is needed for miscellaneous tasks such as managing wastewater containers and sampling equipment.

The following equipment and procedure can be used to collect groundwater samples for chemical and/or radiological analysis.

Equipment
1. Stainless steel Kabis sampler
2. Monofilament line
3. Tripod and reel
4. Water level probe
5. Sample bottles
6. Sample preservatives
7. pH, temperature, conductivity, and turbidity meters
8. Wide-mouth jar
9. Sample labels
10. Cooler packed with Blue Ice (Blue Ice is only required for nonradiological analyses)
11. Trip blank and coolant blank (only required for volatile organic and nonradiological analyses, respectively)

12. Sample logbook
13. Chain-of-custody forms
14. Chain-of-custody seals
15. Permanent ink marker
16. Health and safety and radiological scanning instruments
17. Health and safety clothing
18. Sampling table
19. Waste container
20. Plastic waste bags

Sampling Procedure
1. In preparation for sampling, confirm that all the necessary preparatory work has been completed, including obtaining property access agreements; meeting health and safety, decontamination, and waste disposal requirements; and checking the calibration of all radiological and health and safety scanning instruments.
2. Prior to sampling, a groundwater well must be properly developed (see Section 4.3.3.7.2.1). If a well has been previously developed, there is no need to repeat this procedure unless accumulated sediment is blocking the well screen, or the well was inadequately developed the first time. When using the Kabis sampler, well purging is optional as long as groundwater samples are collected from the screened interval of the well. [*Note*: This is an acceptable practice because the Kabis sampler only begins to fill once it has equilibrated at the sampling depth. Consequently, it passes through any stagnant water that may exist above the well screen.]
3. Collect a water level measurement and record this information on the Well Purging and Sampling Form (see Chapter 5).
4. Unscrew the lid on the sampler and install a sample bottle(s). Screw the lid back onto the sampler until it is snug.
5. Cut a length of monofilament line long enough to reach the well screened interval. Tie one end of the line to the top of the sampler, and the other end to the tripod reel.
6. Lower the sampler steadily down the well to the sampling depth. The lowering rate should not be less than 30 fpm.
7. Once the sampling depth has been reached, allow the sampler to fill for approximately 4 to 6 min. Then retrieve.
8. Unscrew the lid on the sampler and carefully remove the sample bottle(s) and add sample preservatives as required.
9. Cap the sample bottle(s) using a SIBAK ConeCap, attach a sample label and custody seal, then place it into a sample cooler or radiologically shielded container. Samples for chemical analysis should be packed in Blue Ice.
10. See Chapter 5 for details on preparing sample bottles and coolers for sample shipment.
11. Fill a glass jar with sample water and collect and record a pH, temperature, conductivity, and turbidity measurement, in addition to collecting a final water level measurement. Record this information on the Well Purging and Sampling Form.
12. Replace the well cap, lock the well, and containerize any waste in a waste container. Prior to leaving the site all waste containers should be sealed, labeled, and handled appropriately.

4.3.3.7.2.3.3 Bomb sampler method — The bomb sampler method is similar to the Kabis sampler method in that it provides the advantage of being able to collect a grab sample from a specific depth interval. A standard bomb sampler is composed of a sampling tube, center rod, and a support ring (see Figure 4.81). With this method, a support line is used to lower the sampler to the desired sampling depth, while a sampling line is used to open and close the sampler inlet via the center rod. Within the body of the bomb sampler, a spring keeps the center rod in the closed position when lowering the sampler to the desired sampling depth, which prevents water from entering the sampler. When the desired sampling depth is reached, the sampling line is lifted against the pressure of the spring, which allows water to enter the sampling tube. When the sampling line is released, the center rod drops to reseal the sampling tube. After the sampler is retrieved, water is transferred into a sample bottle by placing the bottle beneath the center rod and lifting up on the sampling line. Bomb samplers used to collect water samples for chemical or radiological analysis should be made of Teflon and/or stainless steel.

Some of the major advantages of the bomb sampler are that it is effective for depth interval sampling, easy to operate, portable, available in many sizes, and is relatively inexpensive. The disadvantages are that the sampling procedure is relatively labor-intensive, and aeration of the sample can be a problem when transferring water into sample bottles.

The support and sampling line should be monofilament, such as common fishing line, and should be discarded between wells. The selected line should be cut to a length long enough to reach the desired sampling depth, and it must be strong enough to lift the weight of the sampler when it is full of water. The sampling line should be decontaminated in the same manner as other sampling equipment prior to sampling (see Chapter 9).

The most effective means of lowering a bomb sampler down a well is using the tripod and reel method. To implement this method, one end of the sampling line is tied to the top of the sampler, and the other end is tied to the reel. To lower the sampler down the well, the line is allowed to unwind from the reel. The handle on the reel is then used to rewind the line when retrieving the sampler.

For most sampling programs, four people are sufficient for the purging and sampling procedure. Two are needed for field testing, sample collection, labeling, and documentation; a third is needed for health and safety; and a fourth is needed for miscellaneous tasks such as managing wastewater containers and equipment decontamination.

The following equipment and procedure can be used to collect groundwater samples for chemical and/or radiological analysis.

Equipment
1. Teflon and/or stainless steel bomb sampler
2. Monofilament line
3. Tripod and reel
4. Water level probe
5. Sample bottles
6. Sample preservatives

7. pH, temperature, conductivity, and turbidity meters
8. Wide-mouth glass jar
9. Sample labels
10. Cooler packed with Blue Ice (Blue Ice is only required for nonradiological analyses)
11. Trip blank and coolant blank (only required for volatile organic and nonradiological analyses, respectively)
12. Sample logbook
13. Chain-of-custody forms
14. Chain-of-custody seals
15. Permanent ink marker
16. Health and safety and radiological scanning instruments
17. Health and safety clothing
18. Sampling table
19. Waste container
20. Plastic waste bags

Sampling Procedure

1. In preparation for sampling, confirm that all the necessary preparatory work has been completed, including obtaining property access agreements; meeting health and safety, decontamination, and waste disposal requirements; and checking the calibration of all radiological and health and safety screening instruments.

2. Prior to sampling, a groundwater well must be properly developed and purged (see Sections 4.3.3.7.2.1 and 4.3.3.7.2.2). If a well has been previously developed, there is no need to repeat this procedure unless accumulated sediment is blocking the well screen, or the well was inadequately developed the first time. A well must be purged each time it is sampled.

3. When well purging is complete, collect a final water level measurement, and record this information on the Well Purging and Sampling Form (see Chapter 5).

4. Cut two lengths of monofilament line long enough to reach the sampling interval. Tie one end of one line to the top of the bomb sampler, and the other end to the sampling reel. The second line is tied to the top of the center rod.

5. The bomb sampler can be quickly lowered down the well to a depth just above the water table, then slowly lowered through the water to the sampling interval.

6. When the sampling interval is reached, hold the support line steady while lifting up on the sampling line. When the sampler is full of water, release the sampling line. Using the support line, slowly raise the sampler just above the water surface, then retrieve it quickly to the ground surface. Collecting a groundwater sample in this manner creates little disturbance of the water column, which in turn reduces the loss of volatile organics and minimizes the turbidity of the water sample.

7. Transfer water from the sampler carefully into sample bottles by placing the bottle beneath the center rod and lifting up on the center rod. If analyses to be performed require the sample to be preserved, do so prior to filling the sample bottle.

8. After the bottle is capped, attach a sample label and custody seal to the bottle and immediately place it into a sample cooler or radiologically shielded container. Samples for chemical analysis should be packed in Blue Ice.

9. See Chapter 5 for details on preparing sample bottles and coolers for sample shipment.

10. Fill a glass jar with sample water and collect and record a final pH, temperature, conductivity, and turbidity measurement, in addition to collecting a final water level measurement. Record this information on a Well Purging and Sampling Form.

11. Replace the well cap, lock the well, and containerize any waste in a waste container. Prior to leaving the site, all waste containers should be sealed, labeled, and handled appropriately.

4.3.3.7.2.3.4 Bladder pump method — A bladder pump combined with a source of compressed air, and an electronic controller, can be used as a very effective well purging and sampling tool. The bladder pump is composed of a stainless steel pump body, bottom and top check ball, fill tube, Teflon bladder, outer sleeve, and discharge and air supply line (Figure 4.94). The advantages of this sampling tool are that it can be used effectively as either a dedicated or portable pump. It is a very clean system where the compressed air does not contact the sample water, it is effective at collecting samples as deep as 250 ft, and it is regarded highly by the EPA as an effective tool to collect samples for all parameters including volatile organic compounds. The disadvantages of this method are the cost for the pump and low pumping rate.

This pump operates on an alternating fill and discharge cycle. During the fill cycle, the bottom check ball allows the bladder to fill with water, while the upper check ball prevents any liquid in the discharge line from dropping back into the pump. During the discharge cycle, the bottom check ball seats as compressed gas squeezes the bladder, which in turn forces water up the discharge line. Both the air supply and discharge lines should be made of Teflon or should be Teflon lined.

The bladder pump is powered by a source of compressed air such as a gasoline-powered or electric-powered compressor, or a nitrogen gas cylinder. As a general rule, 0.5 psi is required per foot of well depth, plus 10 psi for pressure that is lost in the pressure line. If a gasoline-powered compressor is used, extreme care must be taken not to contaminate the water sample with gasoline or exhaust derived from the compressor. An electronic controller is used to control the timing of the fill and discharge cycle of the bladder pump. The controller also monitors and controls the air pressure.

A 1.6-in.-diameter bladder pump is able to pump water at a rate between 1.0 and 1.5 gpm at a depth of 25 ft. At a depth of 100 ft the pumping rate typically drops to around 0.5 gpm. These pumps are ideal for use in 2-in.-diameter wells, but can also be used in larger-diameter wells. Bladder pumps are also available in a 3.0-in.-diameter size, which is ideal for larger-diameter wells. This larger pump is able to pump water at a rate between 3 and 4 gpm at a depth of 25 ft, and 1 to 2 gpm at 100 ft.

When the wells at a site are to be sampled regularly for an extended period of time, it is recommended that one bladder pump be installed in each groundwater well. This is an expensive recommendation for any site; however, a significant amount of sampling time and money will be saved in the long run by not having to

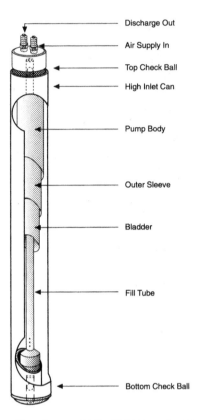

Figure 4.94 Bladder pump. (From Byrnes, M.E., *Field Sampling Methods for Remedial Investigations*, Lewis Publishers, Boca Raton, FL, 1994. With permission.)

decontaminate the pump and discharge line between wells. Also, equipment blank samples are not needed if a dedicated sampling system is used.

For most sampling programs, four people are sufficient for the purging and sampling procedure. Two are needed for field testing, sample collection, labeling, and documentation; a third is needed for health and safety; and a fourth is needed for miscellaneous tasks such as managing wastewater containers, and equipment decontamination.

The following equipment and procedure can be used to collect a groundwater sample for chemical and/or radiological analysis.

Equipment
1. Bladder pump
2. Teflon discharge and air supply line
3. Air compressor or compressed nitrogen gas bottle
4. Electronic controller
5. Water level probe
6. Sample bottles
7. Sample preservatives

8. pH, temperature, conductivity, and turbidity meters
9. Wide-mouth glass jar
10. Sample labels
11. Cooler packed with Blue Ice (Blue Ice is only required for nonradiological analyses)
12. Trip blank and coolant blank (only required for volatile organic and nonradiological analyses, respectively)
13. Sample logbook
14. Chain-of-custody forms
15. Chain-of-custody seals
16. Permanent ink marker
17. Health and safety and radiological scanning instruments
18. Health and safety clothing
19. Sampling table
20. Waste container
21. Plastic waste bags

Sampling Procedure
1. In preparation for sampling, confirm that all the necessary preparatory work has been completed, including obtaining property access agreements; meeting health and safety, decontamination, and waste disposal requirements; and checking the calibration of all radiological and health and safety scanning instruments.
2. Prior to sampling, a groundwater well must be properly developed and purged (see Sections 4.3.3.7.2.1 and 4.3.3.7.2.2). If a well was previously developed as part of an earlier sampling effort, there is no need to repeat this procedure unless accumulated sediment is blocking the well screen, or the well was inadequately developed the first time. A well must be purged each time it is sampled.
3. When well purging is complete, leave the pump running and collect a final water level measurement, and record this information on the Well Purging and Sampling Form (see Chapter 5).
4. Reposition the pump intake to the desired sampling depth and reduce the pumping rate so that a slow but steady flow of water flows from the discharge line.
5. Begin sampling by filling bottles for volatile organic analysis first. Bottles for radiological and other analyses should be filled in an order consistent with their relative importance to the sampling program.
6. If the analyses to be performed require the sample to be preserved, this should be performed prior to filling the sample bottle.
7. After the jar is capped, attach a sample label and custody seal to the bottle and immediately place it into a sample cooler or radiologically shielded container. Samples for chemical analysis should be packed in Blue Ice.
8. See Chapter 5 for details on preparing sample bottles and coolers for sample shipment.
9. Fill a glass jar with sample water and collect and record final pH, temperature, conductivity, and turbidity measurement, in addition to collecting a final water level measurement. Record this information on a Well Purging and Sampling Form.
10. Replace the well cap, lock the well, and containerize any waste in a waste container. Prior to leaving the site, all waste containers should be sealed, labeled, and handled appropriately.

4.3.3.7.2.3.5 Submersible pump method — In the past, submersible pumps have been used primarily for well development and to collect groundwater samples from depths that exceed the limitations of other sampling methods. Currently, more-sophisticated submersible pump systems are designed for purging and sampling. These systems utilize a pump made of all stainless steel and Teflon components, a Teflon discharge line, an electronic flow rate control system, and a power generator (Figure 4.95).

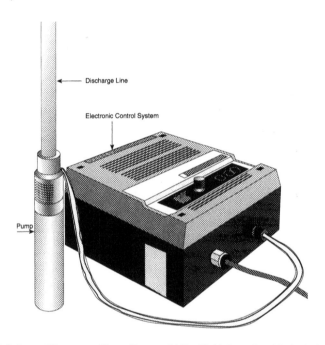

Figure 4.95 Submersible pump. (From Byrnes, M.E., *Field Sampling Methods for Remedial Investigations*, Lewis Publishers, Boca Raton, FL, 1994. With permission.)

The submersible pump utilizes an electric motor which rotates a number of impellers, which in turn force water up the discharge line. These pumps are effective in collecting samples hundreds of feet in depth, and depending upon the size of pump, they are able to pump at rates as low as 100 mL/min to 100 gpm or more. The relatively light weight of the smaller pumps allows the flexibility of using the pump in a dedicated or nondedicated mode.

The advantage of the electronic flow rate control system is that the sampler can control the discharge rate of the pump without throttling the discharge line. Throttling is not desired when sampling, since it facilitates the loss of volatile organics. Other advantages include pumps being available in sizes small enough for use in 2-in.-diameter wells, and the fact that EPA guidance considers this method suitable for collecting groundwater samples for all sampling parameters.

When the monitoring wells at a site are to be sampled regularly, it is recommended that dedicated pumps be installed in each well. This is an expensive recommendation for a large site; however, a significant amount of sampling time and money will be saved in the long run by not having to decontaminate the pump and discharge line between wells, nor will equipment blanks be needed each time the well is sampled.

For most sampling programs, four people are sufficient for the purging and sampling procedure. Two are needed for field testing, sample collection, labeling, and documentation; a third is needed for health and safety; and a fourth is needed for miscellaneous tasks such as managing wastewater drums, and equipment decontamination.

The following equipment and procedure can be used to collect a groundwater sample for chemical and/or radiological analysis.

Equipment
1. Submersible pump
2. Teflon discharge line
3. Generator or electrical power source
4. Electronic flow rate controller
5. Water level probe
6. Sample bottles
7. Sample preservatives
8. pH, temperature, conductivity, and turbidity meters
9. Wide-mouth glass jar
10. Sample labels
11. Cooler packed with Blue Ice (Blue Ice is only required for nonradiological analyses)
12. Trip blank and coolant blank (only required for volatile organic and nonradiological analyses, respectively)
13. Sample logbook
14. Chain-of-custody forms
15. Chain-of-custody seals
16. Permanent ink marker
17. Health and safety and radiological scanning instruments
18. Health and safety clothing
19. Sampling table
20. Waste container
21. Plastic waste bags

Sampling Procedure
1. In preparation for sampling, confirm that all the necessary preparatory work has been completed, including obtaining property access agreements; meeting health and safety, decontamination, and waste disposal requirements; and checking the calibration of all health and safety scanning instruments.
2. Prior to sampling, a groundwater well must be properly developed and purged (see Sections 4.3.3.7.2.1 and 4.3.3.7.2.2). If a well was previously developed as part of an earlier sampling effort, there is no need to repeat this procedure unless accumulated sediment is blocking the well screen, or the well was inadequately developed the first time. A well must be purged each time it is sampled.

3. When well purging is complete, leave the pump running and collect a final water level measurement, and record this information on the Well Purging and Sampling Form (see Chapter 5).
4. Reposition the pump intake to the desired sampling depth, and reduce the pumping rate so that a slow but steady flow of water flows from the discharge line.
5. Begin sampling by filling bottles for volatile organics first. Bottles for radiological and other analyses should be filled in an order consistent with their relative importance to the sampling program.
6. If the analyses to be performed require the sample to be preserved, do so prior to filling the sample bottle.
7. After the bottle is capped, attach a sample label and custody seal to the bottle and immediately place it into a sample cooler or radiologically shielded container. Samples for chemical analysis should be packed in Blue Ice.
8. See Chapter 5 for details on preparing sample bottles and coolers for sample shipment.
9. Fill a glass jar with sample water and collect and record a final pH, temperature, conductivity, and turbidity measurement, in addition to collecting a final water level measurement. Record this information on a Well Purging and Sampling Form.
10. Replace the well cap, lock the well, and containerize any waste in a waste container. Prior to leaving the site, all waste containers should be sealed, labeled, and handled appropriately.

4.3.3.7.2.3.6 Suction-lift method

— The suction-lift method utilizes a surface-mounted pump, and is effective in collecting groundwater samples from wells where the water level is less than 25 ft below the ground surface. This system consists of a vacuum pump, collector flask, and sample collection tube. The method works by using the suction-lift pump to create a vacuum inside of the sampling tube. When the end of the sampling tube is positioned below the water table, air pressure on the water outside of the tube forces water to rise inside the tube. As water rises to the top of the sampling line, it empties into the collector flask. The three most common types of suction-lift systems utilize a centrifugal, peristaltic, or diaphragm pump.

Some of the advantages that the suction-lift method provides are that the equipment is relatively inexpensive, very portable, and it can be used in small-diameter wells. The disadvantages of the method are that sampling is limited to shallow subsurface aquifers less than 25 ft in depth; pumping rates are generally low; and degassing, loss of volatiles, and pH modification are possible.

Since the disadvantages of this method outweigh the advantages, the author does not recommend that this method be used. For shallow aquifers, the bailer or bladder pump method is easier to implement and provides higher-quality samples (see Sections 4.3.3.7.2.3.1 and 4.3.3.7.2.3.4). EPA guidance varies from "do not use this method at all," to "do not use the method to collect samples for volatile organic analysis."

4.3.3.7.2.3.7 Air-lift method

— The air-lift method utilizes a high-pressure hand pump or a small air compressor to force air into a well, which in turn forces water out of the discharge line. This method is not recommended since it introduces air into the sample, which can cause oxidation, gas stripping, and other negative effects on the sample. This method can also damage the integrity of the filter pack around the well screen if high-pressure evacuation is used.

4.3.3.8 Drum and Waste Container Sampling

In an effort to reduce the spread of contamination at a radiological site, soil cuttings, decontamination water, well development and purging water, personal protective equipment, and other types of investigation-derived waste material should be containerized, labeled, and handled as radiological and chemical waste. The following section provides the reader with guidance for collecting soil, sludge, and water samples from waste drums and other containers to support waste disposition.

Waste containers of known or unknown origin must be sampled to determine their content prior to disposition. The only exception to this rule is when there is thoroughly documented process knowledge of the container contents. In this case, the container is designated based on process knowledge and appropriate references are cited. For those containers that require sampling, whatever process knowledge is available is used to assist in identifying which analyses need to be run on the waste samples, and which (if any) RCRA-listed waste codes apply to the waste. The analyses that must be addressed either through process knowledge or sampling and analysis include RCRA characteristics (toxicity, corrosivity, ignitability, and reactivity), applicable RCRA listed waste codes, TSCA PCB and asbestos content, and radiological composition. Radiological analyses should focus on differentiating between different waste classifications (e.g., TRU, high-level waste, low-activity waste).

Much greater health and safety concerns are related to the sampling of containers of unknown origin since the contents may be highly radioactive, toxic, corrosive, reactive, and/or explosive. To address health and safety concerns related to highly radioactive containers, the TRU Drum Monitor (Section 4.2.2.3.2.1), RTR-3 Radiography System (Section 4.2.2.3.2.4), or other similar methods should be considered to survey the radiological content of these containers as they do not require that containers be opened. From a chemical standpoint, it is common practice to pierce the lid of containers of unknown origin with a spike by means of an unmanned mechanical device. By piercing the drum, the volatile organics in the head space will be released, which in turn reduces the chances for explosion. The piercing device and other tools used to open these drums should be made from a "nonsparking" material. Using robotic devices, supplied air, chemical safety suits, or other health and safety precautions is essential whenever sampling these types of containers.

4.3.3.8.1 Soil Sampling from Drum and Waste Container

The methods recommended for collecting soil samples from drums or waste containers for chemical and radiological analyses are the hand auger and slide-hammer methods discussed earlier in Sections 4.3.3.4.1.2 and 4.3.3.4.1.3. [*Note:* These methods are only intended to be used for drums or containers with low radiation levels.] These methods are recommended over other soil sampling methods since they can be used effectively to collect composite soil samples throughout the depth of one or more drums. When more than just a few containers need to be sampled for waste disposition, a statistical sampling approach should be considered.

Refer to the DQO process (Section 4.1.1) to assist in defining the required number of samples.

4.3.3.8.2 Sludge Sampling from Drum and Waste Container

The tool recommended for collecting sludge samples from drums or containers for chemical and radiological analyses is the AMS Sludge Sampler (Figure 4.96). [*Note:* This method is only intended to be used for drums or containers with low radiation levels.] The AMS Sludge Sampler is recommended over other sampling methods since it is designed to prevent the sludge from falling out of the sampler as it is being removed from the drum. When more than just a few containers need to be sampled for waste disposition, a statistical sampling approach should be considered. Refer to the DQO process (Section 4.1.1) to assist in defining the required number of samples. Note that the Coliwasa method described in Section 4.3.3.8.3 may also be used to collect sludge samples from drums or other waste containers. For more information on the AMS Sludge Sampler, see www.ams-samplers.com.

Figure 4.96 AMS Sludge Sampler.

4.3.3.8.3 Liquid Sampling from Drum and Waste Container

The Coliwasa method is effective for collecting grab and composite samples of water or product from drums or other waste containers. This sampler is composed of a vertical sampling tube, piston suction plug, and handle (Figure 4.97). The inlet for the sampler is located at the base of the sample tube. To collect a grab sample, the bottom of the sampler is lowered to the desired sampling depth. The suction plug is then raised to draw the sample into the sampling tube. When the sampler is full, it is removed from the drum and its contents are transferred into a sample bottle. On the other hand, to collect a composite sample, the bottom of the sampler is positioned at more than one sampling depth. At each depth, the suction plug is raised to fill a portion of the sampling tube. When the sampler is full, it is removed from the drum and its contents are transferred into a sample bottle.

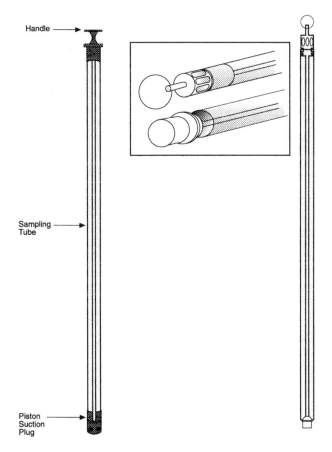

Handle

Sampling
Tube

Piston
Suction
Plug

Figure 4.97 Coliwasa sampler. (From Byrnes, M.E., *Field Sampling Methods for Remedial Investigations*, Lewis Publishers, Boca Raton, FL, 1994. With permission.)

4.4 AIR SAMPLING

The Clean Air Act, discussed in Section 2.1.8, is designed to ensure the quality of the air that we breathe. While air quality sampling is an important component to environmental investigations performed in radiological environments, it is beyond the scope of the book. For guidance on the types of air sampling instrumentation to use, and criteria/methodologies for monitoring ambient air quality, refer to 40 CFR Part 50 (appendices) and 40 CRF Part 58.

4.5 DEFINING BACKGROUND CONDITIONS

When performing soil and/or groundwater remediation, remedial action objectives may require a comparison of radionuclides (naturally occurring) and/or metals concentrations from the site under investigation against background concentrations, as opposed to comparison against a regulatory threshold/standard or a risk-based action level.

Background soil samples should be collected from the same soil formation and approximately the same depths as the soil samples collected from the site. For this reason, background soil samples should preferably be collected from locations within approximately 0.5 miles of the study area. Background soil samples must be collected from locations where the soil has not been disturbed using the same type of sampling equipment used to collect the samples for the remediation site. In an urban environment, background sampling may need to be performed at a local park, or private property (will require written authorization) that has not been developed. Background soil samples should be analyzed for all of the parameters as the samples collected from the remediation site. It is recommended that a minimum of 10 background samples be collected for analysis to assure that a reliable estimate of the population mean and variance can be calculated.

Background groundwater samples should be collected from a groundwater monitoring well located upgradient from the site under investigation and should be distant from other potential contamination sources. This background well must be screening the same aquifer as the wells located downgradient from the site, and the screening should be positioned at the same depth interval within the aquifer as the downgradient wells. Groundwater samples should be collected from the background well at the same frequency as samples collected from the downgradient wells, and should be analyzed for the same contaminants of concern.

4.6 REGULATORY INTERFACE

One of the keys to implementing a successful radiological sampling program is developing a positive team relationship with the involved federal, state, and/or local regulatory agencies. Too often in the past, DQOs and sampling and analysis plans have been developed in isolation, then turned over to the regulators for evaluation once they were complete. The results from this approach often led to major document revisions, which wasted both time and budget.

The process outlined in Section 4.1 is designed to get early regulator involvement before the seven-step DQO process is implemented. Regulator input is first obtained from personal interviews that are intended to identify their specific requirements and concerns related to a particular project. If two regulatory agencies provide conflicting requirements during the interviews, or the specified requirements cannot be met for one reason or another, a Global Issues Meeting is held to resolve these discrepancies. The regulators should then be encouraged to participate in the seven-step DQO process using the workbook provide in Appendix A and the accompanying CD-ROM to keep discussions on track. This approach leads to a lot fewer surprises, lower cost, and a more pleasant working environment.

REFERENCES

Bowen and Bennett, Statistical methods for Nuclear Material Management, Pacific Northwest Laboratory, Prepared for Office of Nuclear Regulatory Research, U.S. Nuclear Regulatory Commission, Washington, D.C., NRC FIN B2420, NUREG/CR-4604, PNL-5849, TK9152.S72, 1988, ISBN 0-87079-588-0.

Byrnes, M.E., Complementary investigative techniques for site assessment with low-level contaminants, *Groundwater Monit. Rev.*, Fall, 1990.

Byrnes, M.E., *Field Sampling Methods for Remedial Investigations*, Lewis Publishers, Boca Raton, FL, 1994.

Davidson, J.R., ELIPGRID-PC Program for Calculating Hot Spot Probabilities, ORNL/TM-12774, Oak Ridge National Laboratory, Oak Ridge, TN, 1994.

Davidson, J.R., Monte Carlo Tests of the ELIPGRID-PC Algorithm, ORNL/TM-12899, Oak Ridge National Laboratory, Oak Ridge, TN, 1995.

DOE (U.S. Department of Energy), Cone Pentrometer, Office of Science and Technology, DOE/EM-0309, April, 1996a.

DOE (U.S. Department of Energy), Field Screening Technology Demonstration Evaluation Report, DOE/OR/21950-1012, FUSRAP, March, 1996b.

DOE (U.S. Department of Energy), Raman Probe, DOE/EM-0442, Office of Science and Technology, July, 1999a.

DOE (U.S. Department of Energy), Near-Infrared Spectroscopy, DOE/EM-0446, Tank Focus Area, Office of Science and Technology, July, 1999b.

DOE (U.S. Department of Energy), Corrosion Probe, DOE/EM-0430, Tank Focus Area, Office of Science and Technology, May, 1999c.

DOE (U.S. Department of Energy), Comparative Testing of Pipeline Slurry Monitors, DOE/EM-0490, Tank Focus Area, Office of Science and Technology, September, 1999d.

DOE (U.S. Department of Energy), Fluidic Sampler, DOE/EM-0485, Tank Focus Area, Office of Science and Technology, September, 1999e.

Driscoll, F.G., *Groundwater and Wells*, Johnson Division, St. Paul, Minnesota, 1986, 438–446.

EPA (Environmental Protection Agency), Interim methods for the Sampling and Analysis of Priority Pollutants in Sediment and Fish Tissue, Cincinnati, OH, ESML, October 1980.

EPA (Environmental Protection Agency), Handbook for Sampling and Sample Preservation of Water and Wastewater, EPA/600/4-82-029, 1982.

EPA (Environmental Protection Agency), Drum Handling Practices at Hazardous Waste Sites, EPA/600/S2-86/013, 1-4, 1986.

EPA (Environmental Protection Agency), A Compendium of Superfund Field Operations methods: Vol. 1, EPA/540/P-87/001a, 1987a.

EPA (Environmental Protection Agency), A Compendium of Superfund Field Operations methods, 540/P-87/001A, Vol. 2, 10-41-10-50, 10-61, 1987b.

EPA (Environmental Protection Agency), RCRA Facility Investigation (RFI) Guidance Volume I, PB89-200299, 7-19-7-23, 1989a.

EPA (Environmental Protection Agency), RCRA Facility Investigation (RFI) Guidance Volume II, PB89-200299, 9-66-9-69, 11-11, 1989b.

EPA (Environmental Protection Agency), RCRA Facility Investigation (RFI) Guidance Volume II and III, PB89-200299, 9-64-9-78, 1989c.

EPA (Environmental Protection Agency), Soil Sampling Quality Assurance User's Guide, EPA/600/8-89/046, 1989d.

EPA (Environmental Protection Agency), Handbook of Suggested Practices for the Design and Installation of Ground-Water Monitoring Wells, EPA/600/4-89/034, 1-123, 1991.

EPA (Environmental Protection Agency), Statistical methods for Evaluating the Attainment of Cleanup Standards, Vol. 3: Reference-Based Standards for Soils and Solid Media, EPA 230-R-94-004, PB 94-176831, December, 1992a.

EPA (Environmental Protection Agency), RCRA Ground-Water Monitoring: Draft Technical Guidance, PB93-139350, 1992b.

EPA (Environmental Protection Agency) Region VIII, Draft Standard Operation Procedures for Field Samplers, Rev. 4, 1992c.

EPA (Environmental Protection Agency), Guidance for the Data Quality Objectives Process, QA/G-4, EPA/600/R-96/055, September, 1994a.

EPA (Environmental Protection Agency), Data Quality Objectives Decision Error Feasibility Trials, DEFT—Version 4.0, U.S. EPA [Quality Assurance Management Staff (QAMS)] Contract #68D10152, Work Assignment 94-2, August, 1994b.

EPA (Environmental Protection Agency), Guidance for Data Quality Assessment, Practical Methods for Data Analysis, QA/G-9, EPA/600/R-96/084, July, 1996.

EPA (Environmental Protection Agency), Multi-Agency Radiation Survey and Site Investigation Manual (MARSSIM), EPA 402-R-97-016, NUREG-1575, December, 1997.

EPA (Environmental Protection Agency), Guidance for Quality Assurance Project Plans, EPA QA/G-5, 1998.

Ford, P.J. et al., Characterization of Hazardous Waste Site—A Methods Manual, Vol. II, Available Sampling Methods, NTIS PB85-168771, EPA 600/4-84-076, Las Vegas, NV, 1984.

Gilbert, R.O., *Statistical Methods for Environmental Pollution Monitoring*, Van Nostrand Reinhold, New York, 1987.

IAEA (International Atomic Energy Agency), Airborne Gamma Ray Spectrometer Surveying, Technical Report Series No. 323, Vienna, Austria, 1991.

Isaaks and Srivastava, *An Introduction to Applied Geostatistics*, Oxford University Press, New York, 1989.

Mason, B.J., Preparation of a Soil Sampling Protocol: Techniques and Strategies, NTIS PB83-206979, EPA, Las Vegas, NV, 1983.

Personal communication with representatives from SAIC, Oak Ridge, TN; Bechtel Hanford, Inc., Richland, WA; Pacific Northwest National Laboratory, Richland, WA; ThermoRetec Inc., Richland, WA; BNFL Instruments Ltd., NM; R.J. Electronics, Turner, OR; and RSI Research Ltd., Sidney, British Columbia, Canada.

Singer, D.A., ELIPGRID, a FORTRAN IV program for calculating the probability of success in locating targets with square, rectangular, and hexagonal grids, Geocom Programs, 4: 1–16, 1972.

Singer, D.A. and F.E. Wickman, *Probability Tables for Locating Elliptical Targets with Square, Rectangular and Hexagonal Point Nets*, Pennsylvania State University, University Park, PA, 1969.

BIBLIOGRAPHY (CHAPTER 4)

Acker, W.L. III, *Basic Procedures for Soil Sampling and Core Drilling*, Acker Drilling Company, Inc., Scranton, PA, 1974, 56–57.

Anderson, M.P., Movement of contaminants in ground water: ground water transport—advection and dispersion; ground-water contamination, in *Studies in Geophysics*, National Academy Press, Washington, D.C., 1984, 179.

Barcelona, M.J., J.P. Gibb, J.A. Helfrich, and E.E. Garske, Practical Guide for Ground-Water Sampling, Illinois State Water Survey, SWS Contract Report 374, Champaign, IL, 1985, 93.

Beck, B.F., A common pitfall in the design of rcra ground-water monitoring programs, *Ground Water*, 21(4), 488–489, 1983.

Brobst, R.D. and P.M. Buszka, The effect of three drilling fluids on ground-water sample chemistry, *Groundwater Monit. Rev.*, 6(1), 62–70, 1986.

Bryden, G.W., W.R. Mabey, and K.M. Robine, Sampling for toxic contaminants in ground water, *Groundwater Monit. Rev.*, 6(2), 67–72, 1986.

Cherry, J.A., R.W. Gillham, and J.F. Barker, Contaminants in ground water: chemical processes, ground-water contamination, in *Studies in Geophysics*, National Academy Press, Washington, D.C., 1984, 179.

DeVera, E.R., B.P. Simmons, R.D. Stephens, and D.L. Storm, Samplers and Sampling Procedures for Hazardous Waste Streams, EPA 600/2-80-810, January 1980.

Environmental Monitoring System Laboratory (EMSL), ORD, EPA (Environmental Protection Agency), Characterization of Hazardous Waste Sites—A Method Manual, Vol. II: Available Sampling methods, Las Vegas, NV, 1983.

Environmental Protection Agency, Soil Gas Monitoring Techniques Videotape, National Audio Visual Center, Capital Heights, MD, 1987.

Environmental Protection Agency, Final Comprehensive State Ground Water Protection Program Guidance, PB93-163087, 1992.

Environmental Protection Agency, Guidance for Planning for Data Collection in Support of Environmental Decision Making Using the Data Quality Objective Process, EPA QA/G-4, 1993.

Heath, R.C., Ground-Water Regions of the United States, United States Geological Survey Water Supply Paper 2242, Superintendent of Documents, U.S. Government Printing Office, Washington, D.C., 1984, 78.

Hinchee, R.E. and H.J. Reisinger, Multi-phase transport of petroleum hydrocarbons in the subsurface environment: theory and practical application, in *Proceedings of the NWWA/APA Conference on Petroleum Hydrocarbons and Organic Chemicals in Ground Water: Prevention, Detection and Restoration*, National Water Well Association, Dublin, OH, 1985, 58–76.

Huber, W.F., The use of downhole television in monitoring applications, in *Proceedings of the Second National Symposium on Aquifer Restoration and Ground-Water Monitoring*, National Water Well Association, Dublin, OH, 1982, 285–286.

Kovski, J.R., Physical transport process for hydrocarbons in the subsurface, in *Proceedings of the Second International Conference on Ground Water Quality Research*, Oklahoma State University Printing Services, Stillwater, OK, 1984, 127–128.

Marrin, D.L. and G.M. Thompson, Remote detection of volatile organic contaminants in ground water via shallow soil gas sampling, in *Proceedings of the NWWA/API Conference on Petroleum Hydrocarbons and Organic Chemicals in Ground Water: Prevention, Detection and Restoration*, National Water Well Association, Dublin, OH, 1984, 172–187.

Microseeps, Methods and Procedures for Use of Microseeps Soil Gas Sampling System, 1992, 1–8.

Morahan, T. and R.C. Doorier, The application of television borehole logging to ground-water monitoring programs, *Groundwater Monit. Rev.*, 4(4), 172–175, 1984.

Nielsen, D.M. and G.L. Yeates, A comparison of sampling mechanisms available for small-diameter ground-water monitoring wells, in *Proceedings of the Fifth National Symposium and Exposition on Aquifer Restoration and Ground-Water Monitoring*, National Water Well Association, Dublin, OH, 1985, 237–270.

Norman, W.R., An effective and inexpensive gas-drive ground-water sampling device, *Groundwater Monit. Rev.*, 6(2), 56–60, 1986.

Northeast Research Institute, Inc., Guide to Petrex Environmental Survey Field Procedures, 20, 1992.

Pettyjohn, W.A., Cause and effect of cyclic changes in ground-water quality, *Groundwater Monit. Rev.*, 2(1), 43–49, 1982.

Schwarzenbach, R.P. and W. Giger, *Behavior and Fate of Halogenated Hydrocarbons in Ground Water*, John Wiley & Sons, New York, 1985, 446–471.

Smith, R., and G.V. James, *The Sampling of Bulk Materials*, the Royal Society of Chemistry, London, 1981.

Target Environmental Services, Site Investigation and Remediation with "Direct Push Sampling Technology," Columbia, MD, 1993, 11.

Villaume, J.F., Investigations at sites contaminated with dense, non-aqueous phase liquids (DNAPLs), *Groundwater Monit. Rev.*, 5(2), 60–74, 1985.

Wyatt, D.E., Pirkle, R.J., and Masdea, D.J., Soil Gas Investigations at the Sanitary Landfill (U), Westinghouse Savannah River Company, WSRC-RP-92-878, 1992, Appendix I.

CHAPTER 5

Sample Preparation, Documentation, and Shipment

After the sample collection procedure is complete, sample containers must be preserved (if required), capped, custody-sealed, and transported along with appropriate documentation to the on-site or fixed analytical laboratory for analysis. Great care should be taken when preparing samples for shipment since an error in this procedure has the potential of invalidating the samples and subsequent data.

5.1 SAMPLE PREPARATION

Immediately after a sample bottle has been filled, it must be preserved as specified in the Sampling and Analysis Plan. Sample preservation requirements vary, based on the sample matrix and the analyses being performed. Radiological analyses run on soil, sediment, or solid waste samples rarely require any sample preservation since the radiological composition and activity levels are not influenced by temperature or other factors as the chemical composition is. Water or liquid waste samples for radiological analysis are often preserved with nitric acid (HNO_3) to prevent isotopes from adhering to the walls of the sample container. Enough nitric acid is added to lower the pH to < 2. For chemical analysis, the only preservation typically required for soil or sediment samples is cooling the sample to 4°C. For water samples, some analyses only require cooling to 4°C, whereas others also require a chemical preservative such as HNO_3, sulfuric acid (H_2SO_4), hydrochloric acid (HCl), or sodium hydroxide (NaOH). Enough acid or base is added to the sample bottle either to lower the pH to < 2 or to raise the pH to >10. The laboratory running the analyses will specify which preservative is required for a particular analysis. The chemicals used to preserve a sample must be of analytical grade to avoid the potential for contaminating the sample. Cooling samples to 4°C is particularly important for samples to be analyzed for volatile organic compounds since cooling the sample slows the rate of chemical degradation.

To avoid any difficulties associated with adding chemical preservatives to sample containers in the field, it is recommended that these preservatives be added to sample bottles in a controlled setting prior to entering the field. This alternative reduces the chances of improperly preserving sample bottles or introducing field contaminants into a sample bottle while adding the preservative.

The preservative should be transferred from the chemical bottle to the sample container using either a disposable polyethylene pipette or a standard glass pipette. A glass eye dropper with rubber bulb is not recommended since the rubber has a potential of introducing contaminants into the sample.

The disposable pipette is made of polyethylene, and should be used only once, and then discarded. This pipette is more convenient than the standard glass pipette method and provides the least opportunity for the cross-contamination of samples. The standard glass pipette is preferred over the disposable pipette when bottles for volatile organic analysis need to be preserved, since polyethylene has the potential of providing trace volatile organics to the sample.

After a sample container has been filled, a Teflon-lined cap or lid is screwed on tightly to prevent the container from leaking. The sample label is filled out, noting the sampling time and date, sample identification number, sampling depth, analyses to be performed, sampler's initials, etc. (see Section 5.2.5). A custody seal is then placed over the cap or lid just prior to placing the sample bottle into the sample cooler. The custody seal is used to detect any tampering with the sample prior to analysis.

5.2 DOCUMENTATION

Accurate documentation is essential for the success of a sampling program. It is only through documentation that a sample can be tied into a particular sampling time, date, location, and depth. Consequently, field logbooks must be kept by every member of the field team, and should be used to record information ranging from weather conditions to the time the driller stubbed his right toe. To assist the documentation effort, standardized forms are commonly used to outline the information that needs to be collected. Some of the more commonly used forms include:

- Scanning instrument quality control check form
- Borehole log forms
- Well completion forms
- Well development forms
- Well purging/sampling forms
- Water level measurement forms

Other documentation needs associated with sample identification and shipment include:

- Sample labels
- Chain-of-custody forms

- Custody seals
- Shipping airbills

In addition to the above documentation requirements, a careful file must be kept to track important information, such as:

- Field variances
- Equipment shipping invoices
- Sample bottle lot numbers
- Documented purity specifications for preservatives, distilled water, and calibration standards
- Instrument serial numbers
- Copies of shipping paperwork
- Quality assurance nonconformance notices

5.2.1 Field Logbooks

Field logbooks are intended to provide sufficient data and observations to enable participants to reconstruct events that occurred during projects and to refresh the memory of the field personnel if called upon to give testimony during legal proceedings. In a legal proceeding, logbooks are admissible as evidence, and consequently must be factual, detailed, and objective.

Field logbooks must be permanently bound, the pages must be numbered, and all entries must be written with permanent ink, signed, and dated. If an error is made in the logbook, corrections can be made by the person who made the entry. A correction is made by crossing out the error with a single line, so as not to obliterate the original entry, and then entering the correct information. All corrections must be initialed and dated.

Observations or measurements that are taken in an area where the logbook may be contaminated can be recorded in a separate bound and numbered logbook before being transferred into the master field notebook. All logbooks must be kept on file as permanent records, even if they are illegible or contain inaccuracies that require a replacement document.

The first page of the logbook should be used as a Table of Contents to facilitate the location of pertinent data. As the logbook is being completed, the page numbers where important events can be found should be recorded. The very next page should begin recording daily events. The first daily event entry should always be the date, followed by a detailed description of the weather conditions. All of the following entries should begin with the time that the entry was made. Any space remaining on the last line of the entry should be lined out to prevent additional information being added in the future. At the end of the day, any unused space between the last entry and the bottom of the page should be lined out, signed, and dated, to prevent additional entries being made at a later date.

To assure that a comprehensive record of all important events is recorded, each team member should keep a daily log. The field manager's logbook should record information at the project level, such as:

- Time when team members, subcontractors, and the client arrive or leave the site;
- Names and company affiliation of all people who visit the site;
- Summary of all discussions and agreements made with team members, subcontractors, and the client;
- Summary of all telephone conversations;
- Detailed explanations of any deviations from the sampling and analysis plan, noting who gave the authorization, and what paperwork was completed to document the change;
- Detailed description of any mechanical problem that occurred at the site, noting when and how it occurred, and how it is being addressed;
- Detailed description of any accidents that occurred, noting who received the injury, how it occurred, how serious the injury was, how the person was treated, and who was notified;
- Other general information such as when and how equipment was decontaminated, what boreholes were drilled, and what samples were collected that day.

The team member's logbook should record information more at the task level. Examples of the types of information that should be recorded in these logbooks include:

- Time when radiological surveys began and ended on a particular site;
- Details on the instruments used to collect radiological measurements;
- Results from instrument calibration checks;
- Details on remedial activities performed at the site;
- Radiological measurement data;
- Level of personal protective equipment used at the site;
- Sample collection times for all samples collected;
- Total depth of any boreholes drilled;
- Detailed description of materials used to build monitoring wells, including type of casing material used; screen slot size; length of screen; screened interval; brand name, lot number, and size of sand used for the sand pack; brand name, lot number, and size of bentonite pellets used for the bentonite seal; brand name and lot number of bentonite powder and cement used for grout; well identification number;
- Details on when, how, and where equipment was decontaminated, and what was done with the wastewater;
- Description of any mechanical problems that occurred at the site, noting when and how it occurred, and how it was addressed;
- Summary of all discussions and agreements made with other team members, subcontractors, and the client;
- Summary of all telephone conversations;
- Detailed description of any accidents that occur, noting who received the injury, how it occurred, how serious the injury was, how the person was treated, and who was notified.

It is essential that field team members record as much information as possible in their logbooks, since this generates a written record of the project. Years after the project is over, these notebooks will be the only means of reconstructing events that occurred. With each team member recording information, it is not uncommon for one member to record information that another member missed.

5.2.2 Photographic Logbook

A photographic logbook should be used to record all photographs taken at a site. This log should record the date, time, subject, frame, roll number (or disk number for a digital camera), and the name of the photographer. For "instant photos," the date, time, subject, and name of the photographer should be recorded directly on the developed picture. Clear photographs of field activities can be very useful in reconciling any later discrepancies.

5.2.3 Field Sampling Forms

It is recommended that standardized field sampling/measurement forms be used to assist the sampler in a number of field activities. Some commonly used forms are presented in Figures 5.1 through 5.6. Forms are most often used to reduce the amount of documentation required in the field logbook. Forms are also effective in reminding the sampler of what information needs to be collected, and they make it more obvious when the necessary information was not collected.

When forms are used, they should be permanently bound in a notebook, the pages should be numbered, and all entries must be written with permanent ink. If an error is made in the notebook, corrections can be made by the person who made the entry. A correction is made by crossing out the error with a single line so as not to obliterate the original entry and then entering the correct information. All corrections must be initialed and dated. The person who completed the form should sign and date the form at the bottom of the page. It is recommended that the field manager also sign the form to confirm that it is complete and accurate.

5.2.4 Identification and Shipping Documentation

The essential documents for sample identification and shipment include the sample label, custody seal, chain-of-custody form, shipping manifest, and shipping airbill. Together, these documents allow radioactive samples to be shipped and/or transported to an analytical laboratory under custody. If custody seals are broken when the laboratory receives the samples, the assumption must be made that the samples were tampered with during shipment. Consequently, the samples may need to be collected over again. When shipping radioactive samples off site, it is essential that 49 CFR 170 through 180 requirements for shipping container inspection and surveying be carefully observed. If shipping radioactive samples by air, the requirements provided in the International Air Transportation Association (IATA), Dangerous Goods Regulations, must be followed.

Before shipment, and upon receipt of a radioactive shipment, a visual inspection of packages should be performed to ensure that packages are not damaged. Prior to shipment, gross alpha and gross beta/gamma measurements should be collected from the outside surfaces of the individual sample containers and the shipping container and the results recorded on the shipping paperwork. Care should be taken prior to shipping multiphased radioactive samples, as settling of hot particulate or layering may occur, resulting in an increase in the radiation levels that are measured for the container. The radiation levels should be measured after settling has occurred and the highest value used for preparing the shipment. When the shipping container is received by the laboratory, gross alpha and gross beta/gamma measurements should once again be collected from the outside surfaces of the shipping container and should be compared against the readings reported on the shipping papers. This practice ensures accountability. Any differences should be reconciled prior to accepting the sample shipment.

SAIC _An Employee-Owned Company_						
Scanning Instrument Quality Control Check Form						
Project:						
Instrument Number:						
Operating Voltage:						
Date When Instrument was Last Calibrated:						
Date When Re-Calibrated is Required:						
Calibration Check Source No:						
Acceptable Range for Source Check:						
Date	Time	Initials	Measured Activity (cpm)	Background Activity Measurement (cpm)	Acceptable (Y/N)	Comment
Approval Signature:			Date:			

Figure 5.1 Example of scanning instrument quality control check form.

Drivers of motor vehicles transporting radioactive samples should have a copy of their emergency response plan or the emergency response information with them as required by 49 CFR 172.600.

SAIC
An Employee-Owned Company

Borehole # _____

Page _____ of _____

Borehole Log Form

	START	FINISH

Project_____
Location_____
Geologic Log by_____
Driller_____
Geophysics by_____
Weather_____

Total Depth_____
Borehole Diameter_____
Depth to Water_____
Rig_____
Bit(s)_____
Drilling Fluid_____

Date
Time
How Left_____

Depth	Pene. Rate/ Blow Cts	Circu- lation Q (gpm)	CPM	Sample		Geologic and Hydrologic Description		
				#	Interval	Lithology Symbol		% Core Recovery
0								

Approval Signature:		Date:	

Figure 5.2 Example of a borehole log form used to record borehole lithology.

SAIC
An Employee-Owned Company

Well Completion Form

Location: Personnel:	Elevation: Ground Level Top of Casing

DRILLING SUMMARY:

Total Depth
Borehole Diameter
Driller
Rig
Bit(s)
Drilling Fluid
Surface Casing

WELL DESIGN:
Basis:
Geologic Log
Casing String(s): C=Casing S=Screen

CONSTRUCTION TIME LOG:

Task	Start Date	Start Time	Finish Date	Finish Time
Drilling:				
Geophys. Logging:				
Casing:				
Filter Placement:				
Cementing:				
Development:				
Other:				

Casing: C1	Comments:
C2	
C3	
C4	
Screen: S1	
S2	
S3	
S4	
Centralizers	
Filter Material	
Other	
Approval Signature:	Date:

Figure 5.3 Example of a well completion form.

5.2.5 Sample Labels

The primary objective of the sample label is to link a sample bottle to a sample number, sampling date and time, and analyses to be performed. The sample label in combination with the chain-of-custody form is used to inform the laboratory what the sample is to be analyzed for. At a minimum, a sample label should contain the following information (Figure 5.7):

- Sample identification number
- Sampling time and date

SAIC
An Employee-Owned Company

Well Development Form

Project Name and Number:

Well Number and Location:

Development Crew: Driller (if applicable):

Water Levels/Time: Initial: Pumping: Final:

Total Well depth: Initial: Final:

Date and Time: Begin: Completed:

Development: Method(s):

Total Quantity of Water Removed: gallons

Date/Time and Pump Setting	Discharge Rate and Measurement Method	Field Measurements				Remarks
		Temp (°C)	Specific Conductivity (umhos/cm).	PH (Standard Units)	Turbidity	

Approval Signature: Date:

Figure 5.4 Example of a well development form.

- Analyses to be performed
- Preservatives used
- Sampler's initials
- Name of the company collecting the sample
- Name and address of the laboratory performing the analysis.

To save time in the field, and to avoid the potential for errors, all of the above information should be added to the sample label before going into the field, with the exception of the sampling time and date, and sampler's initials. This information

SAIC
An Employee-Owned Company

Well Purging and Sampling Form

Site: _____   Well No.: _____

Date(s): _____   Geologist: _____

Purging Bailer: _____   Equipment Used: _____

Sampling Bailer: _____   Measurement Reference Datum: _____

DATA FROM IMMEDIATELY BEFORE AND AFTER DEVELOPMENT:

Depth to water measured from TOC (ft): Before   After Purging: _____
Purging: _____

                                      After Sampling: _____

Total purging time (min): _____

Depth to sediment in well (ft): Before Purging: _____   After Purging: _____

	Time Since Purging Started (min)	Time	Cumulative Volume Removed	Water Temp °C	pH of Water	Conductivity (μmhos/cm)	Turbidity (NTUs)	Water Appearance	Date
Before									
During									
During									
During									
During									
During									
During									
After									

      *CL=clear   CO = cloudy   TU = turbid

Comments:
_____
_____
_____
_____
_____
_____
_____
_____
_____
_____

Approval Signature: _____   Date: _____

Figure 5.5 Example of a well purging and sampling form.

should be added to the label following the capping of the sample bottle, immediately after sample collection, and should reflect the time that sampling began, as opposed to the time sampling was completed.

An effective sample numbering system is a key component of any field sampling program since it serves to tie the sample to its sampling location. The problem with using a simple numbering system such as 1, 2, 3, … is that the number tells you nothing regarding the location, depth, or sample media.

The number of digits used in a sample numbering scheme should be discussed with the laboratory performing the analyses since it may have limitations on the

SAIC
An Employee-Owned Company

Water Level Measurements

Measurement Team:

Project Number and Location:

Measuring Instrument:

Well No.	Date	Time	Tape Reading		Depth to Water (ft)	Initials	Remarks
			Measure Pt.	Water Level			

Measuring Point: Point where measurement was taken. Top of casing (TOC); Top of Protective Steel Casing (TOPSC); Land Surface (LS), etc.
Depth of Water: Measurements should be recorded to the nearest 0.01 ft.
Remarks: Any conditions that may influence the water level measurements.

Approval Signature: Date:

Figure 5.6 Example of water level measurement form.

number of digits used in a sample number. If only six digits are available, an effective numbering system can be developed with a little imagination. As an example, assume that a project has a total of nine buildings where swipe, paint, dust, concrete, and shallow soil samples are to be collected. In this example, each of the six sample number digits could be used to represent the following:

First digit: Site Number (1, 2, 3, … 9)

1 = 103-R Reactor Building
2 = 106-D Biological Laboratory

SAIC An Employee-Owned Company	SAMPLE LABEL		
Client:			Date/Time:
Lab No.:		Sample ID:	
Initials:		Time:	
Analyze for:			
Preservatives:			
H₂SO₄	HNO₃	NaOH	Other:
(Laboratory Name and Address)			

CUSTODY SEAL	CUSTODY SEAL
Date	Date
Signature	Signature
SAIC An Employee-Owned Company	SAIC An Employee-Owned Company

Figure 5.7 Example of sample label and custody seal.

3 =   110-A Pump House
4 =   121-L Treatment Plant
5 =   158-A Testing Laboratory
6 =   185-B Pump House
7 =   205-D Storage Building
8 =   242-S Reactor Building
9 =   251-D Transfer Station

Second digit: Media Number (1, 2, 3, ... 5)

1 =   swipe
2 =   paint
3 =   dust
4 =   concrete
5 =   shallow soil

Third, fourth, and fifth digits: Sampling Location Number

S99 = swipe (1,2,3,. . .99)
P99 = paint (1, 2, 3,. . .99)
D99 = dust (1, 2, 3,. . .99)
C99 = corehole (1, 2, 3,. . .99)
B99 = borehole (1, 2, 3,. . .99)

Sixth digit: Sample number

1 =   first sampling interval

2 =   second sampling interval
3 =   third sampling interval
4 =   fourth sampling interval
5 =   fifth sampling interval
6 =   sixth sampling interval
7 =   seventh sampling interval
8 =   eighth sampling interval
9 =   ninth sampling interval

The sample number "85B032" would therefore indicate that the sample was collected from the 242-S Reactor Building, it is a shallow soil sample, it was collected from Borehole No. 3, and it was collected from the second sampling interval. When using both numbers and letters in a sample number, one should try to minimize the use of letters such as S, l, and O, which can easily be misinterpreted as the numbers 5, 1, and 0.

Some laboratories will allow sample numbers as long as 10 or 12 characters in length. These additional characters should be taken advantage of, since the easier it is to interpret the sample identification number, the easier it will be to interpret the data reported by the laboratory.

5.2.6 Chain-of-Custody Forms and Seals

Chain-of-custody is the procedure used to document who has the responsibility for ensuring the proper handling of a sample from the time it is collected to the time the resulting analytical data are reported by the laboratory to the customer. After a sample bottle has been filled, preserved, and labeled, a custody seal is signed and dated, then placed over the bottle cap to assure that the sample is not tampered with (see Figure 5.7). The custody seal is a fragile piece of tape designed to break if the bottle cap is turned or tampered with.

The person who signs his or her name to the custody seal automatically assumes ownership of the sample until the time it is packaged and custody is signed over to the laboratory, carrier, or to an overnight mail service. This transfer of the custody is recorded on a chain-of-custody form (Figure 5.8). The chain-of-custody form is also the document used by the laboratory to identify which analyses to perform on which samples. When the laboratory receives a sample shipment, it assumes custody of the samples by signing the chain-of-custody form. The sample bottles in the shipment are then counted, and the requested analyses on the sample bottles are compared against the requested analyses on the chain-of-custody form. If there are any inconsistencies, the laboratory should contact the project manager for clarification. These communications should all be carefully documented.

An overnight shipping airbill can be used to document the transfer of custody from the field to the mail service. If this option is selected, the shipper's airbill number should be recorded on the chain-of-custody form, which can be taped inside the sample container. When the laboratory signs the shipper's delivery form and the chain-of-custody form inside the sample container, it has assumed custody of the sample.

SAIC An Employee-Owned Company

Chain of Custody Record

Laboratory Log Number ________

Client Name	Job Number or Purchase Order Number
Project Name	
Project Manager	Sampler(s)

Sample Number	Date Sampled	Time Sampled	Matrix Type	Sample Description	Number of containers	Analyses required							Special handling required	Remarks

Signature	Company	Date	Time
Relinquished by:			
Received by:			
Relinquished by:			
Received by:			
Relinquished by:			
Received by:			

Figure 5.8 Example of a chain-of-custody record.

Although most analytical laboratories will dispose of chemically contaminated samples once the analyses have been completed, they often return radiologically contaminated samples to the customer, who must then assume responsibility for disposal. Laboratories must be licensed by the Nuclear Regulatory Commission and/or state to receive and handle radioactive materials. The license specifies the maximum quantity of radioactive material that the laboratory can receive and store.

5.2.7 Other Important Documentation

Careful documentation should also be maintained for all incoming shipments of materials and supplies used to support sampling activities. The most critical documents to keep on file include:

- Shipping invoices
- Sample bottle lot numbers
- Purity specification for preservatives, distilled water, and calibration liquids and gases
- Instrument serial numbers and calibration logs.

Copies of shipping invoices can be used by the project manager to keep track of equipment costs, and can expedite the return of malfunctioning equipment. It is critical to keep track of sample bottle lot numbers and purity specifications for all chemicals used, since improperly decontaminated bottles and low-quality preservation of decontamination chemicals can contaminate samples. Instrument serial numbers are recorded to assist in tracking which instruments are working well, and which must undergo repair.

REFERENCES

49 CFR 173, Shippers—General Requirements for Shipments and Packagings, Code of Federal Regulations (as amended).

DOE (U.S. Department of Energy), Radiological Control Manual, Chapter 4—Radioactive Materials. Part 2—Release and Transportation of Radioactive Material, DOE/EH-0256T, Article 423, April, 1994.

EPA (Environmental Protection Agency), Guidance on Remedial Investigations under CERCLA, EPA/540/G-85/002, 1985, 4–2, 4–4.

EPA (Environmental Protection Agency), A Compendium of Superfund Field Operations Methods, EPA/540/P-87/001a, 1987, 4–13.

EPA (Environmental Protection Agency), RCRA Facility Investigation (RFI) Guidance, PB89-200299, 1989.

IATA (International Air Transportation Association), Dangerous Goods Regulations.

CHAPTER 6

Data Verification and Validation

Data verification and validation are the first steps in assessing project data quality and usefulness. They represent a standardized review process for determining the analytical quality of discrete sets of chemical data. The primary purpose is to summarize the technical defensibility of the analytical data for the end users and decision makers.

Direction is provided by the EPA under the Contract Laboratory Program (CLP) in the form of the National Functional Guidelines for Organic Data Review (EPA-540/R-94/012, February 1994) and the National Functional Guidelines for Inorganic Data Review (EPA-540/R-94/013, February 1994). Interpretation of this guidance and its application to individual programs and projects needs to be made at the operational level and incorporated into the Sampling and Analysis Plan for a given investigation. Verification and validation must be consistent with the project data quality objectives, laboratory scope of work, and designated analytical methods.

Data verification/validation represent the decision process by which established quality control criteria are applied to the data. Individual sample results are accepted, rejected, or qualified as estimated. Data that meet all criteria are acceptable and can be used as needed by the project. Data not meeting critical criteria are "rejected" (R) and should not be used for any project objectives. Some data meet most criteria and meet all critical measures; however, they have quality control deficiencies that question their degree of accuracy, precision, or sensitivity. These data are qualified as "estimated" (J) to indicate that certain validation criteria were not met. Estimated data may or may not be usable depending on the intended data use, and all estimated data should be followed by a rationale for their estimation, so that the data user is capable of making an intelligent decision regarding their employment in the decision process.

EPA Region I has provided the environmental community with a useful "Tiered Approach" to validation that allows a program or project to establish the level of intensity and depth of review applicable to its needs. Guidance to this approach appears in "Region I, EPA-New England Data Validation Functional Guidelines for Evaluating Environmental Analyses," July 1996, revised December 1996. This document and its appendices may prove useful during project data validation development.

Data verification is the first step in a tiered or graded approach (Tier I). It is performed for the purpose of "verifying" that the laboratory provided the reporting information required by the Sampling and Analysis Plan and met a few minimum requirements that ensure the data are usable. Data verification involves evaluating laboratory analytical data packages to confirm that:

- The data packages are complete and contain all of the information specified in the Sampling and Analysis Plan, e.g., All samples and analyses requested, case narrative, summary data report, completed chain-of-custody form, analytical quality control data (blanks, matrix spikes, matrix spike duplicates, etc.), Date and time when each analysis was performed;
- The laboratory ran the correct analytical methods specified in the Sampling and Analysis Plan;
- Samples did not exceed the maximum analytical holding times specified in the Sampling and Analysis Plan;
- Sample chain-of-custody was not broken from the time the sample was collected, analyzed, and the data reported; and
- The laboratory reported analytical results for each analytical method and each analyte required by the laboratory statement of work and the project Sampling and Analysis Plan.

Data verification does not require an extensive effort and all analytical data packages should undergo this level of evaluation prior to use. Minimal qualification of the data would result from this level of verification and the applicability to objectives such as risk assessment would be questionable. However, for site characterization studies and definition of contamination nature and extent for engineering planning purposes Tier I may be all that is needed, since the consequences of decision error are relatively minor.

Basic data validation is a more extensive evaluation than data verification and represents the next tier in the review process (Tier II). Basic data validation implements an evaluation of laboratory quality control data and analytical procedures. This ensures the analytical process and instrumentation used to perform the analyses met all of the data quality requirements specified in the DQOs and Sampling and Analysis Plan. Focus is given to laboratory/instrument performance criteria, sample preparation and matrix effects evaluation, and field quality control measures. In addition to Tier I verification, basic data validation involves evaluating laboratory analytical data packages to confirm that:

Laboratory/instrument performance criteria
- Laboratory case narrative documentation is clear and accurate.
- Analytical preparation procedures are acceptable and documented.
- Instrument operational and method calibration criteria have been achieved.
- Laboratory calibration blank contamination is under control.
- Laboratory control standard criteria are being met.

Sample preparation and matrix effects criteria
- Laboratory method blank contamination is under control.
- Sample surrogate compound recovery, tracer recovery, and internal standard criteria have been achieved.
- Sample matrix spike recoveries meet minimum accuracy requirements specified in the DQOs and Sampling and Analysis Plan.
- Sample matrix spike duplicate or duplicate comparisons meet minimum precision requirements specified in the DQOs and Sampling and Analysis Plan.
- Sample dilution review and reanalyses are performed.

Field quality control measures
- Field source water blank, equipment rinsate blank, and sample trip blank contents have not impacted the project data results.
- Field duplicate comparisons meet minimum precision requirements specified in the DQOs and Sampling and Analysis Plan.

Tier II validation does require more effort; however, this level enables meaningful qualification of the data and is considered acceptable for documenting data uses, such as risk assessment, transport modeling, cleanup confirmation, and site closing. This degree of review conforms with that identified by the EPA in "Guidance for Data Usability in Risk Assessment" EPA540/G90/009, October 1990. Tier II validation should also be acceptable documentation to confirm legal defensibility.

Complete data validation encompasses both Tier I and Tier II information and adds a detailed examination of the analytical raw data. This level of review requires all information generated by the laboratory to be presented as part of the data deliverable. This would include copies of all chromatograms, spectral printouts, quantification details, preparation logbooks, standard logbooks, calculation programs, etc., produced by the laboratory. In addition to the previously reviewed material, comprehensive data validation will include:

- A detailed examination of the raw data analyte identification;
- A check of calculations used to quantify analyte results, normally a minimum of 10% of the reported concentrations are checked by recalculation from original raw data information; and
- Raw data results are verified against final reported concentrations to preclude transcription errors.

Complete data validation does require an extensive effort, and unless there are special project goals requiring this level of review, it is probably not necessary to implement such intensive inspection of the data. This level of validation is considered acceptable for all data uses including legal actions. When Tier I and Tier II reviews show persistent laboratory errors or indicate a possibility of laboratory impropriety, complete validation is warranted to justify possible contractual recourse relative to laboratory performance.

Regardless of which level is implemented, validation must be documented formally for project records. The principal mechanisms utilized to establish this documentation are qualification flags (U, J, UJ, R) attached to the data values and relevant checklists targeted at verification and method-specific validation. The qualification flags U, J, UJ, and R represent below detection, estimated value, below detection–estimated value, and rejected, respectively. Validation flags should be placed on both hardcopy "Form 1" results and in electronic databases. More sophisticated and useful documentation is implemented by several organizations in the form of reason codes applied to the traditional qualification flags. These codes express in detail why the primary qualification was made. This information can be employed by data users to ascertain potential concentration bias and concentration utility to the project. Additional documentation may be written in validation summary reports (per sample delivery group or per project), Quality Control Summary Reports, or Data Quality Assessments.

Various combinations of data review, verification, and validation can be imposed on the project data depending on the intended use of the information. However, the end result of the verification/validation process must provide a consistent and defensible data set.

REFERENCES

EPA (Environmental Protection Agency), *Data Useability in Risk Assessment*, EPA-540/G90/009, October, 1990.

EPA (Environmental Protection Agency), National Functional Guidelines for Organic Data Review, EPA-540/R-94/012, February, 1994.

EPA (Environmental Protection Agency), National Functional Guidelines for Inorganic Data Review, EPA-540/R-94/013, February, 1994.

EPA (Environmental Protection Agency), Region I, New England Data Validation Functional Guidelines for Evaluating Environmental Analyses, July 1996, (revised December 1996).

CHAPTER **7**

Radiological Data Management

The purpose of sampling and surveying in radiological environments is to generate data. Sampling and Analysis Plans focus on specifying how data will be generated, and to a lesser extent how they will be used. Less attention is typically paid to how data will be managed during fieldwork and after fieldwork is completed. The lack of a data management system and a detailed data management plan can severely impact the effectiveness of sampling and surveying efforts. For complex settings that involve multiple sampling programs across a site (or over time), integrating, managing, and preserving information garnered from each of these sampling programs may be crucial to the overall success of the characterization and remediation effort.

Environmental data management systems comprise the hardware, software, and protocols necessary to integrate, organize, manage, and disseminate data generated by a sampling and analysis program. Data management plans describe how an environmental data management system will be used to address the data management needs of sampling and analysis programs. This chapter reviews the objectives of data management programs, describes the components of a typical data management system, and discusses data management planning for successful radiological data management programs.

7.1 DATA MANAGEMENT OBJECTIVES

Radiological data management serves two basic objectives. The first is to ensure that data collected as part of sampling and surveying programs are readily available to support the site-specific decisions that must be made. The second is to preserve information in a manner that ensures its usefulness in the future and satisfies regulatory requirements for maintaining a complete administrative record of activities at a site. These two objectives impose very specific and at times contradictory requirements on data management plans and systems.

7.1.1 Decision Support

Data management for decision making has the following requirements. Decision making requires the integration of information from sources that will likely go beyond just the data generated by a sampling and survey program. Decision making often focuses on derived information rather than the raw data itself. Decision making demands timely, decentralized, and readily accessible data sets.

For radiologically contaminated sites, decision making requires integrating spatial data from a wide variety of data sources. Examples of these data include:

- Maps that show surface infrastructure, topography, and hydrology. These maps would likely come from a variety of sources, such as the U.S. Geological Survey (USGS) or site facility management departments.
- Historical and recent aerial photography. This may include flyover gross gamma measurement results for larger sites.
- Borehole logs from soil bores and monitoring wells in the area. These logs might include a wide variety of information, including soil or formation type, moisture content, depth to the water table, and results from soil core or downhole radiological scans.
- Nonintrusive geophysical survey data from techniques such as resistivity, ground penetrating radar, magnetics, etc.
- Surface gross gamma activity scans collected by walk- or driveovers. These data may have matching coordinate information obtained from a Global Positioning System, or perhaps may only be assigned to general areas.
- Data from direct *in situ* surface measurements using systems such as high-purity germanium gamma spectroscopy.
- Results from traditional soil and water sample analyses.

These data are likely to come in a variety of formats, including simple ASCII files, databases, spreadsheets, raster image files, electronic mapping layers, hard-copy field notebooks, and hard-copy maps. All of these data contribute pieces to the overall characterization puzzle.

Decision making often focuses on information that is a derived from the basic data collected as part of sampling and survey programs. For example, in estimating contaminated soil volumes, the decision maker may be primarily interested in the results of interpolations derived from soil sampling results, and not in the original soil sampling results themselves. For final status survey purposes, the statistics derived from a MARSSIM-style analysis may be as important as the original data used to calculate those statistics. In the case of nonintrusive geophysical surveys, the final product is often a map that depicts a technician's interpretation of the raw data that were collected. Decision making may also require that basic information be manipulated. For example, outlier or suspect values may be removed from an analysis to evaluate their impacts on conclusions. Results from alpha spectroscopy might be adjusted to make them more directly comparable to gamma spectroscopy results.

Timely and efficient decision making presumes that sampling and survey results are quickly available to decision makers wherever those decision makers may be located. Off-site analyses for soil samples often include a several-week turnaround time. When quality assurance and quality control (QA/QC) requirements are imposed as well, complete final data sets produced by a sampling and survey program might not be available for months after the program has completed its fieldwork. In many cases, decisions are required before these final data sets are available. For example, in a sequential or adaptive sampling program, additional sample collection and the placement of those samples are based on results from prior samples. In a soil excavation program, back-filling requirements may demand that final status survey conclusions be drawn long before final status closure documentation is available. Site decision makers may be physically distributed as well. For example, the decision-making team might include staff on site, program management staff in home offices, off-site technical support contractors, and off-site regulators.

7.1.2 Preserving Information

Preserving, or archiving, sampling and survey results imposes a completely different set of requirements on data management. Data preservation emphasizes completeness of data sets and documentation. Data preservation focuses primarily on raw data and not derived results. Data preservation presumes a centralized repository for information, controlled access, and a very limited ability to manipulate the information that is stored. Environmental data archiving systems can become relatively complex, including a combination of sophisticated database software, relational database designs, and QA/QC protocols governing data entry and maintenance.

7.2 RADIOLOGICAL DATA MANAGEMENT SYSTEMS

Radiological data management systems include the hardware, software, and protocols necessary to integrate, organize, manage, and disseminate data generated by a sampling and analysis program. The particulars of any given data management system are highly site and program specific. However, there are common components that appear in almost all systems. These components include relational databases for storing information and software for analyzing and visualizing environmental data stored in databases.

7.2.1 Relational Databases

Relational databases are the most common means for storing large volumes of radiological site characterization data. Examples of commercially available relational database systems include Microsoft's Access™, SQL Server™, and Oracle™. The principal differences between commercial products are the presumed complexity of the application that is being developed and the number of users that will be supported. For example, Access is primarily a single-user database package that is relatively easy to configure and use on a personal computer. In contrast, Oracle is an enterprise system demanding highly trained staff to implement and maintain, but capable of supporting large amounts of information and a large number of concurrent users within a secure environment.

Relational databases store information in tables, with tables linked together by common attributes. For example, there may be one table dedicated to sampling station (locations where samples are collected) data, one to sample information, and one to sample results. Individual rows of information are commonly known as records, while columns are often referred to as data fields. For example, each record in a sampling stations table would correspond with one sampling station. The fields associated with this record might include the station identifier, easting, northing, and elevation. In the samples table, each record would correspond to one sample. Common fields associated with a sample record might include station identifier, sample identifier, depth from sampling station elevation, date of sample, and type of sample. The results table would contain one record for each result returned. Common fields associated with a result record might include sample identifier, analyte, method, result, error, detection limits, QA/QC flag, and date of analysis. Records in the results table would be linked to the sample table by sample identifier. The sample table would be linked back to the stations table by a station identifier.

7.2.2 Radiological Data Analysis and Visualization Software

While relational databases are very efficient and effective at managing and preserving large volumes of environmental data, they do not lend themselves to data analysis or visualization. Consequently, radiological data management systems also include software that allow data to be analyzed and visualized. In most cases, relational databases are most aligned with the goals of preserving information and so tend to be centralized systems with limited access. In contrast, data analysis and visualization is most commonly associated with decision support activities. Consequently, these software are usually available on each user's computer, with the exact choice and combination of software user specific.

Data analysis software includes a wide variety of packages. For example, spreadsheets can be used for reviewing data and performing simple calculations or statistics on data sets. Geostatistical analyses would demand specialized and more-sophisticated software such as the EPA GeoEAS, the Stanford GSLIB, or similar packages. Analysis involving fate-and-transport calculations could require fate-and-transport modeling codes. The U.S. Army Corps of Engineers Groundwater Modeling System (GMS) software is an example of a fate-and-transport modeling environment.

There are also a wide variety of data visualization packages that can be applied to radiological data. Since most environmental data are spatial (i.e., have coordinates associated with them), Geographical Information Systems (GIS) can be used. Examples of commercial GIS software include ArcInfo™, ArcView™, MapInfo™, and Intergraph™ products. GIS systems are particularly effective at handling large volumes of gamma walkover data. Most GIS systems focus on two-dimensional maps of spatial information. However, radiological contamination often includes the subsurface or vertical dimension as well. Specialized packages such as Dynamic Graphic's EarthVision™ product, SiteView™, and GISKey™ allow some capabilities for three-dimensional modeling and visualization as well. Most of these packages require a significant amount of experience to use them effectively, but are invaluable for making sense out of diverse sets of data associated with a radiological characterization program.

7.3 DATA MANAGEMENT PLANNING

Sampling and survey data collection programs should include a formal data management plan. For large sites, an overarching data management plan may already be in place, with individual sampling and analysis plans simply referencing the master data management plan, calling out project-specific requirements where necessary. For smaller sites, a project-specific data management plan may be an important supporting document to the sampling and analysis plan.

The data management plan should include the following components:

- Identify decisions that will use information garnered from the sampling and surveying effort.
- Identify data sources expected to be producing information:
  — Link to decision that must be made;
  — Define meta-data requirements for each of these data sources;
  — Develop data delivery specifications;
  — Specify QA/QC requirements;
  — Establish standard attribute tables;
  — Identify preservation requirements.
- Specify how disparate data sets will be integrated:
  — Specify master coordinate system for the site;
  — Identify organizational scheme for tying data sets together.
- Specify data organization and storage approaches, including points of access and key software components.
- Provide data flowcharts for overall data collection, review, analysis, and preservation.

7.3.1 Identify Decisions

One goal of data collection is to support decisions that must be made. If the EPA Data Quality Objective (DQO) approach was used for designing data collection (see Section 4.1.1), these decisions should have already been explicitly identified, and the decision points identified in the data management plan should be consistent with these. Avoid general decision statements. The identification of decision points should be as detailed and complete as possible. This is necessary to guarantee that the overall data management strategy will support the data needs of each decision point. Each decision point should have its unique data needs. Again, if the EPA DQO process is followed, these data needs should already be identified and the data management plan need only be consistent with these.

7.3.2 Identify Sources of Information

The purpose of sampling and surveying data collection programs is to provide the information that will feed the decision-making process. The data management plan needs to identify each of the sources of information. The obvious sources of data are results from physical samples that are collected and analyzed. Less obvious but often just as important are secondary sources of information directly generated by field activities. These may include results from nonintrusive geophysical surveys, gross activity screens conducted over surfaces, civil surveys, stratigraphic information from soil cores and bore logs, downhole or soil core scans, and air-monitoring results. These may also include tertiary sources of information, data that already exist. Examples of these data include USGS maps, facility maps of infrastructure and utilities, and aerial photographs. Finally, sources of information may include derived data sets. Examples of derived data sets include flow and transport modeling results, interpretations of nonintrusive geophysical data, rolled-up statistical summaries of raw data, and results from interpolations.

For each source of data, the data management plan should specify what meta-data need to accompany raw results and who has responsibility for maintaining meta-data. Meta-data are data that describe the source, quality, and lineage of raw data. Meta-data allow the user to identify the ultimate source of information, and provide some indication of the accuracy and completeness of the information. Meta-data provide a means for tracking individual sets of information, including modifications, additions, or deletions. Meta-data are particularly important for information from secondary or tertiary sources, or derived data sets.

The data management plan should explicitly define the formats in which data should be delivered. This is particularly true for results from actual data collection activities that are part of the sampling/surveying process. Clearly defined electronic deliverable specifications can greatly streamline and simplify the process of integrating and managing newly obtained data sets. Conversely, without clearly specifying the form of electronic data submission, data management can quickly become chaotic.

QA/QC protocols should also be clearly stated in the data management plan for each data source. QA/QC protocols are typically associated with the process of ensuring that laboratory data meet comparability, accuracy, and precision standards. In the context of data management, however, QA/QC protocols are meant to ensure that data sets are complete and free from egregious errors. QA/QC protocols will vary widely depending on the data source. With gamma walkover data, for example, the QA/QC process may include mapping data to verify completeness of coverage, to identify coordinate concerns (e.g., points that map outside the surveyed area) and instrumentation issues (e.g., sets of sequential readings that are consistently high or consistently low), or unexpected and unexplainable results. For laboratory sample results, the QA/QC process might include ensuring that the sample can be tied back to a location, determining that all analytical results requested were returned, ensuring that all data codes and qualifiers are from preapproved lists. For any particular source of data, there are likely to be two sets of QA/QC conditions that must be met. The first are QA/QC requirements that must be satisfied before data can be used for decision-making purposes. The second are more formal and complete QA/QC requirements that must be satisfied before data can become part of the administrative record. In the latter case, the presumption is that no further modification to a data set is expected.

Standardized attribute tables are particularly important for guaranteeing that later users of information will be able to interpret data correctly. Standardized attribute tables refer to preapproved lists of acceptable values or entries for certain pieces of information. One common example is a list of acceptable analyte names. Another is a list of acceptable laboratory qualifiers. Still another is soil type classification. Standardized attribute tables help avoid the situation where the same entity is referred to by slightly different names. For example, radium-226 might be written as Ra226, or Ra-226, or Radium226. While all four names are readily recognized by the human eye as referring to the same isotope, such variations cause havoc within electronic relational databases. The end result can be lost records. The data management plan should clearly specify for each data source which data fields require a standardized entry, and should specify where those standardized entries can be found. Ensuring that standardized attribute names have been used is one common data management QA/QC task.

The data management plan should specify the preservation requirements of each of the data sets. Not all data sets will have the same preservation requirements. For example, data collected in field notebooks during monitoring well installation will have preservation requirements that are significantly different from samples collected to satisfy site closure requirements. The data management plan must identify what format a particular data set must be in for preservation purposes, what QA/QC requirements must be met before a data set is ready for preservation, and where the point of storage is.

7.3.3 Identify How Data Sets Will Be Integrated

Because environmental decision making routinely relies on disparate data sets, it is important that the data management plan describe how data sets will be integrated so that effective decisions can be made. This integration typically occurs in two ways, through locational integration and/or through relational integration.

Locational integration relies on the coordinates associated with spatial data to integrate different data sets. GIS software excel at using location integration to tie disparate data sets together. For example, GIS packages allow spatial queries of multiple data sets where the point of commonality is that data lie in the same region of space. In contrast, relational database systems are not efficient at all in using coordinate data to organize information. Locational integration requires that all spatial data be based on the same coordinate system. All sampling and survey data collection programs rely on locational integration to some degree. For this reason it is extremely important that the data management plan specify the default coordinate system for all data collection. In some cases, local coordinate systems are used to facilitate data collection. In these instances, the data management plan needs to specify explicitly the transformation that should be used to bring data with local coordinates into the default coordinate system for the site.

Spatial coordinates, however, are not always completely effective in organizing and integrating different data sets. For example, one might want to compare surface soil sample results with gross gamma activity information for the same location. Unless the coordinate information for both pieces of information is exactly the same, it may be difficult to identify which set of gross activity information is most pertinent to the soil sample of concern. There may be data that are lacking exact coordinate information, but that are clearly tied to an object of interest. An example of this type of data are smear samples from piping or pieces of furniture. In these cases coordinates cannot be used at all for linking data sets. Finally, a site may be organized by investigation area, solid waste management unit, or some similar type of logical grouping. In these cases one would want to be able to organize and integrate different data sets based on these groupings.

For these reasons, relationships are also a common tool for facilitating the integration of data. The most common relational organization of sampling data is the paradigm of sampling station, sample, and sample results. A sampling station is tied to a physical location with definitive coordinates. Samples can refer to any type of data collection that took place at that station, including physical samples of media, direct measurements, observed information, etc. Samples inherit their location from the sampling station. For sampling stations that include soil bores, samples may include a depth from surface to identify their vertical location. One sampling station may have dozens of samples, but each sample is assigned to only one sampling station. Individual samples yield results. Any one sample may yield dozens of results (e.g., a complete suite of gamma spectroscopy data), but each result is tied to one sample. Another example of relationship-based organization is the MARSSIM concept of a final status survey unit. Sampling locations may be assigned to a final status unit, as well as direct measurement data from systems such as *in situ* high-purity germanium gamma spectroscopy and scanning data from mobile NaI data collection. With the proper well-defined relationships, decision makers should be able to select a final status survey unit and have access to all pertinent information for that unit.

When relationships are used for data integration, the data management plan must clearly define the relationships that will be used, as well as the naming nomenclature for ensuring that proper relational connections are maintained.

7.3.4 Data Organization, Storage, Access, and Key Software Components

The data management plan should include schematics for how data from each of the data sets will be stored. Some data such as sampling results may best be handled by a relational database system. Other data, such as gamma walkover data, might better be stored in simple ASCII format. Still other data, such as aerial photographs or interpreted results from nonintrusive geophysical surveys, might best be stored in a raster format. The data management plan must identify these formats, as well as specify software versions if commercial software will be used for these purposes.

Data are collected for decision-making purposes. Consequently, the data management plan should describe how data will be made accessible to decision makers. Decision makers tend to have very individualistic software demands based on past experience and training. A key challenge for the developer of a data management plan is to identify the specific requirements of data users, and then to design a process that can accommodate the various format needs of key decision makers. Recent advances in Internet technologies have the potential for greatly simplifying data accessibility.

7.3.5 Data Flowcharts

A complete data management plan will include data flowcharts that map the flow of information from the point of acquisition through to final preservation. These flowcharts should identify where QA/QC takes place, the points of access for decision makers, and the ultimate repository for data archiving. Data flowcharts should also identify where and how linkages will be made among disparate data sets that require integration. This is essential for determining critical path items that might interrupt the decision-making process. For example, if sample station coordinate identification relies on civil surveys, the merging of survey information with other sample station data may well be a critical juncture for data usability. Staff responsibility for key steps in the data flow process should be assigned in the flowcharts.

7.4 THE PAINESVILLE EXAMPLE

The Painesville, Ohio, site provides an example of how data management can be integrated into a radiological site characterization program. Issues at the Painesville site included surface and subsurface soil contaminated with Th-232, U-238, Th-230, and Ra-226. In addition there was the potential for mixed wastes because of volatile organic compounds and metals contamination in soils. The characterization work planned for the site was intended to expedite the cleanup process within an Engineering Evaluation/Cost Analysis framework. The principal goals of the characterization work were to identify areas with contamination above presumed cleanup goals, delineate the extent of those areas, determine the potential for off-site migration of contamination either through surficial or subsurface pathways, evaluate RCRA characteristic waste concerns, and provide sufficient data to perform a site-specific baseline risk assessment and to allow for an evaluation of potential remedial actions through a feasibility study.

The characterization work was conducted using an Expedited Site Characterization approach that integrated Adaptive Sampling and Analysis Program techniques. In this context, the characterization program fielded a variety of real-time data collection technologies and on-site analytical capabilities. These included nonintrusive geophysics covering selected portions of the site, complete gamma walkover/GPS surveys with two different sensors, an on-site gamma spectrometry laboratory, and gamma screens for soil bore activities. Select subsets of samples were sent off site for a broader suite of analyses. The characterization program was designed so that the selection of sampling locations and the evolution of the data collection would be driven by on-site results. The Painesville characterization work imposed particularly harsh demands on data management. Large amounts of data were generated daily. Timely analysis and presentation of these data were important to keep additional characterization work focused and on-track. The work involved four different contractors on site, with off-site technical support from offices in Tennessee, Illinois, and California. In addition, regulatory staff located elsewhere in Ohio needed to be kept informed of progress, results, and issues.

The data management system devised for the site consisted of several key components. An Oracle environmental data-archiving system was maintained by one of the contractors for long-term data preservation. A second contractor handled data visualization using SitePlanner™ and ArcView™ and organized, presented, and disseminated results using a secure (login and password protected) Web site. Contractors on site had access to the outside world (including the data-archiving system and the Web site) via modem connections. Additional on-site data capabilities included mapping and data analysis with AutoCad™ and Excel™. A detailed data management plan for the characterization work specified roles and responsibilities of various contractors, identified data sources and associated data flow paths, determined levels of QA/QC required for each data set, and specified software and hardware standards for the program.

Gamma walkover data and on-site gamma spectrometry results were screened on site for completeness and egregious errors. These data were then forwarded via modem for a more complete review and analysis to contractors off site. The results of this analysis (including maps, graphics, and tables of data) were made available via the Web site. Maps of gamma walkover data were available for viewing and downloading from the Web site within 24 h of the collection of data. On- and off-site laboratory results were loaded into temporary tables with the Oracle data-archiving system. Every night the contents of these tables were automatically transferred to the Web site so that these data would be available to all project staff. Once formal QA/QC procedures had been completed, data were copied from the temporary tables into permanent data tables for long-term storage.

The Web site served a variety of purposes from a data management perspective. In addition to maps and links to data tables, the Web site also tracked data collection status, served as a posting point for electronic photographs of site activities, summarized results, and provided interim conclusions. The Web site included a secure FTP directory for sharing large project files. The Web site ensured that all project staff worked with the same data sets, whether they were on site or off site. The Web site also allowed regulators to track progress without having to be physically present at the site.

Coordinated, rapid, and reliable access to characterization results provided the characterization program with several key advantages. First, it allowed adjustments to the data collection program to take place "on-the-fly," keeping the data collection as focused and efficient as possible. Second, it forced at least a preliminary review of the quality of all data. This review was able to identify problems quickly and correct them before they became significant liabilities to the program. Examples of problems encountered and corrected at Painesville were malfunctioning gamma sensors and issues with survey control. Third, it allowed additional outside technical support to be brought in at key points without requiring expensive staff to be assigned to the site full-time. Off-site technical support had access to all data via the Web site. Finally, by providing regulators with a "window" into the work being done, regulatory concerns could be quickly identified and addressed.

REFERENCE

EPA (Environmental Protection Agency), Multi-Agency Radiation Survey and Site Investigation Manual (MARSSIM), EPA 402-R-97-016, NUREG-1575, December 1997.

CHAPTER 8

Data Quality Assessment

The term *data quality assessment* (DQA) refers to the five-step EPA process (EPA, 1998) that provides a comparison of the implemented sampling approach and resulting analytical data against the sampling, data quality, and error tolerance requirements specified by the DQOs (Section 4.1.1). Figure 8.1 identifies each of the five steps that make up the DQA process. The results from the DQA are used to determine whether or not the null hypothesis (site is assumed to be contaminated until shown to be clean) can be rejected so that the site or facility can be considered "clean" (having met the remedial action goals). Note that rejecting the null hypothesis provides evidence (not proof) that the site meets the remedial action goals.

The DQA process is designed to evaluate statistically based sample designs (simple random, stratified random, systematic, sequential, etc.). DQA Steps 1 and 2 should be implemented by an analytical chemist (radiochemist), while DQA Steps 3 through 5 should be implemented by a statistician.

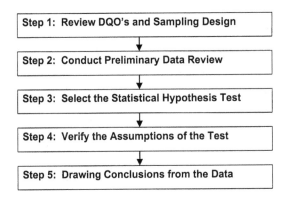

Figure 8.1 Five steps that comprise the DQA process.

8.1 DQA STEP 1: REVIEW DQOs AND SAMPLING DESIGN

Step 1 of the DQA process identifies any discrepancies that exist between the sampling and analytical requirements specified in the DQO and sampling and analysis plan and what was actually performed in the field. The DQA checklist presented in Table 8.1 should be used to assist in performing this evaluation.

This step requires the implementation of the following activities:

- Obtain a copy of the DQA checklist (Table 8.1), project DQO summary report, sampling and analysis plan, data verification/validation packages, maps showing final sampling locations, and any design change notices.
- Review the project DQO summary report and sampling and analysis plan to become familiar with the project data requirements that must be compared with the collected analytical data set.
- Review the data verification/validation packages, maps showing final sampling locations, and design change notices with the intent of identifying any discrepancies that exist between the sampling and analytical requirements specified in the sampling and analysis plan and what was actually performed.
- Complete the DQA checklist presented in Figure 8.1.

8.2 DQA STEP 2: CONDUCT PRELIMINARY DATA REVIEW

This step requires review of the analytical data set, as well as any related quality assurance/quality control reports that are relevant to the project.

As part of this step, the following activities should be performed:

- Review the data verification/validation package and available quality control reports, laboratory audit reports, and any other relevant quality assurance reports that describe the data collection and reporting process as it was actually implemented. Remove all invalid data from the data set. Clearly document the rationale for removing any data from the data set.
- Calculate basic statistical quantities (i.e., summary statistics) from the data set. Examples of statistical quantities include mean, median, percentile, range, standard deviation, and coefficient of variation. Use a spreadsheet to present the results.
- Graph the analytical data to identify distribution patterns and trends and to identify potential problems with the data set. Graphical representations that should be considered include frequency plots, histograms, ranked data plots, normal probability plots, scatter plots, and time plots.

8.3 DQA STEP 3: SELECT THE STATISTICAL HYPOTHESIS TEST

This step requires the selection of the most appropriate statistical hypothesis test for drawing conclusions from the data set. All statistical hypothesis tests make assumptions about the data set. Parametric tests (e.g., one sample t-test) assume that the data have some distributional form (e.g., normal, lognormal), whereas

Table 8.1 DQA Checklist

Task	Completed Yes	Completed No	Name	Date	Explanation
DQO Workbook					
1. Reviewed the project-specific DQO workbook					
1a. Reviewed decision statements					
1b. Reviewed decision rules					
1c. Reviewed the null hypothesis					
1d. Reviewed the gray region and tolerable limits on decision error					
1e. Reviewed the sampling design rationale					
Sampling and Analysis Plan					
2. Reviewed the project-specific sampling and analysis plan					
2a. Reviewed maps showing proposed sampling locations					
2b. Reviewed analytical method, detection limit, and precision and accuracy requirements					
2c. Reviewed field and laboratory quality control sampling requirements (i.e., duplicates, rinsate blanks, matrix spikes)					
2d. Reviewed sample bottle and preservation requirements					
2e. Reviewed field and laboratory quality assurance requirements					
Maps Showing Actual Sampling Locations					
3. Reviewed project-specific maps showing actual sampling locations, and compare against the requirements specified in the DQA report and sampling and analysis plan					
Other					
4a. Laboratory analytical reports					
4b. Field screening data					
4c. Field logbooks					
4d. Chain-of-custody forms					
4e. Maps showing final sampling locations					
4f. Design Change Notices					

nonparametric tests (e.g., Wilcoxon Signed Rank Test) make no distributional assumptions. Table 8.2 presents some of the more common statistical hypothesis tests that are recommended by EPA (1998). The statistical hypothesis tests about a single population are designed for a comparison against a fixed threshold (e.g., a regulatory cleanup guideline), while the statistical hypothesis tests about two populations are designed for comparison between two populations (e.g., investigation site and background).

Table 8.2   List of Statistical Hypothesis Tests for Consideration

Type of Test	Test Name[a]
Tests of Hypotheses about a Single Population	
Test for a mean	One-sample t-test (parametric test)
	Wilcoxon Signed Rank (one-sample) test for the mean (nonparametric test)
Tests for a proportion or percentile	One-sample proportion test
Tests for a median	One-Sample proportion test
	Wilcoxon Signed Rank (one-sample) test for the median
Tests of Hypotheses between Two Populations	
Test for two means	Two-sample t-test
	Satterthwaite's two-sample t-test
Test for two proportions/ two percentiles	Two-sample test for proportions
Nonparametric comparison of two populations	Wilcoxon Rank Sum Test
	Quantile test

[a] Refer to EPA (1998) and Gilbert (1987) for formulas and specific details on these statistical hypothesis tests.

When selecting a statistical hypothesis test, it is important to consider the sensitivity of each test to departures from the assumptions. When small sample populations (i.e., fewer than ten samples) are being assessed, it is recommended that a nonparametric statistical hypothesis test be selected to draw conclusions from the data. This selection will avoid incorrectly assuming that the data are normally distributed when there is simply not enough information to test this assumption. In all cases, the rationale for the selected statistical hypothesis test should be clearly documented.

This step requires the implementation of the following activities:

- Review the statistical quantities and graphical data plots generated in DQA Step 2.
- Select the appropriate statistical hypothesis test and document all of the assumptions made about the data set to justify the selection.
- Note any sensitive assumptions where relatively small deviations could jeopardize the validity of the test results.

8.4 DQA STEP 4: VERIFY THE ASSUMPTIONS OF THE STATISTICAL HYPOTHESIS TEST

This step is performed to assess the validity of the statistical hypothesis test chosen in DQA Step 3. DQA Step 4 is used to determine whether the data support the underlying assumptions necessary for the selected test, or whether the data set must be transformed before further statistical analysis, or whether another statistical hypothesis test must be chosen. The graphical representations of the data developed in DQA Step 2 (Section 8.2) should be used to provide important qualitative information about the reasonableness of the assumptions. Table 8.3 presents a list of the statistical analyses that should be considered.

Table 8.3   Statistical Analyses for Verifying Assumptions

Type of Test	Name of Test[a]
Tests for distributional assumptions	Shapiro Wilk W Test
	Filliben's statistic
	Coefficient of variation test
	Skewness and Kurtosis tests
	Geary's test
	Range test
	Chi-Square test
	Lilliefors Kolmogorov-Smirnoff test
Tests for trends	Regression-based methods:
	• Estimation of a trend using slope of regression line
	• Testing for trends using regression methods
	General trend estimation methods:
	• Sen's slope estimator
	• Seasonal Kendall slope estimator
	Hypothesis tests for detection trends:
	• One observation per time period for one sampling location
	• Multiple observations per time period for one sampling location
	• Multiple sampling locations with multiple observations
	• One observation for one station with multiple seasons
Outliers	Extreme value test
	Discordance test
	Extreme value test (Dixon's test)
	Rosner's test
	Walsh's test
	Multivariate outliers
Test for dispersions	Confidence intervals for a single variance
	The F-test for the equality of two variances
	Bartlett's test for the equality of two or more variances
	Levene's test for the equality of two or more variances

[a] Refer to EPA (1998) and Gilbert (1987) for specific details on these statistical analyses.

If the results from this statistical analysis support the key assumptions of the statistical hypothesis test, the DQA process continues on to DQA Step 5 (Section 8.5), where conclusions are drawn from the data. However, if one or more of the assumptions are questioned, then one must return to DQA Step 3 (Section 8.3) and reevaluate the selection of the most appropriate statistical hypothesis test.

This step requires the implementation of the following activities:

- Review the assumptions about the data set used to justify the statistical hypothesis test selection.
- Use the graphical representations of the data set developed in DQA Step 2 (Section 8.2) to provide an initial determination of the reasonableness of the assumptions.
- Perform a statistical analysis of the data set to confirm or reject the assumptions of the statistical hypothesis test selected in DQA Step 3 (Section 8.3).
- If the results from this assessment support the key assumptions of the statistical hypothesis test, proceed to DQA Step 5 (Section 8.5); otherwise, return to DQA Step 3 (Section 8.3) and reevaluate the most appropriate statistical hypothesis test.

8.5 DQA STEP 5: DRAWING CONCLUSIONS FROM DATA

In this step, the selected statistical hypothesis test is performed and conclusions are drawn from the results. The results from the statistical hypothesis test shall either (a) *reject the null hypothesis* (site is assumed to be contaminated until shown to be clean), or (b) *fail to reject the null hypothesis*. In case (a), the data have provided the evidence needed to reject the null hypothesis, so the decision can be made that the site is now "clean" (having met remedial action goals) with sufficient confidence and without further analysis. This is acceptable because the statistical hypothesis test inherently controls the false-positive decision error rate within the data user's tolerable limits.

In case (b), the data do not provide sufficient evidence to reject the null hypothesis. Therefore, the data shall be analyzed further to determine whether the data user's tolerable limits on false-negative decision errors have been satisfied (see Figure 8.2).

The overall performance of the sampling shall be evaluated by performing a statistical power calculation on the statistical hypothesis test over the range of possible parameter values. The power of a statistical test is defined as the probability of rejecting the null hypothesis when the null hypothesis is false. A power analysis helps evaluate the adequacy of the sampling design when the true parameter value lies near the action level.

This step requires the implementation of the following activities:

- Perform the selected statistical hypothesis test.
- Use the flowchart presented in Figure 8.2 to identify the activities to be performed based on the results from the statistical hypothesis test.
- Summarize the results from DQA Steps 1 through 5 in the DQA summary report.

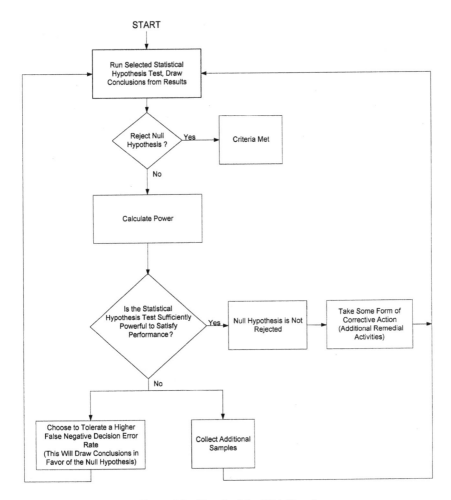

Figure 8.2 Flowchart for DQA Step 5.

REFERENCES

EPA (Environmental Protection Agency), Guidance Document on the Statistical Analysis of Groundwater Monitoring Data at RCRA Facilities, EPA/530/R-93/003, U.S. Environmental Protection Agency, Washington, D.C., 1992.

EPA (Environmental Protection Agency), Guidance for the Data Quality Objectives Process, EPA QA/G-4, U.S. Environmental Protection Agency, Washington, D.C., 1994.

EPA (Environmental Protection Agency), Guidance for Data Quality Assessment—Practical Methods for Data Analysis, EPA QA/G-9, U.S. Environmental Protection Agency, Washington, D.C., 1998.

Gilbert, R.O., *Statistical Methods for Environmental Pollution Monitoring*, Van Nostrand Reinhold, New York, 1987.

CHAPTER 9

Equipment Decontamination

When performing environmental investigations, all sampling equipment must be treated as if it is contaminated, and therefore should be thoroughly decontaminated between sampling points. Decontamination procedures for radiologically and chemically contaminated equipment are presented in Sections 9.1 and 9.2, respectively.

Decontamination is defined as the process of removing, neutralizing, washing, and rinsing surfaces of equipment and personal protective clothing to minimize the potential for contaminant migration. Decontamination is performed to assure the collection of representative environmental samples. The only way to eliminate decontamination is by using disposable equipment.

It is critical to test the effectiveness of any decontamination procedure so that the credibility of environmental samples cannot be questioned. For radiological contaminants, this is accomplished by scanning the decontaminated equipment and by collecting swipe samples for radiological counting. If the results from the scanning measurements and swipe sampling exceed radiological release limits, additional decontamination steps must be taken to remove the contamination. If the results are below the release limit, the equipment may be released for use in collecting additional samples.

For highly radioactive environments, the equipment may not be able to be decontaminated sufficiently to meet the release limits. Sampling equipment that cannot meet the release limits should be disposed of. Large equipment (used to support sampling activities) that cannot meet the release limits may in some instances still be used as long as it does not come in direct contact with the samples being collected. In this situation, every effort should be made to reduce the amount of removable contamination.

For chemical contaminants, testing the effectiveness of the decontamination procedure is accomplished through the preparation of equipment rinsate blank samples which are prepared by pouring ASTM Type II reagent-grade (or equivalent) water over decontaminated sampling equipment, and collecting the rinsate in sample bottles. The rinsate sample is then shipped to the laboratory and is analyzed for the same parameters as the environmental sample. If the results from the analysis of equipment rinsate samples identify contamination, the chemical decontamination procedure was proved to be ineffective. In this case, the analytical data for any samples collected with the contaminated equipment should not be used.

The decontamination of large equipment and sampling equipment should be performed as close to the study area as possible to prevent the possible spread of contamination.

9.1 RADIOLOGICAL DECONTAMINATION PROCEDURE

The following four methods are available to support the decontamination of radiologically contaminated equipment, and are presented in order for increasing levels of contamination. One should begin with the first decontamination method that is practical. If the first method used is not effective in removing the contamination, proceed to the next method.

9.1.1 Tape Method

1. Apply masking tape or duct tape to the surface of the equipment.
2. Remove the tape.
3. Repeat Steps 1 and 2 until the entire surface of the equipment has been covered.
4. Scan the surfaces of the equipment for gross alpha and gross beta/gamma activity to determine if equipment meets radiological release limits.
5. Collect swipe samples from the areas showing the highest activity (based on the scanning measurements) to determine if equipment meets radiological release limits for removable gross alpha and gross beta/gamma activity.
6. If radiological release limits have been met, release the equipment for reuse. If radiological release limits have not been met, proceed to Section 9.1.2 and perform the next level of decontamination.
7. Document the results from the scanning surveys and the swipe samples.
8. When practical, wrap the equipment in plastic sheeting or aluminum foil to keep it clean prior to use.

9.1.2 Manual Cleaning Method

1. Remove soil adhering to equipment by scraping, brushing (wire brush), or wiping.
2. Wash thoroughly with a strong nonphosphate detergent/soap wash.
3. Rinse equipment with tap water.
4. Rinse equipment with ASTM Type II (or equivalent) water. Note, this step is only required when decontaminating sampling equipment.
5. Allow equipment to air dry.
6. Scan the surfaces of the equipment for gross alpha and gross beta/gamma activity to determine if equipment meets radiological release limits.
7. Collect swipe samples from the areas showing the highest activity (based on the scanning measurements) to determine if equipment meets radiological release limits.
8. If radiological release limits have been met, release the equipment for reuse. If radiological release limits have not been met, proceed to Section 9.1.3.
9. Document the results from the scanning surveys and the swipe samples.

10. When practical, wrap the equipment in plastic sheeting or aluminum foil to keep it clean prior to use.

9.1.3 HEPA Vacuum Method

1. Remove soil adhering to equipment by scraping, brushing (wire brush), or wiping.
2. Use a HEPA Vacuum system to vacuum surfaces of the equipment.
3. Rinse equipment with tap water.
4. Rinse equipment with ASTM Type II (or equivalent) water. Note, this step is only required when decontaminating sampling equipment.
5. Allow equipment to air dry.
6. Scan the surfaces of the equipment for gross alpha and gross beta/gamma activity to determine if equipment meets radiological release limits.
7. Collect swipe samples from the areas showing the highest activity (based on the scanning measurements) to determine if equipment meets radiological release limits.
8. If radiological release limits have been met, release the equipment for reuse. If radiological release limits have not been met, proceed to Section 9.1.4.
9. Document the results from the scanning surveys and the swipe samples.
10. When practical, wrap the equipment in plastic sheeting or aluminum foil to keep it clean prior to use.

9.1.4 High-Pressure Wash Method

1. Remove soil adhering to equipment by scraping, brushing (wire brush), or wiping.
2. Use a pressure washer to steam clean equipment thoroughly with potable water and a nonphosphatic laboratory-grade detergent. The pressure washer should produce a minimum water pressure of 80 psi and a minimum temperature of 180°F.
3. Use a pressure washer to rinse equipment with tap water.
4. Rinse equipment with ASTM Type II (or equivalent) water. Note, this step is only required when decontaminating sampling equipment.
5. Allow equipment to air dry.
6. Scan the surfaces of the equipment for gross alpha and gross beta/gamma activity to determine if equipment meets radiological release limits.
7. Collect swipe samples from the areas showing the highest activity (based on the scanning measurements) to determine if equipment meets radiological release limits.
8. If radiological release limits have been met, release the equipment for reuse. If radiological release limits have not been met for sampling equipment, dispose of equipment. If radiological release limits have not been met for large equipment, it may be considered for reuse only if does not come in direct contact with the samples being collected.
9. Document the results from the scanning surveys and the swipe samples.
10. When practical, wrap the equipment in plastic sheeting or aluminum foil to keep it clean prior to use.

9.2 CHEMICAL DECONTAMINATION PROCEDURE

9.2.1 Large Equipment

The following method should be used to decontaminate all types of large equipment used to support chemical sampling activities (e.g., drilling augers, drill bit, A-rod).

1. Remove soil adhering to equipment by scraping, brushing (wire brush), or wiping.
2. Use a pressure washer to steam clean equipment thoroughly with potable water and a nonphosphatic laboratory-grade detergent. The pressure washer should produce a minimum water pressure of 80 psi and a minimum temperature of 180°F.
3. Use a pressure washer to rinse equipment with tap water.
4. Allow equipment to air dry.
5. When practical, wrap the equipment in plastic sheeting or aluminum foil to keep it clean prior to use.

9.2.2 Sampling Equipment

The following methods should be used to decontaminate all types of equipment used to perform chemical sampling activities. This equipment includes but is not limited to:

- Deep soil sampling equipment (e.g., split-spoon, thin-walled tube samplers)
- Shallow soil sampling equipment (e.g., hand augers, core sampler)
- Sediment sampling equipment (e.g., box sampler, core sampler)
- Surface water/groundwater sampling equipment (e.g., dipper, bailer, bladder pump)
- Concrete/asphalt sampling equipment (e.g., core barrel)
- Paint sampling equipment (e.g., spatula)
- Air sampling equipment
- Support equipment coming in contact with sample (e.g., discharge lines, bowls, spoons)

When samples are to be analyzed for inorganics, a dilute hydrochloric or nitric acid solution rinse must be used in the decontamination procedure. Dilute hydrochloric acid is generally preferred over nitric acid when cleaning stainless steel because nitric acid may oxidize this material.

When samples are to be analyzed for organic contaminants, the equipment decontamination procedure should include a rinse with pesticide-grade acetone and hexane, in that order. Pesticide-grade methanol may be used instead of acetone. When samples are to be analyzed for both inorganics and organics, a dilute hydrochloric or nitric acid solution rinse must be used in combination with a pesticide-grade acetone and hexane rinse.

Below are three equipment decontamination procedures. The first procedure should be used when samples are to be analyzed for inorganic parameters only. The second procedure should be used when samples are to be analyzed for organic parameters only, and the third should be used if samples are to be analyzed for both organic and inorganic parameters.

Inorganic Procedure
1. Remove soil adhering to equipment by scraping, brushing (wire brush), or wiping.
2. Wash thoroughly with a strong nonphosphate detergent/soap wash.
3. Rinse thoroughly with tap water.
4. Rinse with ASTM Type II (or equivalent) water.
5. Rinse with dilute hydrochloric or nitric acid solution.
6. Rinse with ASTM Type II (or equivalent) water.
7. Allow equipment to air dry.
8. Wrap equipment in aluminum foil (shiny side out) to keep it clean prior to use.

Organic Procedure
1. Remove soil adhering to equipment by scraping, brushing (wire brush), or wiping.
2. Wash thoroughly with a strong nonphosphate detergent/soap wash.
3. Rinse thoroughly with tap water.
4. Rinse with ASTM Type II (or equivalent) water.
5. Rinse with pesticide-grade acetone (or methanol).
6. Rinse with pesticide-grade hexane.
7. Allow equipment to air dry.
8. Wrap equipment in aluminum foil (shiny side out) to keep it clean prior to use.

Combined Inorganic/Organic Procedure
1. Remove soil adhering to equipment by scraping, brushing (wire brush), or wiping.
2. Wash thoroughly with a strong nonphosphate detergent/soap wash.
3. Rinse thoroughly with tap water.
4. Rinse with ASTM Type II (or equivalent) water.
5. Rinse with dilute hydrochloric or nitric acid solution.
6. Rinse with ASTM Type II (or equivalent) water.
7. Rinse with pesticide-grade acetone (or methanol).
8. Rinse with pesticide-grade hexane.
9. Allow equipment to air dry.
10. Wrap equipment in aluminum foil (shiny side out) to keep it clean prior to use.

The dilute acid solution, acetone, hexane, and distilled/deionized water are most easily handled when they are contained within Teflon squirt bottles. Squirt bottles made of substances other than Teflon should not be used, since they have the potential to contaminate the solutions. When using these solutions, the drippings should be caught in a bucket or tub. The acetone and hexane drippings should be allowed to volatilize into the air, while the acid solution drippings should be neutralized using baking soda.

Decontaminating pumps is more difficult than other sampling equipment since many of them are not easy to disassemble. The procedure to use for these tools is to place the pump in a large decontamination tub full of nonphosphate detergent and tap water. The pump is turned on, which forces the wash water through the pump and discharge line. The same procedure is repeated in a tub full of clean tap water. The outside of the pump and outside of the pressure and discharge lines are then decontaminated using the sampling decontamination methods described above.

All of the water generated from this decontamination procedure must be containerized and then analytically tested to determine the appropriate method for disposal. Discharge permits must be acquired from the city and/or county prior to discharging decontamination water into any sanitary or storm sewer system.

REFERENCES

EPA (Environmental Protection Agency), A Compendium of Superfund Field Operations Methods, EPA/540/P-87/001a, 1987.

EPA (Environmental Protection Agency), RCRA Facility Investigation (RFI) Guidance, PB89-200299, 1989.

Data Quality Objectives Summary
Report Template*

Data Quality Objectives Summary Report for

[Add Project Description]

Prepared for the *[Note: Insert name of organization]*

Submitted by *[Note: Add company name]*

* Appendix A appears on the CD accompanying this book.

Data Quality Objectives Summary
Report for *[Add Project Description]*

Author(s)

[Add Author Name(s)]
[Add Company Name(s)]

Date Published

[Add Publication Date]

Executive Summary

[Insert executive summary text here.]

Contents

Tables

Acronyms

[List all of the acronyms used in the workbook. The following list is only an example.]

COC   contaminant of concern
COPC   contaminant of potential concern
D&D   decontamination and decommissioning
DQO   data quality objective
DR   decision rule
DS   decision statement
EPA   U.S. Environmental Protection Agency
LBGR   lower bound of the gray region
PQL   practical quantitation limit
PSQ   principal study question
SAP   sampling and analysis plan
UCL   upper confidence level
WS   waste stream

Metric Conversion Chart

	Into Metric Units			Out of Metric Units		
If You Know	*Multiply By*	*To Get*		*If You Know*	*Multiply By*	*To Get*
Length				**Length**		
inches	25.4	millimeters		millimeters	0.039	inches
inches	2.54	centimeters		centimeters	0.394	inches
feet	0.305	meters		meters	3.281	feet
yards	0.914	meters		meters	1.094	yards
miles	1.609	kilometers		kilometers	0.621	miles
Area				**Area**		
sq. inches	6.452	sq. centimeters		sq. centimeters	0.155	sq. inches
sq. feet	0.093	sq. meters		sq. meters	10.76	sq. feet
sq. yards	0.0836	sq. meters		sq. meters	1.196	sq. yards
sq. miles	2.6	sq. kilometers		sq. kilometers	0.4	sq. miles
acres	0.405	hectares		hectares	2.47	acres
Mass (weight)				**Mass (weight)**		
ounces	28.35	grams		grams	0.035	ounces
pounds	0.454	kilograms		kilograms	2.205	pounds
ton	0.907	metric ton		metric ton	1.102	ton
Volume				**Volume**		
teaspoons	5	milliliters		milliliters	0.033	fluid ounces
tablespoons	15	milliliters		liters	2.1	pints
fluid ounces	30	milliliters		liters	1.057	quarts
cups	0.24	liters		liters	0.264	gallons
pints	0.47	liters		cubic meters	35.315	cubic feet
quarts	0.95	liters		cubic meters	1.308	cubic yards
gallons	3.8	liters				
cubic feet	0.028	cubic meters				
cubic yards	0.765	cubic meters				
Temperature				**Temperature**		
Fahrenheit	Subtract 32, then multiply by 5/9	Celsius		Celsius	Multiply by 9/5, then add 32	Fahrenheit
Radioactivity				**Radioactivity**		
picocuries	37	millibecquerel		millibecquerel	0.027	picocuries

1.0 Step 1—State the Problem

The purpose of this data quality objective (DQO) process is to support decision-making activities as they pertain to the remediation of *[Note: Add the name of the site to be remediated]*. The objective of DQO Step 1 is to use the information gathered from the DQO scoping process, as well as other relevant information to state the problem to be resolved clearly and concisely. The free-form text sections included in this step are intended to define the task objectives and assumptions, present the task issues, summarize the site background information, and provide a concise statement of the problem. The tables provided in this section are designed to document the personnel involved in the DQO process, identify the contaminants of concern, and summarize the key information needed to support the writing of the problem statement.

1.1 PROJECT OBJECTIVES

[Note on completion of Section 1.1: Clearly state the task objectives as they pertain to remediation activities. Begin discussing the task objectives on a large scale, then focus the discussion on the site-specific objectives.]

1.2 PROJECT ASSUMPTIONS

[Note on completion of Section 1.2: Clearly state all of the task-specific assumptions that have been made based on DQO Team discussions and interviews with the regulators.]

1.3 PROJECT ISSUES

1.3.1 Global Issues

[Note on completion of Section 1.3.1: Present the date when the global issues meeting was held, and note the organizations that were represented at the meeting. List each of the global issues that were discussed and the resolutions that were agreed upon.]

1.4 EXISTING REFERENCES

Table 1-1 presents a list of all of the references that were reviewed as part of the scoping process, as well as a summary of the pertinent information contained within each reference. These references are the primary source for the background information presented in Section 1.5.

Table 1-1 Existing References

Reference	Summary

1.5 SITE BACKGROUND INFORMATION

[*Note on completion of Section 1.5:* In this section, provide background information about the site under investigation. This information will be used to support the development of the problem statement. This section should address the following key areas:

- Site description
- Site history
- Specific areas within the site to be investigated
- Summary of all recorded spills and releases
- Summary of historical analytical data.]

1.6 DQO TEAM MEMBERS AND REGULATORS

Individual members of the DQO Team were carefully selected to participate in the seven-step DQO process based on their technical background to provide expertise in all of the technical areas needed to meet the task objectives. The regulators included representatives from the [*Note:* Insert the name of the regulatory agencies]. The role of the regulators was to make final decisions related to the sampling design.

Tables 1-2 and 1-3 identify each of the individual members of the DQO team and the regulators. These tables also identify the organization that each DQO team member or regulator represents, as well as their technical area of expertise and telephone number.

Table 1-2 DQO Team Members

Name	Organization	Role and Responsibility	Telephone Number

Table 1-3 Regulators

Name	Organization	Role and Responsibility	Telephone Number

1.7 PROJECT BUDGET AND CONTRACTUAL VEHICLES

Table 1-4 presents the budget for all of the task activities associated with the development and implementation of the sampling program, the performance of laboratory analyses, the performance of the data quality assessment, and the evaluation and reporting of investigation results. For those activities that need to be subcontracted, Table 1-4 presents the available contractual vehicles.

Table 1-4 Task Budget and Contractual Vehicles

Task Activities	Budget	Contractual Vehicle
DQO development		
Sampling and analysis plan development		
Field implementation		
Laboratory analyses		
Data quality assessment		
Documentation of investigation results		

1.8 MILESTONE DATES

Table 1-5 presents the milestone dates for the completion of all of the task activities associated with the development and implementation of the sampling program, the performance of laboratory analyses, the performance of a data quality assessment, and the evaluation and reporting of investigation results.

Table 1-5 Milestone Dates

Task Activities	Milestone Date
DQO workbook development	
Sampling and analysis plan development	
Field implementation	
Laboratory analyses	
Data quality assessment	
Documentation of investigation results	

1.9 CONTAMINANTS OF CONCERN

A list of the contaminants of concern (COCs) for the site under investigation was generated by initially listing all of the contaminants of potential concern (COPCs) based on historical process operations. Certain COPCs were then removed from the list if they were being addressed under either a separate sampling and analysis plan (SAP) or a waste management plan. Certain COPCs may also have been removed if they have a short half-life, are not regulated, pose no risk, or are nontoxic, or if process knowledge/analytical data confirms that insignificant releases have occurred.

1.9.1 Total List of Contaminants of Potential Concern

Table 1-6 identifies all of the COPCs for each of the types of media to undergo remediation.

1.9.2 Contaminants of Potential Concern Addressed by Concurrent Remediation Activities

The scope of a DQO prepared in support of remediation activities typically assumes the responsibility for all media at the site. However, if certain media and associated COPCs are already being addressed by concurrent activities (i.e., under a separate SAP or waste management plan), these media/associated COPCs may be excluded from further consideration from this DQO. Table 1-7 presents a list of the COPCs that are being removed from the total list of COPCs for this reason.

Table 1-6 Total List of COPCs for Each Media Type

Media	Known or Suspected Source of Contamination	Type of Contamination (General)	COPCs (Specific)

Table 1-7 COPCs Addressed by Concurrent Remediation Activities

Media	COPCs	Remediation Activity

1.9.3   Other Contaminant of Potential Concern Exclusions

Table 1-8 presents a list of all other COPCs to be excluded from the investigation. These exclusions are in addition to the exclusions identified in Table 1-7 and are based on physical laws, process knowledge, task focus, or other mitigating factors. Table 1-8 provides the specific rationale for the exclusion of each of the identified COPCs.

Table 1-8 Rationale for COPC Exclusions

Media	COPCs	Rationale for Exclusion

1.9.4   Final List of Contaminants of Concern

Table 1-9 presents the final list of COCs for each media to be carried through the remainder of the DQO process.

Table 1-9 Final List of COCs

Media	COCs

1.9.5   Distribution of Contaminants of Concern

Table 1-10 identifies the best understanding of how each of the COCs arrived at the site and the fate and transport mechanisms (e.g., wind or water) that may have impacted the distribution (e.g., layering or lateral homogeneity) of each of the COCs.

Table 1-10 Distribution of COCs

Media	COCs	How COC Arrived at Site	Fate and Transport Mechanisms	Expected Distribution (Heterogeneous/ Homogeneous)

1.10   CURRENT AND POTENTIAL FUTURE LAND USE

The current and potential future uses for the land in the immediate vicinity of the site under investigation are summarized in Table 1-11. This information is needed later in the DQO process to support the evaluation of decision error consequences.

1.11 CONCEPTUAL SITE MODEL

The goal of the DQO process is to develop a sampling design that will either confirm or reject the conceptual site model. The conceptual site model is continuously being refined as more data become available. Table 1-12 presents a tabular

Table 1-11 Current and Potential Future Land Use

Current Land Use	Potential Future Land Use

depiction of the conceptual site model, identifying the sources, release mechanisms, migration pathways, and potential receptors for each of the COCs. This table also summarizes the exposure scenarios.

Table 1-12 Tabular Depiction of the Conceptual Site Model

Media	COCs	Source	Release Mechanism	Migration Pathways	Potential Receptors
Exposure Scenario:					
Exposure Scenario:					

1.12 STATEMENT OF THE PROBLEM

[*Note on completion of Section 1.12:* Combine the relevant background information into a concise statement of the problem to be resolved in a descriptive text format. The problem description should be concise but comprehensive. This discussion should be a comprehensive summary of the problem to be resolved and should provide a textual description of the information provided in the previous tables, particularly Table 1-12.]

2.0 Step 2—Identify the Decision

The purpose of DQO Step 2 is to define the principal study questions (PSQs) that need to be resolved to address the problem identified in DQO Step 1 and the alternative actions that would result from the resolution of the PSQs. The PSQs and alternative actions are then combined into decision statements that express a choice among alternative actions. Table 2-1 presents the task-specific PSQs, alternative actions, and resulting decision statements. This table also provides a qualitative assessment of the severity of the consequences of taking an alternative action if it is incorrect. This assessment takes into consideration human health, environment (flora/fauna), political, economic, and legal ramifications. The severity of the consequences is expressed as low, moderate, or severe.

Table 2-1 Summary of DQO Step 2 Information

PRINCIPAL STUDY QUESTION #1—			
PSQ -AA#	Alternative Action	Description of Consequences of Implementing the Wrong Alternative Action	Severity of Consequences (Low/Moderate/Severe)

Decision Statement #1—

PRINCIPAL STUDY QUESTION #2—			
PSQ -AA#	Alternative Action	Description of Consequences of Implementing the Wrong Alternative Action	Severity of Consequences (Low/Moderate/Severe)

Decision Statement #2—

PRINCIPAL STUDY QUESTION #3—			
PSQ -AA#	Alternative Action	Description of Consequences of Implementing the Wrong Alternative Action	Severity of Consequences (Low/Moderate/Severe)

Decision Statement #3—

3.0 Step 3 – Identify Inputs to the Decision

The purpose of DQO Step 3 is to identify the type of data needed to resolve each of the decision statements identified in DQO Step 2. This data may already exist or may be derived from computational or surveying/sampling and analysis methods. Analytical performance requirements (e.g., practical quantitation limit, or PQL, requirement, precision, and accuracy) are also provided in this step for any new data that need to be collected.

3.1 INFORMATION REQUIRED TO RESOLVE DECISION STATEMENTS

Table 3-1 specifies the information (data) required to resolve each of the decision statements identified in Table 2-1 and identifies whether these data already exist. For the data that are identified as existing, the source references for the data have been provided with a qualitative assessment as to whether or not the data are of sufficient quality to resolve the corresponding decision statement. The qualitative assessment of the existing data was based on the evaluation of the corresponding quality control data (e.g., spikes, duplicates, blanks), detection limits, data collection methods, etc.

Table 3-1 Required Information and Reference Sources

DS #	Remediation Variable	Required Data	Does Data Exist? (Y/N)	Source Reference	Sufficient Quality (Y/N)	Additional Information Required? (Y/N)

3.2 BASIS FOR SETTING THE ACTION LEVEL

The action level is the threshold value that provides the criterion for choosing between alternative actions. Table 3-2 identifies the basis (i.e., regulatory threshold or risk-based) for establishing the action level for each of the COCs. The numerical value for the action level is defined in DQO Step 5.

Table 3-2 Basis for Setting Action Level

DS #	Remediation Variable	COCs	Basis for Setting Action Level

3.3 COMPUTATIONAL AND SURVEY/ANALYTICAL METHODS

Table 3-3 identifies the decision statements where existing data either do not exist or are of insufficient quality to resolve the decision statements. For these decision statements, Table 3-3 presents computational and/or surveying/sampling methods that could be used to obtain the required data.

Table 3-3 Information Required to Resolve the Decision Statements

DS #	Remediation Variable	Required Data	Computational Methods	Survey/Analytical Methods

Table 3-4 presents details on the computational methods identified in Table 3-3. These details include the source and/or author of the computational method and information on how the method could be applied to this study.

Table 3-4 Details on Identified Computational Methods

DS #	Computational Method	Source/Author	Application to Study

Table 3-5 identifies each of the survey and/or analytical methods that may be used to provide the required information needed to resolve each of the decision statements. The possible limitations associated with each of these methods are also provided with the estimated cost.

Table 3-5 Potentially Appropriate Survey/Analytical Methods

DS #	Remediation Variable	Potentially Appropriate Survey/Analytical Method	Possible Limitations	Cost

3.4 ANALYTICAL PERFORMANCE REQUIREMENTS

Table 3-6 defines the analytical performance requirements for the data that need to be collected to resolve each of the decision statements. These performance requirements include the PQL and precision and accuracy requirements for each of the COCs.

Table 3-6 Analytical Performance Requirements

DS #	COCs	Survey/ Analytical Method	Preliminary Action Level	PQL	Precision Requirement	Accuracy Requirement

4.0 Step 4—Define the Boundaries of the Study

The primary objective of DQO Step 4 is to identify the population of interest, define the spatial and temporal boundaries that apply to each decision statement, define the scale of decision making, and identify any practical constraints (hindrances or obstacles) that must be taken into consideration in the sampling design. Implementing this step ensures that the sampling design will result in the collection of data that accurately reflect the true condition of the site under investigation.

4.1 POPULATION OF INTEREST

Prior to defining the spatial and temporal boundaries of the site under investigation, it is first necessary to define clearly the populations of interest that apply for each decision statement (Table 4-1). The intent of Table 4-1 is to clearly define the attributes that make up each population of interest by stating them in a way that makes the focus of the study unambiguous.

Table 4-1 Characteristics that Define the Population of Interest

DS #	Population of Interest	Unit Measurement Size	Total Number of Potential Measurement Units within the Population

4.2 GEOGRAPHIC BOUNDARIES

Table 4-2 identifies the geographic boundaries that apply to each decision statement. Limiting the geographic boundaries of the study area ensures that the investigation does not expand beyond the original scope of the task.

Table 4-2 Geographic Boundaries of the Investigation

DS #	Geographic Boundaries of the Investigation

4.3 ZONES WITH HOMOGENEOUS CHARACTERISTICS

Table 4-3 defines the zones within the site under investigation that have relatively homogeneous characteristics. These zones were identified by using existing information to segregate the elements of the population into subsets that exhibit relatively homogeneous characteristics (e.g., types of contaminants). Dividing the site into separate zones reduces the overall complexity of the problem by breaking the site into more manageable pieces.

Table 4-3 Zones with Homogeneous Characteristics

DS #	Population of Interest	Zone	Homogeneous Characteristic Logic

4.4 TEMPORAL BOUNDARIES

Table 4-4 identifies temporal boundaries that may apply to each decision statement. The temporal boundary refers to both the time frame over which each decision statement applies (e.g., number of years) and when (e.g., season, time of day, weather conditions) the data should optimally be collected.

Table 4-4 Temporal Boundaries of the Investigation

DS #	Time Frame	When to Collect Data

4.5 SCALE OF DECISION MAKING

In Table 4-5, the scale of decision making has been defined for each decision statement. The scale of decision making is defined by joining the population of interest and the geographic and temporal boundaries of the area under investigation.

Table 4-5 Scale of Decision Making

| DS # | Population of Interest | Geographic Boundary | Temporal Boundary | | Scale of Decision |
			Time Frame	When to Collect Data	

4.6 PRACTICAL CONSTRAINTS

Table 4-6 identifies all of the practical constraints that may impact the data collection effort. These constraints include physical barriers, difficult sample matrices, high radiation areas, or any other condition that will need to be taken into consideration in the design and scheduling of the sampling program.

Table 4-6. Practical Constraints on Data Collection

5.0 Step 5—Develop a Decision Rule

The purpose of DQO Step 5 is initially to define the statistical parameter of interest (i.e., mean, median) that will be used for comparison against the action level. The statistical parameter of interest specifies the characteristic or attribute that the regulator would like to know about the population. The final action level for each of the COCs is also identified in DQO Step 5. When this is established, a decision rule is developed for each decision statement in the form of an "if ... then ..." statement that incorporates the parameter of interest, the scale of decision making, the action level, and the alternative action(s) that would result from resolution of the decision. Note that the scale of decision making and alternative actions were identified earlier in DQO Steps 4 and 2, respectively.

5.1 INPUTS NEEDED TO DEVELOP DECISION RULES

Tables 5-1 and 5-2 present all of the information needed to formulate the decision rules identified in Section 5.2. This information includes the decision statements and alternative actions identified earlier in DQO Step 2, the scale of decision making identified in DQO Step 4, the statistical parameter of interest, and the final action levels for each of the COCs.

Table 5-1 Decision Statements

DS #	Decision Statement

Table 5-2 Inputs Needed to Develop Decision Rules

DS#	COCs	Statistical Parameter of Interest	Scale of Decision Making	Final Action Level	Alternative Actions

5.2 DECISION RULES

Table 5-3 presents decision rules that correspond to each of the decision statements identified in Table 5-1.

Table 5-3 Decision Rules

DS #	DR #	Decision Rule

6.0 Step 6—Specify Tolerable Limits on Decision Errors

Since analytical data can only estimate the true condition of the site under investigation, decisions that are made based on measurement data could potentially be in error (i.e., decision error). For this reason, the primary objective of DQO Step 6 is to determine which decision statements (if any) require a statistically based sample design. For those decision statements requiring a statistically based sample design, DQO Step 6 defines tolerable limits on the probability of making a decision error.

6.1 STATISTICAL VS. NONSTATISTICAL SAMPLING DESIGN

Table 6-1 provides a summary of the information used to support the selection between a statistical vs. a nonstatistical sampling design for each decision statement. The factors that were taken into consideration in making this selection included the time frame over which each of the decision statements applies, the qualitative consequences of an inadequate sampling design, and the accessibility of the site if resampling is required.

Table 6-1 Statistical vs. Nonstatistical Sampling Design

DS#	Time Frame (years)	Qualitative Consequences of Inadequate Sampling Design (Low/Moderate/Severe)	Resampling Access After Remediation (Accessible/ Inaccessible)	Proposed Sampling Design (Statistical/ Nonstatistical)

6.2 NONSTATISTICAL DESIGNS

For those decision statements to be resolved using a nonstatistical design, there is no need to define the "gray region" or the tolerable limits on decision error, since these only apply to statistical designs. Refer to Section 7.1 for details on the selected nonstatistical sampling design(s).

6.3 STATISTICAL DESIGNS

An initial step in the process of establishing a statistically based sample design is to define the expected range of the statistical parameter of interest (i.e., mean, median) for each COC. Table 6-2 defines the expected statistical parameter of interest concentration ranges for each COC based on the evaluation of historical analytical data.

Table 6-2 Statistical Parameter of Interest Concentration Ranges

DS #	Media	COCs	Statistical Parameter of Interest	Range Lower Limit	Upper Limit

6.4 DECISION ERRORS

The two types of decision error that could occur are as follows: treating (i.e., managing and disposing of) clean site media as if it were *contaminated*, and treating (i.e., managing and disposing of) contaminated site media as if it were *clean*. The decision error that has the more severe consequence is the latter since the error could result in human health and/or ecological impacts.

6.5 NULL HYPOTHESIS

Table 6-3 identifies the null hypothesis that applies to the site under investigation. The term *null hypothesis* refers to the baseline condition of the site, which has been defined based on the historical data and process knowledge identified in the scoping summary report. The null hypothesis states the opposite of what one hopes to demonstrate.

Table 6-3 Defining the Null Hypothesis

Null Hypothesis Statement	Indicate Selection
Site media is assumed to be contaminated until it is shown to be clean.	
Site media is assumed to be clean until it is shown to be contaminated.	

6.6   TOLERABLE LIMITS FOR DECISION ERROR

For each decision statement, Tables 6-4 and 6-5 present the selected statistical design to be implemented (i.e., simple random, random systematic), the boundaries of the gray region, and the probability values to points above and below the gray region that reflect the regulator's tolerable limits for making an incorrect decision.

Table 6-4 Statistical Designs

DS #	Media	Selected Statistical Design	Boundaries of the Gray Region	Tolerable Limits for Incorrect Decision	
				At Lower Bound of Gray Region	At Action Level

Table 6-5 Tolerable Decision Errors

DS #	Media	COCs	Statistical Parameter of Interest[a]	Statistical Parameter of Interest Range[b]	Final Action Level[a]	Gray Region[c]	Tolerable Decision Error	
							At Lower Bound of Gray Region[c,d] (%)	At Action Level[c,e] (%)

[a   Derived from Table 5-2.

b   Derived from Table 6-2.

c   Derived from Table 6-4.

d   Mistakenly concluding > action level.

e   Mistakenly concluding < action level.]

7.0 Step 7—Optimize the Design

The objective of DQO Step 7 is to present alternative data collection designs that meet the minimum data quality requirements specified in DQO Steps 1 through 6. A selection process is then used to identify the most resource-effective data collection design that satisfies all of the data quality requirements. Table 6-3 differentiates between those decision statements that require a statistical sampling design and those that may be resolved using a nonstatistical design.

7.1 NONSTATISTICAL DESIGN

Tables 7-1 through 7-3 have been completed for those decision statements to be resolved using a nonstatistical approach.

7.1.1 Nonstatistical Screening Method Alternatives

Table 7-1 identifies all of the screening technologies that were considered to resolve each decision statement and the optional methods of implementing each technology. The table also summarizes the limitations associated with each screening technology and/or method of implementation and provides an estimated cost for implementation.

Table 7-1 Potential Nonstatistical Screening Alternatives

DS #	Media	Screening Technology	Potential Implementation Designs	Limitations	Cost

7.1.2 Nonstatistical Sampling Method Alternatives

Table 7-2 identifies the various types of media that need to be sampled to resolve each decision statement and alternative methods for collecting these samples. This table presents alternative implementation designs for each sampling method and identifies any limitations that may be associated with each sampling method and/or design. An estimated cost for the implementation of each sampling design has also been provided for comparison purposes.

Table 7-2 Potential Nonstatistical Sampling Method Alternatives

DS #	Media	Sampling Method	Potential Implementation Designs	Limitations	Cost

7.1.3 Nonstatistical Implementation Design

Table 7-3 presents the selected screening technology(s) and sampling method(s) for resolving each decision statement, along with a summary of the proposed implementation design. The table also provides the basis for the selected implementation design.

7.2 STATISTICAL DESIGN

Tables 7-4 through 7-8 have been completed for those decision statements requiring a statistical approach. For each decision statement, these tables identify the statistical hypothesis test selected for testing the null hypothesis, present the formula for calculating the required number of samples/measurements, identify the total number of samples/measurements to be collected and estimated cost for various Type I (α) and Type II (β) error tolerances, present the results from a trade-off analysis, and summarize the final selected statistical sampling/measurement design.

7.2.1 Data Collection Design Alternatives

Table 7-4 identifies the statistical design alternatives (i.e., simple random, stratified random, systematic) that were evaluated for each decision statement, as well as the selected design and the basis for the selection.

7.2.2 Mathematical Expressions for Solving Design Problems

Table 7-5 identifies the statistical hypothesis test (e.g., Wilcoxon Signed Rank Test, one sample t-test) that has been selected for testing the null hypothesis. The table presents the assumptions that were made about the population distribution (i.e., symmetrical, normal) in the selection process, as well as the formula for calculating the required number of samples/measurements.

Table 7-3 Selected Implementation Design

DS #	Media	Selected Screening Technology(s)	Selected Sampling Method(s)	Potential Implementation Designs
1				
Selected Implementation Design:				
Basis for Selection:				

DS #	Media	Selected Screening Technology(s)	Selected Sampling Method(s)	Potential Implementation Designs
2				
Selected Implementation Design:				
Basis for Selection:				

DS #	Media	Selected Screening Technology(s)	Selected Sampling Method(s)	Potential Implementation Designs
3				
Selected Implementation Design:				
Basis for Selection:				

7.2.3 Select the Optimal Sample/Measurement Size That Satisfies the DQO

Table 7-6 presents the total number of samples/measurements required to be collected for each decision statement with varying error tolerances and varying widths of the gray region. The total number of samples/measurements was calculated using the statistical method identified in Table 7-4. As would be expected, the higher the error tolerances and wider the gray region, the smaller the number of samples/measurements that are required.

Table 7-4 Selected Statistical Design

DS #	Media	Statistical Design Alternatives
1		
Selected Statistical Design:		
Basis for Selection:		
DS #	Media	Statistical Design Alternatives
2		
Selected Statistical Design:		
Basis for Selection:		
DS #	Media	Statistical Design Alternatives
3		
Selected Statistical Design:		
Basis for Selection:		

Table 7-5 Statistical Methods for Testing the Null Hypothesis

DS #	Media	Statistical Method Alternatives	Selected Statistical Method for Testing Null Hypothesis	Assumptions Made in Selecting Statistical Method	Formula for Calculating Number of Samples/ Measurements

Table 7-6 Sample/Measurement Size Based on Varying Error Tolerances and LBGR

		Mistakenly Concluding < Action Level		
		α =	α =	α =
DS #1				
LBGR =				
Mistakenly Concluding > Action Level	β =			
	β =			
	β =			
LBGR =				
Mistakenly Concluding > Action Level	β =			
	β =			
	β =			
LBGR =				
Mistakenly Concluding > Action Level	β =			
	β =			
	β =			
DS #2				
LBGR =				
Mistakenly Concluding > Action Level	β =			
	β =			
	β =			
LBGR =				
Mistakenly Concluding > Action Level	β =			
	β =			
	β =			
LBGR =				
Mistakenly Concluding > Action Level	β =			
	β =			
	β =			

7.2.4   Sampling/Measurement Cost

For varying error tolerances, and varying widths of the gray region, Table 7-7 presents the total cost for sampling and analyzing the number of samples identified in Table 7-6. As would be expected, the higher the error tolerances, and the wider the gray region, the lower the sampling and analysis costs.

7.2.5   Selecting the Most Resource-Effective Data Collection Design

A trade-off analysis was performed for the purpose of identifying the most resource optimal number of samples/measurements for the given budget. It is important to consider trade-offs so contingency plans can be developed and the added value of selecting one set of considerations over another can be quantified. Table 7-8 identifies the sampling/measurement design that provides the best balance between cost (or expected cost) and the ability to meet the DQOs, and a selection was made.

7.2.6   Final Statistical Sampling Design

The results from the trade-off analysis were evaluated for each decision statement. If required, one or more outputs to DQO Steps 1 through 6 were modified to tailor the design to meet all of the DQO constraints most efficiently. For each decision statement, Table 7-9 presents a summary of the final statistical design and the total number of samples/measurements to be collected.

Table 7-7 Sampling Cost Based on Varying Error Tolerances and LBGR

		Mistakenly Concluding < Action Level		
		$\alpha =$	$\alpha =$	$\alpha =$
DS #1				
LBGR =				
Mistakenly Concluding > Action Level	$\beta =$			
	$\beta =$			
	$\beta =$			
LBGR =				
Mistakenly Concluding > Action Level	$\beta =$			
	$\beta =$			
	$\beta =$			
LBGR =				
Mistakenly Concluding > Action Level	$\beta =$			
	$\beta =$			
	$\beta =$			
DS #2				
LBGR =				
Mistakenly Concluding > Action Level	$\beta =$			
	$\beta =$			
	$\beta =$			
LBGR =				
Mistakenly Concluding > Action Level	$\beta =$			
	$\beta =$			
	$\beta =$			
LBGR =				
Mistakenly Concluding > Action Level	$\beta =$			
	$\beta =$			
	$\beta =$			

Table 7-8 Most Resource Effective Data Collection Design

Table 7-9 Final Statistical Sampling/Measurement Design

DS #	Statistical Sampling/Measurement Design	Number of Samples/ Measurements	Total Number of Samples/ Measurements within Population (Table 4-1)

8.0 References

[Insert list of references cited in the summary report.]

APPENDIX **B**

Sampling and Analysis Plan Template*

Sampling and Analysis Plan for *[Add Project Description]*

[Add Company Name]
[Add City, State]

---

* Appendix B appears on the CD accompanying this book.

[Add Document #]

Rev. XX

Sampling and Analysis Plan for *[Add Project Description]*

Date Published
[Add publication date]

[Add Company Name]
[Add City, State]

[Add Document #]

REV:XX

APPROVAL PAGE

Title of Document: Sampling and Analysis Plan For *[Add project description]*

Author(s): *[Add names of authors]*

Approval: *[Add name and job title]*

_____   _____

Signature   Date

> The approval signature on this page indicates that this document has been authorized for information release to the public through appropriate channels. No other forms or signatures are required to document this information release.

Contents

Tables

1.0 INTRODUCTION

This Sampling and Analysis Plan presents the rationale and strategy for the sampling and analysis activities proposed in support of [*Note:* Enter the project-specific activity to be performed (i.e., the characterization of the 100 Valve Pit Area and 200 Waste Pump Overflow Tank). Briefly describe the purpose of the investigation.

Example

> The purpose of the proposed sampling and analysis activities is to determine whether exterior surfaces of the floor, walls, ceiling, overhead lights, duct work, and equipment present within the study areas exceed action levels for fixed and removable radioactivity, as well as to provide data of sufficient quality and quantity to support the selection of the appropriate disposition option for liquid and sediment present at the study areas. The data will also be compared to decontamination and decommissioning (D&D) endpoint acceptance criteria in preparation for turning the facilities over to the D&D project.]

This section provides background information about the project, as well as a discussion of the previous investigations performed at the site, a list of the contaminants of potential concern (COPCs), and a summary of the data quality objectives (DQOs).

1.1 BACKGROUND

[*Note:* Provide site-specific background information to provide a historical perspective on the project. This information should include the following: all process operations or other activities that were historically performed at the site; locations where COPCs were stored/used; estimated volumes of COPCs used and/or disposed of; details on waste disposal practices, spills, etc.]

Example

> The 100 Valve Pit Area and the 200 Waste Pump Overflow Tank are associated with the past operations of the 300 Reactor. These areas are currently in a surveillance and maintenance program and will eventually be subject to D&D.
>
> The 100 Valve Pit Area contains a circular floor that is approximately 10 ft in diameter. The area is located in the very bottom of the pump house silo, which is 24 ft in diameter and 50 ft high. Over 32 ft of the structure is located below ground level. The silo is composed of reinforced concrete. Access to the valve pit is via a narrow, spiral stairway located in the silo.

The 100 Pump House processed an estimated 3000 gallons of decontamination chemicals from the 100 Radioactive Chemical Waste Storage Tank. It is suspected that as many as 80 gal of equipment processing fluids may have leaked from the pump house, resulting in the potential accumulation of the decontamination chemicals in the 100 Valve Pit Area sump. Additionally, the valve pit area received as many as 8500 gallons of water from the 600 Reactor and 700 Building floor drains. Since the 600 and 700 Reactor buildings housed reactor water, treatment and decontamination chemicals, equipment, and the reactor's steam generators, it is expected that equipment leakage and drainage during maintenance were the main contributors to the COPCs that include radioactive chemicals, and heavy metals including lead, chromium, cadmium, arsenic, and mercury. Limited historical characterization data exist for this facility. A single sample was collected from the valve pit in 1994 and was determined to be unusable due to problems with the analytical methods. Radiological exposure measurements were collected inside the pump house in January 1997 and are detailed in Radiological Survey Record 36-SR-103. No records exist of previous efforts to dispose of the waste materials contained within the valve pit area, nor are there any records identifying any historical spills.

The 200 Waste Pump Overflow Tank is a horizontal, cylindrical 1000-gal-capacity tank partially buried below ground level at the 200 Liquid Waste Disposal Station....

1.2 CONTAMINANTS OF CONCERN

The results from previous investigations and/or process knowledge have identified the following radionuclides and/or chemicals as COCs:
[*Note:* Provide a list of the COCs for the site.]

Example

Radiological:

Am-241	Co-60	Sb-125	Cs-134	Cs-137	Eu-152
Eu-154	Eu-155	Ra-226	Ra-228	Ni-63	Sr-90
Tc-99	Th-228	Th-230	Th-232	U-234	U-235
U-238	Pu-238	Pu-239/240			

Metals:

Sb	As	Ba	Be	Cd	Cr
Hg	Pb	Ni	Se	Ag	Th

PCBs: (Aroclors)

1016	1221	1232	1242	1248	1254
1260					

Others:

Ammonia	TOC	Asbestos

1.3 DATA QUALITY OBJECTIVES

The U.S. Environmental Protection Agency DQO procedure was used to support the development of this Sampling and Analysis Plan (EPA, 1994). The DQO procedure is a strategic planning approach that provides a systematic procedure for defining the criteria that a data collection design should satisfy. Using the DQO process ensures that the type, quantity, and quality of environmental data used in decision making will be appropriate for the intended application.

This section presents only a summary of the key outputs resulting from the implementation of the seven-step DQO process. For additional details, the reader should refer to the DQO Summary Report cited in the list of references.

1.3.1 Statement of the Problem

[*Note:* Provide a brief description of the problem. This description should be less than one page in length and should summarize the key elements of the problem statement presented in the DQO Summary Report. Copy the statement of the problem from Step 1 of the DQO Summary Report.]

1.3.2 Decision Rules

The decision rules for each decision statement identified in the DQO Summary Report are summarized below. These "if ... then ..." statements describe what action will be taken based on the results of the data collected.

[*Note:* Copy all of the decision rules from Step 5 of the DQO Summary Report and insert them into this section.]

1.3.3 Decision Consequences and Error Tolerances

[*Note:* Provide a brief summary of the decision consequences that led to the selection of the statistical or nonstatistical sampling design. If a statistical sampling design was selected, identify the error tolerances and the width of the gray region. This discussion should be based on Step 6 of the DQO Summary Report.]

1.3.4 Sample Design Summary

[*Note:* Summarize the optimal sample design presented in Step 7 of the DQO Summary Report. Discuss the assumptions that were made in calculating the total number of samples to be collected.]

2.0 Quality Assurance Project Plan

The following section identifies the individuals or organizations participating in the project and discusses specific roles and responsibilities. This section also discusses the quality objectives for measurement data and discusses the special training requirements for the staff performing the work.

2.1 PROJECT MANAGEMENT

2.1.1 Project/Task Organization

[*Note:* Identify the individuals or organizations participating in the project and discuss their specific roles and responsibilities. Include the Project Manager, Quality Assurance Manager, field implementation team, principal data users, and regulators. The Quality Assurance Manager must be independent of the unit generating the data.]

Example

The project shall be managed through the remedial action and waste disposal project that has an assigned Project Manager. The Field Support Group shall provide project assistance in performing environmental surveys and collecting environmental samples and shall be responsible for sample collection, packaging, and shipping. The Sample and Analytical Group shall arrange for analytical services. The Safety Group shall provide radiological control and safety support as required, while the Environmental Compliance Group shall be responsible for performing independent quality assurance activities.

[*Note:* Provide an organizational chart showing the relationship and lines of communication among the project participants. Include other data users who are outside the organization generating the data but for whom the data are nevertheless intended. The organization chart must also identify any subcontractor relationships relevant to environmental data operations.]

2.1.2 Quality Objectives and Criteria for Measurement Data

The detection limits and precision and accuracy requirements for each of the analyses to be performed are summarized in Table 2-1. These requirements were defined in Step 6 of the DQO Summary Report [*Note:* Reference DQO Summary Report].

[*Note:* Table 2-1 is an example table.]

Table 2-1 Table Summarizing Data Quality Requirements

Media	Analytical Method	Analytical Parameter	Regulatory Limit	Detection Limit	Precision	Accuracy
Sediment	SW-846/8010	TCE	5 ppb	2 ppb	±20%	±20%
Sediment	Gamma spectroscopy	U-238	6 pCi/g	2 pCi/g	±20%	±20%

2.1.3 Special Training Requirements/Certification

Training or certification requirements needed by personnel are described in [*Note:* Cite appropriate company procedures].

[*Note:* Identify any specialized training or certification requirements needed by personnel to complete the project or task successfully.]

Example

Field personnel shall have completed the following training before starting work: Radiation Worker II, Occupational Safety and Health Administration 40-Hour Hazardous Waste Worker Training, etc.).

2.1.4 Documentation and Records

[*Note:* Itemize the information and reporting format for field logs, instrument print-outs, data report packages, results from calibration and quality control checks, etc. Specify the laboratory turnaround time requirements for each requested analytical method. Specify whether a field sampling and/or laboratory analysis case narrative is required to provide a complete description of difficulties encountered during sampling and analysis. Specify the requirements for the final disposition of records and documents, including location and length of retention time.]

2.2 MEASUREMENT/DATA ACQUISITION

The following section presents the requirements for sampling methods, sample handling and custody, analytical methods, and field and laboratory quality control. This section also addresses the requirements for instrument calibration and maintenance, supply inspections, and data management.

2.2.1 Sampling Process Design

A summary of the sampling process design is presented in Section 1.3.4. Chapter 3 presents figures and tables that identify the sampling locations, total number of samples to be collected, sampling procedures to be implemented, analyses to be performed, and sample bottle requirements.

2.2.2   Sampling Methods Requirements

The procedures to be implemented in the field include the following:

[*Note:* List the specific company procedures to be implemented. If company procedures do not exist, provide detailed sampling method requirements in this section.]

2.2.3   Sample Handling, Shipping, and Custody Requirements

All sample handling, shipping, and custody requirements should be performed in accordance with [*Note:* Cite appropriate company procedures]. The sample handling, shipping, and custody requirements shall take Resource Conservation and Recovery Act of 1976–listed waste codes into consideration.

2.2.4   Analytical Methods Requirements

[*Note:* Identify the analytical methods, performance requirements (e.g., detection limits, precision, accuracy) and equipment required to perform the onsite or fixed-laboratory analyses. Identify extraction methods, laboratory decontamination procedures, and waste disposal requirements. This section should also address the actions to be taken when a failure in the analytical system occurs and who is responsible for corrective action.]

2.2.5   Quality Control Requirements

The quality control procedures must be followed in the field and laboratory to ensure that reliable data are obtained. When performing this field sampling effort, care shall be taken to prevent the cross-contamination of sampling equipment, sample bottles, and other equipment that could compromise sample integrity. Tables 2-2 and 2-3 summarize the field and laboratory quality control requirements for supporting the proposed sampling effort.

2.2.6   Instrument/Equipment Testing, Inspection, and Maintenance Requirements

All on-site environmental instruments shall be tested, inspected, and maintained in accordance with [*Note:* Cite appropriate company procedures]. The results from all testing, inspection, and maintenance activities shall be recorded in a bound logbook in accordance with procedures outlined in [*Note:* Cite appropriate company procedures].

[*Note:* The following are example tables.]

Table 2-2 Field Quality Control Requirements

Sample Type	Frequency	Purpose
Duplicate	10%	To check the precision of the laboratory analyses.
Rinsate blank	1/day	To check the effectiveness of the decontamination procedure

Table 2-3 Laboratory Quality Control Requirements

Sample Type	Frequency	Purpose
Blank	1/batch	To determine the existence and magnitude of possible contamination encountered during the sample preparation and analysis process
Matrix spike	1/batch	A sample spiked with known quantities of analytes and subjected to the entire analytical procedure; used as a measure of accuracy
Matrix spike duplicate	1/batch	A second aliquot of the same sample as the matrix spike with the same known quantities of analytes added as the matrix spike; used to estimate method precision

2.2.7 Instrument Calibration and Frequency

All on-site environmental instruments shall be calibrated in accordance with [*Note:* Cite appropriate company procedures]. The results from all instrument calibration activities shall be recorded in a bound logbook in accordance with procedures outlined in [*Note:* Cite appropriate company procedures]. Tags will be attached to all field screening and on-site analytical instruments, noting the date when the instrument was last calibrated, along with the calibration expiration date.

2.2.8 Inspection/Acceptance Requirements for Supplies and Consumables

[*Note:* Describe how and by whom supplies and consumables shall be inspected and accepted for use in the project. Present the acceptance criteria for such supplies and consumables.]

2.2.9 Data Acquisition Requirements (Nondirect Measurements)

The sources for nondirect measurement data, which shall be used to resolve one or more of the DQO decision statements, are presented in Step 3 of the accompanying DQO Summary Report [*Note:* Site DQO Summary Report].

2.2.10 Data Management

Data resulting from the implementation of this Sampling and Analysis Plan will be managed and stored in accordance with [*Note:* Cite appropriate company procedures].

All validated reports and supporting analytical data packages shall be subject to final technical review by qualified reviewers before their submittal to regulatory agencies or inclusion in reports or technical memoranda. Electronic data access, when appropriate, shall be through computerized databases. Where electronic data are not available, hard copies will be used.

2.2.11   Sample Preservation, Containers, and Holding Times

The sample preservation, container, and holding-time requirements for the analyses to be performed are summarized in Table 2-4.

[*Note:* Table 2-4 is an example table.]

Table 2-4 Preservatives, Containers, and Holding Times

Analyte/ Test	Container	Quantity	Preservative	Holding Time
Soil				
U-238	Polyethylene wide-mouth jar	500 mL	None	6 months
PCBs	Amber glass	250 mL	4°C	7 days

2.2.12   Field Documentation

Field documentation shall be kept in accordance with [*Note:* Cite appropriate company procedures], including the following procedures:

[*Note:* List procedures for field logbooks, chain-of-custody, etc.]

2.3   ASSESSMENT/OVERSIGHT

2.3.1   Assessments and Response Actions

Assessment and field activities may be performed in accordance with [*Note:* Cite appropriate company procedures] to verify compliance with the requirements outlined in this Sampling and Analysis Plan, project work packages, procedures, and regulatory requirements.

Deficiencies identified by one of these assessments shall be reported in accordance with [*Note:* Cite appropriate company procedures]. When appropriate, corrective actions will be taken by the Project Manager in accordance with [*Note:* Cite appropriate company procedures] to minimize recurrence.

2.3.2   Reports to Management

Management shall be made aware of all deficiencies identified by the self-assessments and these shall be reported in accordance with [*Note:* Cite appropriate company procedures].

2.4 DATA VALIDATION AND USABILITY

2.4.1 Data Review, Validation, and Verification Requirements

Data verification and validation is performed on analytical data sets, primarily to confirm that sampling and chain-of-custody documentation is complete, sample numbers can be tied to the specific sampling location, samples were analyzed within the required holding times, and analyses met the data quality requirements specified in the sampling and analysis instruction.

2.4.2 Validation and Verification Methods

All data verification and validation shall be performed in accordance with [*Note:* Cite appropriate company procedures].

[*Note:* Specify the percentage of the data that will undergo data verification and/or validation. If no verification or validation is to be performed, provide justification.]

2.4.3 Reconciling Results with DQOs

A data quality assessment (DQA) shall be performed on the resulting analytical data in accordance with EPA (1996). This assessment involves the following activities:

- Identifying any discrepancies that exist between the sampling and analytical requirements specified in the sampling and analysis plan and what was actually performed in the field.
- Reviewing the analytical data set (plotting data on trend diagrams, histograms, etc.) to identify outliers. Reviewing any related quality assurance/quality control reports that are relevant to the project.
- Selecting the most appropriate statistical hypothesis test for drawing conclusions from the data set.
- Performing an assessment of the validity of the chosen statistical hypothesis test.
- Running the selected statistical hypothesis test and drawing conclusions from the results.

3.0 Field Sampling Plan

3.1 SAMPLING OBJECTIVES

The objective of the field sampling plan (FSP) is to identify clearly the sampling and analysis activities needed to resolve the decision rules identified in Step 5 of the DQO Summary Report. The FSP takes the sampling design proposed in Step 7 of the DQO Summary Report and presents this design using figures and tables whenever possible to identify sampling locations, total number of samples to be collected, sampling procedures to be implemented, analyses to be performed, and sample bottle requirements. Flowcharts are also used whenever possible to support the field decision-making process.

3.2 SAMPLING LOCATIONS AND FREQUENCY

[*Note:* Clearly describe the sampling design developed for the project, including:

- Rationale for the design
- Field screening methods to be implemented
- Number and type of samples to be collected
- Sampling locations and frequency
- Parameters of interest.

All preselected sampling locations should be clearly shown on a grid map. If collecting samples from a drum, it is important that both the location of the drum and the origin of the material contained within the drum be carefully documented on a map. When the results from field screening or quick-turnaround laboratory data are to be used to support the field decision-making process, include one or more decision flowcharts to define clearly the alternative actions based on the results from the data.]

3.3 SAMPLING AND ON-SITE ENVIRONMENTAL MEASUREMENT PROCEDURES

The sampling and on-site environmental measurement procedures to be implemented in the field should be consistent with those outlined in [*Note:* Cite appropriate company procedures].

[*Note:* List the specific company procedures to be implemented. Clearly describe any activities to be implemented that do not have procedures for them.]

3.4 SAMPLE MANAGEMENT

Sample management activities shall be performed in accordance with [*Note:* Cite appropriate company procedures].

3.5 MANAGEMENT OF INVESTIGATION-DERIVED WASTE

Investigation-derived waste generated by sampling activities will be managed in accordance with [*Note:* Cite appropriate company procedures]. Unused samples and associated laboratory waste for the analysis will be dispositioned in accordance with the laboratory contract. Pursuant to 40 CFR 300.440, Remedial Project Manager (RPM) approval is required before returning unused samples or waste from off-site laboratories.

4.0 Health and Safety

All field operations will be performed in accordance with health and safety requirements outlined in [*Note:* Cite appropriate company procedures]. In addition, a work control package will be prepared in accordance with [*Note:* Cite appropriate company procedures], which will further control site operations. This package will include an activity hazard analysis, site-specific health and safety plan, and applicable radiological work permits.

The sampling procedures and associated activities will take into consideration exposure reduction and contamination control techniques that will minimize the radiation exposure to the sampling team as required by [*Note:* Cite appropriate company procedures].

5.0 References

EPA, 1994, Guidance for the Data Quality Objectives Process, EAP QA/G-4, U.S. Environmental Protection Agency, Washington, D.C.

EPA, 1996, Guidance for Data Quality Assessment, EPA QA/G-9, U.S. Environmental Protection Agency, Washington, D.C.

[*Note:* Add reference to site-specific DQO Summary Report completed in support of this task, along with any other new references cited in the text.]

Statistical Tables

Table C-1 Quantiles of the Shapiro-Wilk W Test for Normality (values of W such that $100p\%$ of the distribution of W is less than W_p)

n	$w_{0.01}$	$w_{0.02}$	$w_{0.05}$	$w_{0.10}$	$w_{0.50}$
3	0.753	0.756	0.767	0.789	0.959
4	0.687	0.707	0.748	0.792	0.935
5	0.686	0.715	0.762	0.806	0.927
6	0.713	0.743	0.788	0.826	0.927
7	0.730	0.760	0.803	0.838	0.928
8	0.749	0.778	0.818	0.851	0.932
9	0.764	0.791	0.829	0.859	0.935
10	0.781	0.806	0.842	0.869	0.938
11	0.792	0.817	0.850	0.876	0.940
12	0.805	0.828	0.859	0.883	0.943
13	0.814	0.837	0.866	0.889	0.945
14	0.825	0.846	0.874	0.895	0.947
15	0.835	0.855	0.881	0.901	0.950
16	0.844	0.863	0.887	0.906	0.952
17	0.851	0.869	0.892	0.910	0.954
18	0.858	0.874	0.897	0.914	0.956
19	0.863	0.879	0.901	0.917	0.957
20	0.868	0.884	0.905	0.920	0.959
21	0.873	0.888	0.908	0.923	0.960
22	0.878	0.892	0.911	0.926	0.961
23	0.881	0.895	0.914	0.928	0.962
24	0.884	0.898	0.916	0.930	0.963
25	0.888	0.901	0.918	0.931	0.964
26	0.891	0.904	0.920	0.933	0.965
27	0.894	0.906	0.923	0.935	0.965
28	0.896	0.908	0.924	0.936	0.966
29	0.898	0.910	0.926	0.937	0.966
30	0.900	0.912	0.927	0.939	0.967
31	0.902	0.914	0.929	0.940	0.967
32	0.904	0.915	0.930	0.941	0.968
33	0.906	0.917	0.931	0.942	0.968
34	0.908	0.919	0.933	0.943	0.969
35	0.910	0.920	0.934	0.944	0.969
36	0.912	0.922	0.935	0.945	0.970
37	0.914	0.924	0.936	0.946	0.970
38	0.916	0.925	0.938	0.947	0.971
39	0.917	0.927	0.939	0.948	0.971
40	0.919	0.928	0.940	0.949	0.972
41	0.920	0.929	0.941	0.950	0.972
42	0.922	0.930	0.942	0.951	0.972
43	0.923	0.932	0.943	0.951	0.973
44	0.924	0.933	0.944	0.952	0.973
45	0.926	0.934	0.945	0.953	0.973
46	0.927	0.935	0.945	0.953	0.974
47	0.928	0.936	0.946	0.954	0.974
48	0.929	0.937	0.947	0.954	0.974
49	0.929	0.937	0.947	0.955	0.974
50	0.930	0.938	0.947	0.955	0.974

Note: The null hypothesis of a normal distribution is rejected at the α significance level if the calculated W is less than W_α.

Source: After Gilbert, 1987. (With permission.)

Table C-2 Values of $H_{1-\alpha} = H_{0.90}$ for Computing a One-Sided Upper 90% Confidence Limit on a Lognormal Mean

s_y	3	5	7	10	12	15	21	31	51	101
0.10	1.686	1.438	1.381	1.349	1.338	1.328	1.317	1.308	1.301	1.295
0.20	1.885	1.522	1.442	1.396	1.380	1.365	1.348	1.335	1.324	1.314
0.30	2.156	1.627	1.517	1.453	1.432	1.411	1.388	1.370	1.354	1.339
0.40	2.521	1.755	1.607	1.523	1.494	1.467	1.437	1.412	1.390	1.371
0.50	2.990	1.907	1.712	1.604	1.567	1.532	1.494	1.462	1.434	1.409
0.60	3.542	2.084	1.834	1.696	1.650	1.606	1.558	1.519	1.485	1.454
0.70	4.136	2.284	1.970	1.800	1.743	1.690	1.631	1.583	1.541	1.504
0.80	4.742	2.503	2.119	1.914	1.845	1.781	1.710	1.654	1.604	1.560
0.90	5.349	2.736	2.280	2.036	1.955	1.880	1.797	1.731	1.672	1.621
1.00	5.955	2.980	2.450	2.167	2.073	1.985	1.889	1.812	1.745	1.686
1.25	7.466	3.617	2.904	2.518	2.391	2.271	2.141	2.036	1.946	1.866
1.50	8.973	4.276	3.383	2.896	2.733	2.581	2.415	2.282	2.166	2.066
1.75	10.48	4.944	3.877	3.289	3.092	2.907	2.705	2.543	2.402	2.279
2.00	11.98	5.619	4.380	3.693	3.461	3.244	3.005	2.814	2.648	2.503
2.50	14.99	6.979	5.401	4.518	4.220	3.938	3.629	3.380	3.163	2.974
3.00	18.00	8.346	6.434	5.359	4.994	4.650	4.270	3.964	3.697	3.463
3.50	21.00	9.717	7.473	6.208	5.778	5.370	4.921	4.559	4.242	3.965
4.00	24.00	11.09	8.516	7.062	6.566	6.097	5.580	5.161	4.796	4.474
4.50	27.01	12.47	9.562	7.919	7.360	6.829	6.243	5.769	5.354	4.989
5.00	30.01	13.84	10.61	8.779	8.155	7.563	6.909	6.379	5.916	5.508
6.00	36.02	16.60	12.71	10.50	9.751	9.037	8.248	7.607	7.048	6.555
7.00	42.02	19.35	14.81	12.23	11.35	10.52	9.592	8.842	8.186	7.607
8.00	48.03	22.11	16.91	13.96	12.96	12.00	10.94	10.08	9.329	8.665
9.00	54.03	24.87	19.02	15.70	14.56	13.48	12.29	11.32	10.48	9.725
10.00	60.04	27.63	21.12	17.43	16.17	14.97	13.64	12.56	11.62	10.79

Source: After Gilbert, 1987. (With permission.)

Table C-3 Values of $H_\alpha = H_{0.10}$ for Computing a One-Sided Lower 10% Confidence Limit on a Lognormal Mean

s_y	3	5	7	10	12	15	21	31	51	101
0.10	−1.431	−1.320	−1.296	−1.285	−1.281	−1.279	−1.277	−1.277	−1.278	−1.279
0.20	−1.350	−1.281	−1.268	−1.266	−1.266	−1.266	−1.268	−1.272	−1.275	−1.280
0.30	−1.289	−1.252	−1.250	−1.254	−1.257	−1.260	−1.266	−1.272	−1.280	−1.287
0.40	−1.245	−1.233	−1.239	−1.249	−1.254	−1.261	−1.270	−1.279	−1.289	−1.301
0.50	−1.213	−1.221	−1.234	−1.250	−1.257	−1.266	−1.279	−1.291	−1.304	−1.319
0.60	−1.190	−1.215	−1.235	−1.256	−1.266	−1.277	−1.292	−1.307	−1.324	−1.342
0.70	−1.176	−1.215	−1.241	−1.266	−1.278	−1.292	−1.310	−1.329	−1.349	−1.370
0.80	−1.168	−1.219	−1.251	−1.280	−1.294	−1.311	−1.332	−1.354	−1.377	−1.403
0.90	−1.165	−1.227	−1.264	−1.298	−1.314	−1.333	−1.358	−1.383	−1.409	−1.439
1.00	−1.166	−1.239	−1.281	−1.320	−1.337	−1.358	−1.387	−1.414	−1.445	−1.478
1.25	−1.184	−1.280	−1.334	−1.384	−1.407	−1.434	−1.470	−1.507	−1.547	−1.589
1.50	−1.217	−1.334	−1.400	−1.462	−1.491	−1.523	−1.568	−1.613	−1.063	−1.716
1.75	−1.260	−1.398	−1.477	−1.551	−1.585	−1.624	−1.677	−1.732	−1.790	−1.855
2.00	−1.310	−1.470	−1.562	−1.647	−1.688	−1.733	−1.795	−1.859	−1.928	−2.003
2.50	−1.426	−1.634	−1.751	−1.862	−1.913	−1.971	−2.051	−2.133	−2.223	−2.321
3.00	−1.560	−1.817	−1.960	−2.095	−2.157	−2.229	−2.326	−2.427	−2.536	−2.657
3.50	−1.710	−2.014	−2.183	−2.341	−2.415	−2.499	−2.615	−2.733	−2.864	−3.007
4.00	−1.871	−2.221	−2.415	−2.596	−2.681	−2.778	−2.913	−3.050	−3.200	−3.366
4.50	−2.041	−2.435	−2.653	−2.858	−2.955	−3.064	−3.217	−3.372	−3.542	−3.731
5.00	−2.217	−2.654	−2.897	−3.126	−3.233	−3.356	−3.525	−3.698	−3.889	−4.100
6.00	−2.581	−3.104	−3.396	−3.671	−3.800	−3.949	−4.153	−4.363	−4.594	−4.849
7.00	−2.955	−3.564	−3.904	−4.226	−4.377	−4.549	−4.790	−5.037	−5.307	−5.607
8.00	−3.336	−4.030	−4.418	−4.787	−4.960	−5.159	−5.433	−5.715	−6.026	−6.370
9.00	−3.721	−4.500	−4.937	−5.352	−5.547	−5.771	−6.080	−6.399	−6.748	−7.136
10.00	−4.109	−4.973	−5.459	−5.920	−6.137	−6.386	−6.730	−7.085	−7.474	−7.906

Source: After Gilbert, 1987. (With permission.)

Table C-4 Values of $H_{1-\alpha} = H_{0.95}$ for Computing a One-Sided Upper 95% Confidence Limit on a Lognormal Mean

^sy	3	5	7	10	12	15	21	31	51	101
0.10	2.750	2.035	1.886	1.802	1.775	1.749	1.722	1.701	1.684	1.670
0.20	3.295	2.198	1.992	1.881	1.843	1.809	1.771	1.742	1.718	1.697
0.30	4.109	2.402	2.125	1.977	1.927	1.882	1.833	1.793	1.761	1.733
0.40	5.220	2.651	2.282	2.089	2.026	1.968	1.905	1.856	1.813	1.777
0.50	6.495	2.947	2.465	2.220	2.141	2.068	1.989	1.928	1.876	1.830
0.60	7.807	3.287	2.673	2.368	2.271	2.181	2.085	2.010	1.946	1.891
0.70	9.120	3.662	2.904	2.532	2.414	2.306	2.191	2.102	2.025	1.960
0.80	10.43	4.062	3.155	2.710	2.570	2.443	2.307	2.202	2.112	2.035
0.90	11.74	4.478	3.420	2.902	2.738	2.589	2.432	2.310	2.206	2.117
1.00	13.05	4.905	3.698	3.103	2.915	2.744	2.564	2.423	2.306	2.205
1.25	16.33	6.001	4.426	3.639	3.389	3.163	2.923	2.737	2.580	2.447
1.50	19.60	7.120	5.184	4.207	3.896	3.612	3.311	3.077	2.881	2.713
1.75	22.87	8.250	5.960	4.795	4.422	4.081	3.719	3.437	3.200	2.997
2.00	26.14	9.387	6.747	5.396	4.962	4.564	4.141	3.812	3.533	3.295
2.50	32.69	11.67	8.339	6.621	6.067	5.557	5.013	4.588	4.228	3.920
3.00	39.23	13.97	9.945	7.864	7.191	6.570	5.907	5.388	4.947	4.569
3.50	45.77	16.27	11.56	9.118	8.326	7.596	6.815	6.201	5.681	5.233
4.00	52.31	18.58	13.18	10.38	9.469	8.630	7.731	7.024	6.424	5.908
4.50	58.85	20.88	14.80	11.64	10.62	9.669	8.652	7.854	7.174	6.590
5.00	65.39	23.19	16.43	12.91	11.77	10.71	9.579	8.688	7.929	7.277
6.00	78.47	27.81	19.68	15.45	14.08	12.81	11.44	10.36	9.449	8.661
7.00	91.55	32.43	22.94	18.00	16.39	14.90	13.31	12.05	10.98	10.05
8.00	104.6	37.06	26.20	20.55	18.71	17.01	15.18	13.74	12.51	11.45
9.00	117.7	41.68	29.46	23.10	21.03	19.11	17.05	15.43	14.05	12.85
10.00	130.8	46.31	32.73	25.66	23.35	21.22	18.93	17.13	15.59	14.26

Source: After Gilbert, 1987. (With permission.)

Table C-5: Values of $H_{\alpha} = H_{0.05}$ for Computing a One-Sided Lower 5% Confidence Limit on a Lognormal Mean

^sy	3	5	7	10	12	15	21	31	51	101
0.10	−2.130	−1.806	−1.731	−1.690	−1.677	−1.666	−1.655	−1.648	−1.644	−1.642
0.20	−1.949	−1.729	−1.678	−1.653	−1.646	−1.640	−1.636	−1.636	−1.637	−1.641
0.30	−1.816	−1.669	−1.639	−1.627	−1.625	−1.625	−1.627	−1.632	−1.638	−1.648
0.40	−1.717	−1.625	−1.611	−1.611	−1.613	−1.617	−1.625	−1.635	−1.647	−1.662
0.50	−1.644	−1.594	−1.594	−1.603	−1.609	−1.618	−1.631	−1.646	−1.663	−1.683
0.60	−1.589	−1.573	−1.584	−1.602	−1.612	−1.625	−1.643	−1.662	−1.685	−1.711
0.70	−1.549	−1.560	−1.582	−1.608	−1.622	−1.638	−1.661	−1.686	−1.713	−1.744
0.80	−1.521	−1.555	−1.586	−1.620	−1.636	−1.656	−1.685	−1.714	−1.747	−1.783
0.90	−1.502	−1.556	−1.595	−1.637	−1.656	−1.680	−1.713	−1.747	−1.785	−1.826
1.00	−1.490	−1.562	−1.610	−1.658	−1.681	−1.707	−1.745	−1.784	−1.827	−1.874
1.25	−1.486	−1.596	−1.662	−1.727	−1.758	−1.793	−1.842	−1.893	−1.949	−2.012
1.50	−1.508	−1.650	−1.733	−1.814	−1.853	−1.896	−1.958	−2.020	−2.091	−2.169
1.75	−1.547	−1.719	−1.819	−1.916	−1.962	−2.015	−2.088	−2.164	−2.247	−2.341
2.00	−1.598	−1.799	−1.917	−2.029	−2.083	−2.144	−2.230	−2.318	−2.416	−2.526
2.50	−1.727	−1.986	−2.138	−2.283	−2.351	−2.430	−2.540	−2.654	−2.780	−2.921
3.00	−1.880	−2.199	−2.384	−2.560	−2.644	−2.740	−2.874	−3.014	−3.169	−3.342
3.50	−2.051	−2.429	−2.647	−2.855	−2.953	−3.067	−3.226	−3.391	−3.574	−3.780
4.00	−2.237	−2.672	−2.922	−3.161	−3.275	−3.406	−3.589	−3.779	−3.990	−4.228
4.50	−2.434	−2.924	−3.206	−3.476	−3.605	−3.753	−3.960	−4.176	−4.416	−4.685
5.00	−2.638	−3.183	−3.497	−3.798	−3.941	−4.107	−4.338	−4.579	−4.847	−5.148
6.00	−3.062	−3.715	−4.092	−4.455	−4.627	−4.827	−5.106	−5.397	−5.721	−6.086
7.00	−3.499	−4.260	−4.699	−5.123	−5.325	−5.559	−5.886	−6.227	−6.608	−7.036
8.00	−3.945	−4.812	−5.315	−5.800	−6.031	−6.300	−6.674	−7.066	−7.502	−7.992
9.00	−4.397	−5.371	−5.936	−6.482	−6.742	−7.045	−7.468	−7.909	−8.401	−8.953
10.00	−4.852	−5.933	−6.560	−7.168	−7.458	−7.794	−8.264	−8.755	−9.302	−9.918

Source: After Gilbert, 1987. (With permission.)

Table C-6 Cumulative Standard Normal Distribution

z	0.00	0.01	0.02	0.03	0.04	0.05	0.06	0.07	0.08	0.09
0.0	0.50000	0.50399	0.50798	0.51197	0.51595	0.51994	0.52392	0.52790	0.53188	0.53586
0.1	0.53983	0.54379	0.54776	0.55172	0.55567	0.55962	0.56356	0.56749	0.57142	0.57534
0.2	0.57926	0.58317	0.58706	0.59095	0.59483	0.59871	0.60257	0.60642	0.61026	0.61409
0.3	0.61791	0.62172	0.62551	0.62930	0.63307	0.63683	0.64058	0.64431	0.64803	0.65173
0.4	0.65542	0.65910	0.66276	0.66640	0.67003	0.67364	0.67724	0.68082	0.68438	0.68793
0.5	0.69146	0.69497	0.69847	0.70194	0.70540	0.70884	0.71226	0.71566	0.71904	0.72240
0.6	0.72575	0.72907	0.73237	0.73565	0.73891	0.74215	0.74537	0.74857	0.75175	0.75490
0.7	0.75803	0.76115	0.76424	0.76730	0.77035	0.77337	0.77637	0.77935	0.78230	0.78523
0.8	0.78814	0.79103	0.79389	0.79673	0.79954	0.80234	0.80510	0.80785	0.81057	0.81327
0.9	0.81594	0.81859	0.82121	0.82381	0.82639	0.82894	0.83147	0.83397	0.83646	0.83891
1.0	0.84134	0.84375	0.84613	0.84849	0.85083	0.85314	0.85543	0.85769	0.85993	0.86214
1.1	0.86433	0.86650	0.86864	0.87076	0.87285	0.87493	0.87697	0.87900	0.88100	0.88297
1.2	0.88493	0.88686	0.88877	0.89065	0.89251	0.89435	0.89616	0.89796	0.89973	0.90147
1.3	0.90320	0.90490	0.90658	0.90824	0.90988	0.91149	0.91308	0.91465	0.91621	0.91773
1.4	0.91924	0.92073	0.92219	0.92364	0.92506	0.92647	0.92785	0.92922	0.93056	0.93189
1.5	0.93319	0.93448	0.93574	0.93699	0.93822	0.93943	0.94062	0.94179	0.94295	0.94408
1.6	0.94520	0.94630	0.94738	0.94845	0.94950	0.95053	0.95154	0.95254	0.95352	0.95448
1.7	0.95543	0.95637	0.95728	0.95818	0.95907	0.95994	0.96080	0.96164	0.96246	0.96327
1.8	0.96407	0.96485	0.96562	0.96637	0.96711	0.96784	0.96856	0.96926	0.96995	0.97062
1.9	0.97128	0.97193	0.97257	0.97320	0.97381	0.97441	0.97500	0.97558	0.97615	0.97670
2.0	0.97725	0.97778	0.97831	0.97882	0.97932	0.97982	0.98030	0.98077	0.98124	0.98169
2.1	0.98214	0.98257	0.98300	0.98341	0.98382	0.98422	0.98461	0.98500	0.98537	0.98574
2.2	0.98610	0.98645	0.98679	0.98713	0.98745	0.98778	0.98809	0.98840	0.98870	0.98899
2.3	0.98928	0.98956	0.98983	0.99010	0.99036	0.99061	0.99086	0.99111	0.99134	0.99158
2.4	0.99180	0.99202	0.99224	0.99245	0.99266	0.99286	0.99305	0.99324	0.99343	0.99361
2.5	0.99379	0.99396	0.99413	0.99430	0.99446	0.99461	0.99477	0.99492	0.99506	0.99520
2.6	0.99534	0.99547	0.99560	0.99573	0.99585	0.99598	0.99609	0.99621	0.99632	0.99643
2.7	0.99653	0.99664	0.99674	0.99683	0.99693	0.99702	0.99711	0.99720	0.99728	0.99736
2.8	0.99744	0.99752	0.99760	0.99767	0.99774	0.99781	0.99788	0.99795	0.99801	0.99807
2.9	0.99813	0.99819	0.99825	0.99831	0.99836	0.99841	0.99846	0.99851	0.99856	0.99861
3.0	0.99865	0.99869	0.99874	0.99878	0.99882	0.99886	0.99889	0.99893	0.99897	0.99900
3.1	0.99903	0.99906	0.99910	0.99913	0.99916	0.99918	0.99921	0.99924	0.99926	0.99929
3.2	0.99931	0.99934	0.99936	0.99938	0.99940	0.99942	0.99944	0.99946	0.99948	0.99950
3.3	0.99952	0.99953	0.99955	0.99957	0.99958	0.99960	0.99961	0.99962	0.99964	0.99965
3.4	0.99966	0.99968	0.99969	0.99970	0.99971	0.99972	0.99973	0.99974	0.99975	0.99976
3.5	0.99977	0.99978	0.99978	0.99979	0.99980	0.99981	0.99981	0.99982	0.99983	0.99983
3.6	0.99984	0.99985	0.99985	0.99986	0.99986	0.99987	0.99987	0.99988	0.99988	0.99989
3.7	0.99989	0.99990	0.99990	0.99990	0.99991	0.99991	0.99992	0.99992	0.99992	0.99992
3.8	0.99993	0.99993	0.99993	0.99994	0.99994	0.99994	0.99994	0.99995	0.99995	0.99995
3.9	0.99995	0.99995	0.99996	0.99996	0.99996	0.99996	0.99996	0.99996	0.99997	0.99997

Source: After Hines and Montgomery, 1980.

Table C-7 Quantile Values t_p (ν) for the Student's t Distribution with ν Degrees of Freedom

ν	$t_{0.55}$	$t_{0.60}$	$t_{0.70}$	$t_{0.75}$	$t_{0.80}$	$t_{0.90}$	$t_{0.95}$	$t_{0.975}$	$t_{0.99}$	$t_{0.995}$
1	0.158	0.325	0.727	1.000	1.376	3.08	6.31	12.71	31.82	63.66
2	0.142	0.289	0.617	0.816	1.061	1.89	2.92	4.30	6.96	9.92
3	0.137	0.277	0.584	0.765	0.978	1.64	2.35	3.18	4.54	5.84
4	0.134	0.271	0.569	0.741	0.941	1.53	2.13	2.78	3.75	4.60
5	0.132	0.267	0.559	0.727	0.920	1.48	2.02	2.57	3.36	4.03
6	0.131	0.265	0.553	0.718	0.906	1.44	1.94	2.45	3.14	3.71
7	0.130	0.263	0.549	0.711	0.896	1.42	1.90	2.36	3.00	3.50
8	0.130	0.262	0.546	0.706	0.889	1.40	1.86	2.31	2.90	3.36
9	0.129	0.261	0.543	0.703	0.883	1.38	1.83	2.26	2.82	3.25
10	0.129	0.260	0.542	0.700	0.879	1.37	1.81	2.23	2.76	3.17
11	0.129	0.260	0.540	0.697	0.876	1.36	1.80	2.20	2.72	3.11
12	0.128	0.259	0.539	0.695	0.873	1.36	1.78	2.18	2.68	3.06
13	0.128	0.259	0.538	0.694	0.870	1.35	1.77	2.16	2.65	3.01
14	0.128	0.258	0.537	0.692	0.868	1.34	1.76	2.14	2.62	2.98
15	0.128	0.258	0.536	0.691	0.866	1.34	1.75	2.13	2.60	2.95
16	0.128	0.258	0.535	0.690	0.865	1.34	1.75	2.12	2.58	2.92
17	0.128	0.257	0.534	0.689	0.863	1.33	1.74	2.11	2.57	2.90
18	0.127	0.257	0.534	0.688	0.862	1.33	1.73	2.10	2.55	2.88
19	0.127	0.257	0.533	0.688	0.861	1.33	1.73	2.09	2.54	2.86
20	0.127	0.257	0.533	0.687	0.860	1.32	1.72	2.09	2.53	2.84
21	0.127	0.257	0.532	0.686	0.859	1.32	1.72	2.08	2.52	2.83
22	0.127	0.256	0.532	0.686	0.858	1.32	1.72	2.07	2.51	2.82
23	0.127	0.256	0.532	0.685	0.858	1.32	1.71	2.07	2.50	2.81
24	0.127	0.256	0.531	0.685	0.857	1.32	1.71	2.06	2.49	2.80
25	0.127	0.256	0.531	0.684	0.856	1.32	1.71	2.06	2.48	2.79
26	0.127	0.256	0.531	0.684	0.856	1.32	1.71	2.06	2.48	2.78
27	0.127	0.256	0.531	0.684	0.855	1.31	1.70	2.05	2.47	2.77
28	0.127	0.256	0.530	0.683	0.855	1.31	1.70	2.05	2.47	2.76
29	0.127	0.256	0.530	0.683	0.854	1.31	1.70	2.04	2.46	2.76
30	0.127	0.256	0.530	0.683	0.854	1.31	1.70	2.04	2.46	2.75
40	0.126	0.255	0.529	0.681	0.851	1.30	1.68	2.02	2.42	2.70
60	0.126	0.254	0.527	0.679	0.848	1.30	1.67	2.00	2.39	2.66
120	0.126	0.254	0.526	0.677	0.845	1.29	1.66	1.98	2.36	2.62
∞	0.126	0.253	0.524	0.674	0.842	1.28	1.645	1.96	2.33	2.58

Source: After Fisher and Yates, 1974.

Table C-8 Quantiles of the Wilcoxon Signed Ranks Test Statistic

n	$W_{0.005}$	$W_{0.01}$	$W_{0.025}$	$W_{0.05}$	$W_{0.10}$	$W_{0.20}$	$W_{0.30}$	$W_{0.40}$	$W_{0.50}$	$\dfrac{n(n+1)}{2}$
4	0	0	0	0	1	3	3	4	5	10
5	0	0	0	1	3	4	5	6	7.5	15
6	0	0	1	3	4	6	8	9	10.5	21
7	0	1	3	4	6	9	11	12	14	28
8	1	2	4	6	9	12	14	16	18	36
9	2	4	6	9	11	15	18	20	22.5	45
10	4	6	9	11	15	19	22	25	27.5	55
11	6	8	11	14	18	23	27	30	33	66
12	8	10	14	18	22	28	32	36	39	78
13	10	13	18	22	27	33	38	42	45.5	91
14	13	16	22	26	32	39	44	48	52.5	105
15	16	20	26	31	37	45	51	55	60	120
16	20	24	30	36	43	51	58	63	68	136
17	24	28	35	42	49	58	65	71	76.5	153
18	28	33	41	48	56	66	73	80	85.5	171
19	33	38	47	54	63	74	82	89	95	190
20	38	44	53	61	70	83	91	98	105	210
21	44	50	59	68	78	91	100	108	115.5	131
22	49	56	67	76	87	100	110	119	126.5	153
23	55	63	74	84	95	110	120	130	138	176
24	62	70	82	92	105	120	131	141	150	300
25	69	77	90	101	114	131	143	153	162.5	325
26	76	85	99	111	125	142	155	165	175.5	351
27	84	94	108	120	135	154	167	178	189	378
28	92	102	117	131	146	166	180	192	203	406
29	101	111	127	141	158	178	193	206	217.5	435
30	110	121	138	152	170	191	207	220	232.5	465
31	119	131	148	164	182	205	221	235	248	496
32	129	141	160	176	195	219	236	250	264	528
33	139	152	171	188	208	233	251	266	280.5	561
34	149	163	183	201	222	248	266	282	297.5	595
35	160	175	196	214	236	263	283	299	315	630
36	172	187	209	228	251	279	299	317	333	665
37	184	199	222	242	266	295	316	335	351.5	703
38	196	212	236	257	282	312	334	353	370.5	741
39	208	225	250	272	298	329	352	372	390	780
40	221	239	265	287	314	347	371	391	410	820
41	235	253	280	303	331	365	390	411	430.5	861
42	248	267	295	320	349	384	409	431	451.5	903
43	263	282	311	337	366	403	429	452	473	946
44	277	297	328	354	385	422	450	473	495	990
45	292	313	344	372	403	442	471	495	517.5	1035
46	308	329	362	390	423	463	492	517	540.5	1081
47	324	346	379	408	442	484	514	540	564	1128
48	340	363	397	428	463	505	536	563	588	1176
49	357	381	416	447	483	527	559	587	612.5	1225
50§	374	398	435	467	504	550	583	611	637.5	1275

Note: The entries in this table are quantiles w_p of the Wilcoxon Signed Ranks Test statistic W^* for selected values of $p \leq 0.50$. Quantiles w_p for $p > 0.50$ may be computed from the equation

$$w_p = n(n + 1)/2 - w_{1-p}$$

where $n(n + 1)/2$ is given in the right-hand column in the table. Note that $P(W^* < w_p)$ # p and $P(W^* > w_p)$ # $1 - p$ if H_0 is true. Critical regions correspond to values of W^* less than (or greater than) but not including the appropriate quantile.

Table C-8 (continued)   Quantiles of the Wilcoxon Signed Ranks Test Statistic

n	$w_{0.005}$	$w_{0.01}$	$w_{0.025}$	$w_{0.05}$	$w_{0.10}$	$w_{0.20}$	$w_{0.30}$	$w_{0.40}$	$w_{0.50}$	$\dfrac{n(n+1)}{2}$

For n larger than 50, the pth quantile of w_p of the Wilcoxon Signed Ranks Test statistic may be approximated by

$$w_p = [n(n+1)]/4 + x_p\sqrt{n(n+1)(2n+1)/24}$$

where x_p is the pth quantile of a standard normal random variable, obtained from Table C-5.

Source: After Conover, 1980. Adapted from Harter and Owen, 1970. The values in the selected tables had been developed and distributed by the Lederle Laboratories Division of American Cyanamid Co., in cooperation with the Department of Statistics, Florida State University, Tallahassee, FL, and copyrighted in 1963 by American Cyanamid Co. and Florida State University.

REFERENCES

Conover, W.J., 1980, *Practical Nonparametric Statistics*, 2nd ed., John Wiley & Sons, New York.

Fisher, R.A. and Yates, F., 1974, *Statistical Tables for Biological, Agricultural and Medical Research*, 6th ed., Longman Group, London.

Gilbert, R.O., 1987, *Statistical Methods for Environmental Pollution Monitoring*, Van Nostrand Reinhold, New York, Appendix A.

Harter, H.L. and Owen, D.B., 1970, *Selected Tables in Mathematical Statistics*, Vol. 1, Markham Publishing, Chicago, IL.

Hines, W.W. and Montgomery, D.C., 1980, *Probability and Statistics in Engineering and Management Science*, 2nd ed., John Wiley & Sons, New York, Appendix Table II.

Land, C.E., 1975, Tables of confidence limits for linear functions of the normal mean and variance, in *Selected Tables in Mathematical Statistics*, Vol. III, American Mathematical Society, Providence, RI, 385–419.

Shapiro, S.S. and Wilk, M.B., 1965, An analysis of variance test for normality (complete examples), *Biometrika*, 591–611.

APPENDIX **D**

Metric Conversion Chart

Into Metric Units			Out of Metric Units		
If You Know	*Multiply By*	*To Get*	*If You Know*	*Multiply By*	*To Get*
Length			**Length**		
inches	25.4	millimeters	millimeters	0.039	inches
inches	2.54	centimeters	centimeters	0.394	inches
feet	0.305	meters	meters	3.281	feet
yards	0.914	meters	meters	1.094	yards
miles	1.609	kilometers	kilometers	0.621	miles
Area			**Area**		
sq. inches	6.452	sq. centimeters	sq. centimeters	0.155	sq. inches
sq. feet	0.093	sq. meters	sq. meters	10.76	sq. feet
sq. yards	.0836	sq. meters	sq. meters	1.196	sq. yards
sq. miles	2.6	sq. kilometers	sq. kilometers	0.4	sq. miles
acres	0.405	hectares	hectares	2.47	acres
Mass (weight)			**Mass (weight)**		
ounces	28.35	grams	grams	0.035	ounces
pounds	0.454	kilograms	kilograms	2.205	pounds
ton	0.907	metric ton	metric ton	1.102	ton
Volume			**Volume**		
teaspoons	5	milliliters	milliliters	0.033	fluid ounces
tablespoons	15	milliliters	liters	2.1	pints
fluid ounces	30	milliliters	liters	1.057	quarts
cups	0.24	liters	liters	0.264	gallons
pints	0.47	liters	cubic meters	35.315	cubic feet
quarts	0.95	liters	cubic meters	1.308	cubic yards
gallons	3.8	liters			
cubic feet	0.028	cubic meters			
cubic yards	0.765	cubic meters			
Temperature			**Temperature**		
Fahrenheit	Subtract 32, then multiply by 5/9	Celsius	Celsius	Multiply by 9/5, then add 32	Fahrenheit
Radioactivity			**Radioactivity**		
picocuries	37	millibecquerel	millibecquerel	0.027	picocuries

APPENDIX **E**

Radiological Decay Chains

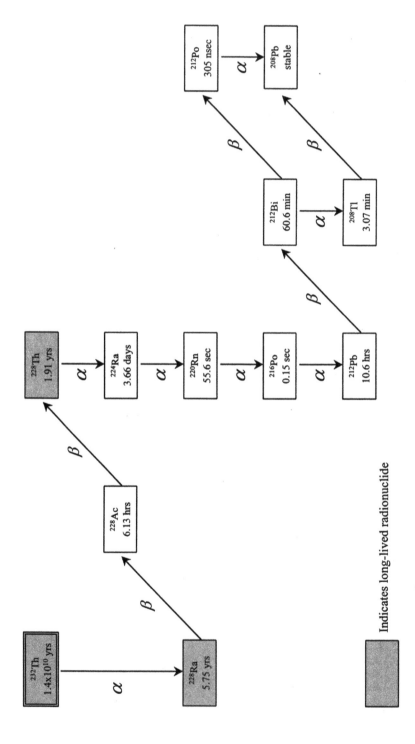

Figure E-1   Th-232 radioactive decay chain. (Modified after Shleien, *The Health Physics and Radiological Health Handbook*, Scinta, Inc., Silver Springs, MD, 1992.

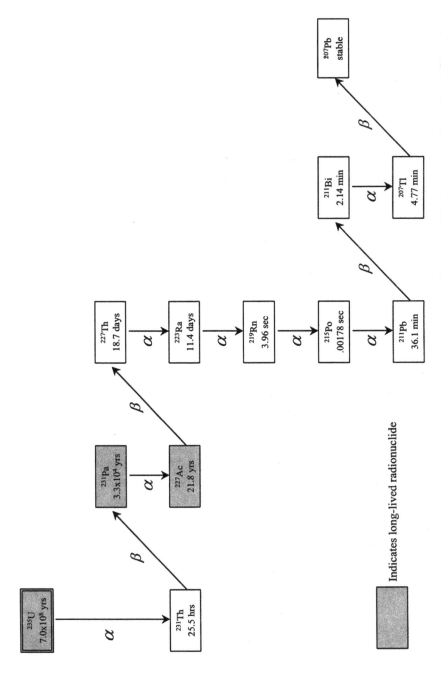

Figure E-2 U-235 radioactive decay chain. (Modified after Shleien, *The Health Physics and Radiological Health Handbook*, Scinta, Inc., Silver Springs, MD, 1992.

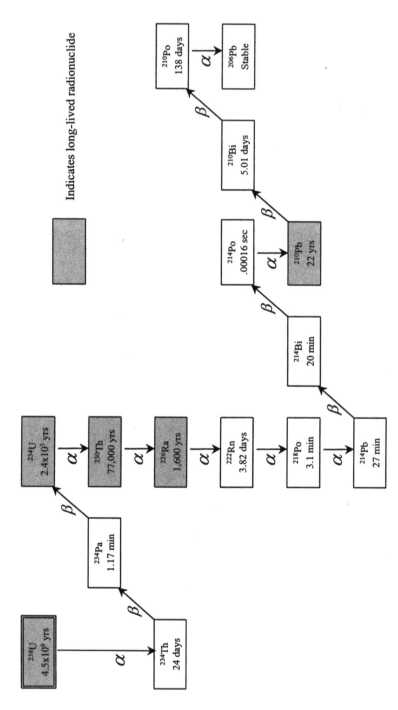

Figure E-3   U-238 radioactive decay chain. (Modified after Shleien, *The Health Physics and Radiological Health Handbook*, Scinta, Inc., Silver Springs, MD, 1992.

APPENDIX **F**

Sample Containers, Preservation, and Holding Times

Analysis	Container	Minimum Sample Size, mL	Preservation	Maximum Holding Time
Activity	P, G(B)	100	Refrigerate	24 h/14 d
Alkalinity	P, G	200	Refrigerate	24 h/14 d
BOD	P, G	1000	Refrigerate	6 h/48 h
Boron	P	100	None required	28 d/6 months
Bromide	P, G	—	None required	28 d/28 d
Carbon, organic, total	G	100	Analyze immediately, or refrigerate and add HCl to pH < 2	7 d/28 d
Carbon dioxide	P, G	100	Analyze immediately	Stat/N.S.
COD	P, G	100	Analyze as soon as possible, or add H_2SO_4 to pH < 2; refrigerate	7 d/28 d
Chlorine, residual	P, G	500	Analyze immediately	0.5 h/stat
Chlorine dioxide	P, G	500	Analyze immediately	0.5 h/N.S.
Chlorophyll	P, G	500	30 d in dark	30 d/N.S.
Color	P, G	500	Refrigerate	48 h/48 h
Conductivity	P, G	500	Refrigerate	28 d/28 d
Cyanide				
Total	P, G	500	Add NaOH to pH > 12, refrigerate in dark	24 h/14 d; 24 h if sulfide present
Amenable to chlorination	P, G	500	Add 100 mg $Na_2S_2O_3$/L	Stat/14 d; 24 h if sulfide present
Fluoride	P	300	None required	28 d/28 d

Analysis	Container	Minimum Sample Size, mL	Preservation	Maximum Holding Time
Hardness	P, G	100	Add HNO_3 to pH < 2	6 months/6 months
Iodine	P, G	500	Analyze immediately	0.5 h/N.S.
Metals, general	P(A), G(A)	—	For dissolved metals filter immediately, add HNO_3 to pH < 2	6 months/6 months
Chromium VI Copper by colorimetry*	P(A), G(A)	300	Refrigerate	24 h/24 h
Mercury	P(A), G(A)	500	Add HNO_3 to pH < 2, 4°C, refrigerate	28 d/28 d
Nitrogen:				
Ammonia	P, G	500	Analyze as soon as possible or add H_2SO_4 to pH < 2, refrigerate	7 d/28 d
Nitrate	P, G	100	Analyze as soon as possible or refrigerate	48 h/48 h (28 d for chlorinated samples)
Nitrate + nitrate	P, G	200	Add H_2SO_4 to pH < 2, refrigerate	none/28 d
Nitrate	P, G	100	Analyze as soon as possible or refrigerate	None/48 h
Organic, Kjeldahl	P, G	500	Refrigerate; add H_2SO_4 to pH < 2	7 d/28 d
Odor	G	500	Analyze as soon as possible; refrigerate	6 h/N.S.
Oil and grease	G, wide-mouth calibrated	1000	Add H_2SO_4 to pH < 2; refrigerate	28 d/28 d
Organic compounds				
Pesticides	G(S), TPE-lined cap	—	Refrigerate; add 1000 mg ascorbic acid/L if residual chlorine present	7 d/7 d until extraction; 40 d after extraction
Phenols	P, G	500	Refrigerate; add H_2SO_4 to pH < 2	[a]/28 d
Purgeables by purge and trap	G, TPE-lined cap	50	Refrigerate; add HCl to pH < 2; add 1000 mg ascorbic acid/L if residual chlorine present	7 d/14 d

Analysis	Container	Minimum Sample Size, mL	Preservation	Maximum Holding Time
Oxygen, dissolved	G, BOD bottle	300		
Electrode			Analyze immediately	0.5 h/stat
Winkler			Titration may be delayed after acidification	8 h/8 h
Ozone	G	1000	Analyze immediately	0.5 h/N.S.
pH	P, G	—	Analyze immediately	2 h/stat
Phosphate	G(A)	100	For dissolved phosphate, filter immediately; refrigerate	48 h/N.S.
Salinity	G, wax seal	240	Analyze immediately or use wax seal	6 months/N.S.
Silica	P	—	Refrigerate—do not freeze	28 d/28 d
Sludge digester gas	G, gas bottle	—	—	N.S.
Solids	P, G	—	Refrigerate	7 d/2–7; see cited reference[b]
Sulfate	P, G	—	Refrigerate	28 d/28 d
Sulfide	P, G	100	Refrigerate; add 4 drops 2 N zinc acetate/100 mL; add NaOH to pH > 9	28 d/7 d
Taste	G	500	Analyze as soon as possible; refrigerate	24 h/N.S.
Temperature	P, G	—	Analyze immediately	Stat/stat
Turbidity	P, G	—	Analyze same day; store in dark up to 24 h; refrigerate	24 h/ 48 h

[a] See text for additional details.

[b] Environmental Protection Agency Rules and Regulations, *Federal Register 49*, No. 209, October 26, 1984. See this citation for possible differences regarding container and preservation requirements.

Note: For analyses not listed, use glass or plastic containers; preferably refrigerate during storage and analyze as soon as possible. Refrigerate = storage at 4°C in the dark. P = plastic (polyethylene or equivalent); G = glass; G(A) or P(A) = rinsed with 1 + 1 HNO_3; G(B) = glass, borosilicate; G(S) = glass, rinsed with organic solvents; N.S. = not stated in cited reference; stat = no storage allowed, analyze immediately.

Source: American Public Health Association, Standard Methods for the Examination of Water and Wastewater, 18th ed., Washington, D.C., 1992.

Index

Milton Keynes UK
Ingram Content Group UK Ltd.
UKHW031536071024
449327UK00024B/1879

9 780367 455477